GW00775926

The New Quantum Age

The New Quantum Age

From Bell's Theorem to Quantum Computation and Teleportation

Andrew Whitaker
Department of Physics
Queen's University Belfast

OXFORD
UNIVERSITY PRESS

Great Clarendon Street, Oxford OX2 6DP

Oxford University Press is a department of the University of Oxford.
It furthers the University's objective of excellence in research, scholarship,
and education by publishing worldwide in

Oxford New York

Auckland Cape Town Dar es Salaam Hong Kong Karachi
Kuala Lumpur Madrid Melbourne Mexico City Nairobi
New Delhi Shanghai Taipei Toronto

With offices in

Argentina Austria Brazil Chile Czech Republic France Greece
Guatemala Hungary Italy Japan Poland Portugal Singapore
South Korea Switzerland Thailand Turkey Ukraine Vietnam

Oxford is a registered trade mark of Oxford University Press
in the UK and in certain other countries

Published in the United States
by Oxford University Press Inc., New York

First published 2012

British Library Cataloguing in Publication Data
Data available

Library of Congress Cataloging in Publication Data
Data available

Typeset by SPI Publisher Services, Pondicherry, India
Printed and bound by
CPI Group (UK) Ltd, Croydon, CR0 4YY

ISBN 978–0–19–958913–5

1 3 5 7 9 10 8 6 4 2

For John and Peter

Preface

Over the past half-century, the approach of the community of physicists to quantum theory has changed quite markedly. At the beginning of the 1960s, the theory was regarded as a superlative tool for the study of physical systems—atoms, molecules, nuclei, radiation, and those properties of solids where their atomic nature was central to the behaviour. However, it was considered totally inappropriate even to think with any novelty about the fundamental nature of quantum theory and its rather surprising properties. That, it was practically universally believed, had been sorted out once and for all by Niels Bohr thirty years before.

Fifty years or so later, it has been fairly generally realized that Bohr's ideas were not necessarily the final word on the topic, that the theory itself certainly merited and still merits much further analysis, and that the theory may inform us, not just of the behaviour of physical systems, but also of fundamental and important aspects of the Universe itself.

Also, and perhaps even more surprisingly, this deeper analysis of the theory has led to the development of the new topic of study of quantum information theory, which may revolutionize several aspects of communication and computation.

This book describes the changes that have taken place, and discusses some of the new ideas current in both quantum theory and quantum information theory. I have also taken the opportunity to give an account of some of those scientists mostly responsible for the change in attitude over this period. I hope that the book will be accessible to anybody interested in the topic; to paraphrase Einstein, it might be said that I have tried to make things as simple as possible, but no simpler, given that I do not wish to avoid even some of the more complex aspects of the topic.

I should state that the initial idea of the book was that it should be written jointly by Dipankar Home and myself. However, after initial planning, Dipankar had unfortunately to drop out because of other commitments. I acknowledge his contribution to the planning of the book.

I would also particularly thank Reinhold Bertlmann and Anton Zeilinger for a particular contribution to this book. About ten years ago I was privileged to speak at *Quantum [Un]Speakables*, a conference they organized in Vienna to celebrate the work of John Bell and his followers. Among those attending this conference were virtually all those who contributed to the rise of the study of quantum foundations, and the early days of quantum information theory. Renate Bertlmann took a unique set of photographs of the participants, and I would thank Renate, Reinhold, and Anton for permission to use many of these in this book. I would also like to thank Gerlinde Fritz who arranged for me to receive these pictures. In the caption they are referred to as [© Zentralbibliothek für Physik, Renate Bertlmann].

I would also like to thank Queen's University Belfast for my appointments as Emeritus Professor and Visiting Senior Research Fellow that have enabled me to complete this book.

At Queen's, I would like to thank Joan Whitaker, Subject Librarian in Mathematics and Physics, for considerable help in tracing books, papers, and pictures.

I would also thank all the publishers and authors who have given permission for the reproduction in this book of copyright figures and pictures. Every effort has been made to contact the holders of copyright in materials reproduced in this book. I apologize for any omissions, which will be rectified in future printings if notice is given to the publisher.

At Oxford University Press, I would thank Sonke Adlung, April Warman, and Clare Charles for their courtesy and helpfulness in various aspects of the writing of this book. I would also like to thank Charles Lauder Jr for highly efficient and helpful copyediting, and Marionne Cronin, Emma Lonie, and Vijayasankar Natesan at Oxford University Press and SPi for all their hard work in the production of this volume.

Lastly I would thank my wife, Joan, and children, John and Peter, for their support and tolerance while I have been writing the book.

Contents

Introduction

The First Quantum Age and the New Quantum Age

The First Quantum Age began with the pioneering work of the founders of quantum theory, in particular Max Planck, Albert Einstein, and Niels Bohr, in the first quarter of the twentieth century. During this period, many crucial results were obtained, but it was not until the work of Werner Heisenberg and Erwin Schrödinger in 1925–6 that the theory was put on a firm comprehensive basis. There then followed an enormous flowering of important results, and within a very few years, a very large number of physicists contributed to the task of using the theory to obtain a remarkably good understanding of atoms and molecules, nuclear physics and the physics of solids. Even as early as the start of the Second World War it was clear that quantum theory was the most successful and powerful physical theory the world had known.

The theory was quite highly mathematical, and it was Niels Bohr in particular who stressed the necessity of constructing an intellectually satisfactory bridge between its mathematical results and the all-important physical predictions. We can say that an *interpretation* of quantum theory was required, and Bohr himself also produced such an interpretation, which is often called the *orthodox* or *standard* interpretation of quantum theory. Since Bohr was based in Copenhagen, it has usually been called the *Copenhagen* interpretation, and Bohr in fact considered his ideas to be applicable far beyond quantum theory, giving them the more general title of *complementarity*.

Whatever the merits of Bohr's work, and they will be discussed later in the book, his approach was certainly successful in its own terms, giving a firm answer to the rather tricky conceptual problems that the theory raised, and so allowing those working on quantum theory to put these matters on one side and concentrate on making use of the theory, with, as we have said, tremendous profit. So the First Quantum Age has undoubtedly been extremely fruitful. The basic ideas of quantum theory and the First Quantum Age are sketched in Part I of this book.

If there was a downside to the Bohr approach it was that its firm answers to conceptual questions often ruled out interesting alternative approaches. A few brave souls did put their heads above the parapet to call attention to such approaches, but they were severely criticized by the powerful group of physicists centred around Bohr, Heisenberg, and Wolfgang Pauli. It was made fairly clear that, to be frank, there was no place in physics—no jobs in physics!—for anybody who dared to question the Copenhagen position.

To a considerable extent the First Quantum Age continues up to the present day. There is still an enormous amount of important work done on extending use of the standard apparatus of quantum theory in novel directions. Powerful computers and increasingly sophisticated experimental techniques provide ample fresh areas for the theory to conquer, and such subjects as astrophysics and the physics of elementary particles demand a great deal of work from quantum theorists. And for many the Copenhagen interpretation still seems perfectly adequate.

Yet, beginning perhaps in the 1960s with the work of John Bell, this standard approach has been challenged with steadily increasing success. In what we may call the New Quantum Age, ways of thinking about quantum theory that were disallowed by Copenhagen have flourished, and physicists have been able to use them to discuss some of the most basic attributes of the Universe. Indeed, and rather amazingly, they have been able to suggest experiments to test their most fundamental ideas—this process has been called *experimental metaphysics*.

Admittedly this work seemed for some time highly esoteric. Yet in recent decades this set of ideas of the New Quantum Age has actually, and again perhaps very surprisingly, born technical fruit in the form of *quantum information theory*—which comprises, among others, the techniques of *quantum computation, quantum cryptography* and *quantum teleportation*—some of which have already been demonstrated, other still requiring a lot of effort to be made to work effectively.

In Part II, we consider the conceptual arguments and crucial experiments of the New Quantum Age, while in Part III we sketch the technical advances of quantum information theory.

PART I
THE FIRST QUANTUM AGE

1
Quantum theory—basic ideas

There is no doubt that quantum theory is a theory of tremendous power and importance; indeed it is probably the most successful physical theory of all time! It tells us a vast amount about the properties of atoms, molecules, nuclei, and solids, and it is also essential for any understanding of astrophysics. In fact we can go further: quantum theory is not only central in virtually every field of physics, but it is also the theoretical background to the whole of modern chemistry, and is becoming of increasing importance in our quest to understand biology at the molecular level. So we can sum up by saying that it is supremely important for any kind of understanding of how the Universe works.

In addition, though, it has generated a large number of technical applications. In particular there are very many specifically quantum devices, starting with the transistor, and moving on through all the related solid state devices that have followed it, right up to the today's famous integrated circuits and 'computer chips'. And we must certainly not forget the ubiquitous laser. Devices based on quantum principles have had an enormous range of applications—in industry, in medicine, in computers, in all the devices we take for granted in home, shop or factory—with the result that, in addition to its intellectual interest, quantum theory manages to contribute a substantial fraction of the economic output of the developed nations.

However, it must be admitted that quantum theory is admittedly difficult—in two rather distinct ways. The first difficulty is that it is mathematically complex; in this book, though, we shall do our best to explain the essential ideas without bringing in too many of these technical details. But over and above the rather advanced mathematics, the second difficulty is that the theory has a number of elements that are very difficult to accept or even understand, at least from the standpoint of the pre-quantum or so-called classical physics that we are all used to, or even just from a commonsense approach to 'physical reality'.

In Part I of this book, we shall describe some of these conceptually challenging aspects of the theory, and also the way in which these challenges have been met in what we call the First Quantum Age. Our approach is not historical, but we do include biographical boxes relating the ideas described to the physicists who discovered or elucidated them. A fuller account of both the ideas and the history is given in [1].

In the rest of this chapter, we make a few introductory remarks about quantum ideas. To start with, we should make the point that nearly everything we discuss follows from the central law of non-relativistic quantum theory, the famous Schrödinger equation, though it must be supplemented by the Born probability rule. Both the Schrödinger equation and the Born rule will be discussed shortly, in Chapters 3 and 4, respectively.

However, two general points on this all-important Schrödinger equation can be mentioned right away. First, we said that it is non-relativistic. This means that the discussion is restricted to the case where the speeds of the bodies being considered are much less than that of light, always called *c*. which has the value 3×10^8 ms^{-1}. In practice this is not really such an important problem as one might expect, since in many of the most important applications of quantum theory the particles do indeed have speeds much less than *c*. It would actually be possible to make the discussion relativistic by going beyond the Schrödinger equation, but this would only be achieved at the expense of considerable extra complication, and as it does not change the conceptual status of the theory to any extent, we shall stick to the non-relativistic case in this book.

What is actually more relevant for our account is that many of the experiments that we shall discuss involve the passage of photons from one point to another. As we shall see later in the following chapter, photons must be regarded as particles of light and therefore obviously travel at speed *c*. Our previous paragraph might seem to rule out consideration of photons, but in fact it turns out that we are able to discuss these experiments in a totally satisfactory way using the Schrödinger equation.

The second point is that the fundamental equation of quantum theory is known as the *time-dependent* Schrödinger equation (TDSE). Shortly we shall meet the *time-independent* Schrödinger equation (TISE). This has no independent status, being obtained directly from the TDSE for situations where the physics does not depend on time. However, the TISE is exceptionally useful in its own right, as we shall see; indeed, although it is less fundamental than the TISE, it is probably used much more.

We now come to a point that is rather subtle but is of very great importance for much of the rest of the book. This is that in quantum theory we must accept that the Schrödinger equation gives much important information about the system in question, but not all the information we are used to in classical (pre-quantum) physics.

In classical physics, for example, we take it for granted that both the position and the momentum of a particle exist and have precise values at all times, and that the central set of equations of classical mechanics, Newton's laws, does indeed use and provide values for both these quantities simultaneously. However, in quantum theory the situation is very different. As we shall see, the Schrödinger equation can give us precise values for one or other of the position and momentum of a particle, but if it gives us a precise value for its position, it can say nothing at all about its momentum, and if it gives us a precise value for its momentum, it can say nothing at all about its position.

Of course to say that the Schrödinger equation cannot provide precise values for both position and momentum does not necessarily mean that the particle itself does not actually *possess* such values. It might indeed seem most natural to assume that it *does* possess the values, and it is merely a limitation of the Schrödinger equation that it cannot provide them.

However, a striking addition to the statement of the Schrödinger equation has often been assumed, virtually universally, in fact, in the First Quantum Age. It is an important part of what is called the Copenhagen interpretation of quantum theory, which we shall meet shortly. This addition says that if the value of a particular quantity is not provided by the Schrödinger equation, such a value actually does not exist in the physical world. This is clearly a stronger statement than saying, for example, that the value of the quantity might exist but the Schrödinger equation cannot provide it (as we suggested in the previous paragraph), or even that it exists but that we are, for some reason, prohibited from knowing it.

Thus, in our example above, it may be that, at a particular time, the Schrödinger equation gives us a value for the position of the particle. Actually we can put that more precisely—we are saying that, at this time, the so-called wave-function of the particle relates to a precise value of position. (We shall study the idea of the wave-function more formally in Chapter 3.) Then, as we have said, the Schrödinger equation, or more specifically the form of the wave-function at this time, can tell us absolutely nothing about the momentum of the particle. That is a central and non-controversial feature of quantum theory.

The additional element we mentioned in the previous paragraph, which is much more controversial, says that we must then assume that the momentum of the particle just does not have a value; it is *indeterminate*. Similarly if the wave-function provides us with a precise value of momentum at a particular time, then we must just take it for granted that the position of the particle just does not have a value at this particular time.

In Einstein's terms, we are talking about *completeness*. We are saying that the Schrödinger equation is *complete* in the sense that nothing exists in the physical universe that cannot be obtained from the equation—if you can't get it from the wave-function, it just doesn't exist! It must be said that Einstein himself strongly disagreed with the idea that the Schrödinger equation was complete. He believed that there were elements of reality lying outside the remit of the Schrödinger equation. This disagreement of Einstein with the Copenhagen interpretation of quantum theory is extremely important and it will be investigated in depth later in the book.

As we have said, this statement of completeness clearly constitutes a very important difference between quantum theory and classical physics. In classical physics it is naturally taken for granted that all physical quantities have particular values at all times, but this is not the case in

Albert Einstein (1879–1955)

Fig. 1.1 Albert Einstein in 1921 [courtesy of Hebrew University of Jerusalem]

Einstein is most famous for his work on relativity, his special theory of relativity, which was published in 1905, and the general theory, which included gravitation and which followed in 1916, but his work on the quantum theory was equally important. Einstein's most important contribution to quantum theory was the argument in 1905 (his *annus mirabilis or* 'wonderful year') that suggested the existence of the *photon*; the fact that light had a particle-like as well as a wavelike nature was the first of a series of conceptual difficulties that were to plague the quantum theory. His idea of the photon emerged from fundamental studies on the theory of radiation, but it was able to predict and explain the essential features of the photoelectric effect, and it was for this work that Einstein was awarded the Nobel Prize for physics in 1921.

Thomas Kuhn has argued [2] that Einstein's work in the first decade of the century was of crucial importance in clarifying the full significance of quantization and discreteness that had been demonstrated but not really explained or even understood by Planck in 1900.

Einstein made other important contributions to quantum theory with his work on the specific heat of gases and solids, and his discovery in 1917 of *stimulated emission*, which was the basis of the *laser*. Then in 1924, Einstein contributed to the development of the ideas of Satyendra Nath Bose that led to the discovery of the Bose–Einstein statistics that are obeyed by particles such as photons, and also by the recently discovered Bose–Einstein condensate, where quantum effects are observable on a large or macroscopic scale rather than a microscopic one.

During the 1920s, Einstein gave crucial support to the ideas of de Broglie and Schrödinger, but his unwillingness to accept Bohr's *complementarity* led to his subsequent ideas being dismissed. Only recently has it gradually been realized that many of his arguments were profound and important [3]. In particular his ideas stimulated the important work of John Bell.

Einstein did not develop strong links to any country. He was born in Germany, but lived and studied in Switzerland from 1895, famously being employed in the Swiss Patent Office in Berne from 1902. Following his exceptionally important early work, he took up academic appointments in Zurich and then Prague from 1909, and then in 1914 he returned to Germany to the supremely important post of Director of the Kaiser Wilhelm Institute of Physics, Professor in the University of Berlin and a member of the Prussian Academy of Sciences. However, he always disliked German militarism, and in 1933 he was driven out by the Nazis, emigrating to the United States and taking up a position at the Institute for Advanced Study in Princeton, where he lived until his death.

quantum theory. We can put this slightly differently. Since quantum theory tells us that physical quantities do not necessarily possess values prior to measurement, we can say that measurement, at least in some sense, actually creates the value measured. Certainly the question of *measurement* is crucial in quantum theory. In contrast we can say that in classical physics it is, of course, taken for granted that physical quantities exist totally independently of whether we measure them, or even whether we *can* measure them.

To return to what we have called this additional feature of the Copenhagen interpretation, the opposing point of view is that there may be additional variables—*hidden variables* or *hidden parameters*—which cannot be obtained from the wave-function, but nevertheless can give us additional information about the physical quantities related to individual systems; in other words, the Schrödinger equation is *not* complete. Thus the orthodox view in the First Quantum Age has been that hidden variables did not exist, but the question of their possible existence became of increasing importance in the New Quantum Age.

References

1. A. Whitaker, *Einstein, Bohr and the Quantum Dilemma: From Quantum Theory to Quantum Information* (Cambridge: Cambridge University Press, 2006).
2. T. S. Kuhn, *Black-Body Theory and the Quantum Discontinuity 1894–1912* (Chicago: University of Chicago Press, 1987).
3. D. Home and A. Whitaker, *Einstein's Struggles with Quantum Theory: A Reappraisal* (New York: Springer, 2008).

2
Quantum theory and discreteness

In this chapter we come to one of the central surprises of quantum theory. Indeed it was the very first surprise, so it actually launched the theory and also gave it its name. This surprise is that, if one measures one of quite a wide range of physical quantities or *physical variables*, for which classical physics would definitely lead us to expect a continuous range of possible results, in fact only certain discrete values are obtained.

A very important example historically is a measurement of the energy of a simple harmonic oscillator or SHO. The SHO is one of most important type of system in physics. Common examples include a pendulum vibrating to and fro, or a weight on the end of a stretched elastic spring vibrating up and down, in both cases with air resistance ignored so that there is no dissipation of energy and the oscillatory motion continues indefinitely. In quantum theory, it is important that a mode of electromagnetic vibration behaves in a way mathematically exactly analogous to that of an SHO and so it will obey the same rules.

We shall first consider the energy of an SHO from the classical point of view. We shall define the energy for the pendulum as zero when it is stationary, and then when it is moving the energy must be positive, as both its kinetic and its gravitational potential energy will be positive. That though is the *only* restriction on the energy—that the energy cannot be negative. Classically it would be taken for granted that the energy can take *any* positive value (or, of course, zero). To put it another way, the possible results of a measurement are *continuous*.

In contrast, and most surprisingly for anyone used to classical physics, in quantum theory the possible values obtained in the measurement of the energy of an SHO, or we can say a little more formally, a particle subject to an SHO potential, are restricted to a very few possibilities; the values are *discrete* rather than *continuous*, and are shown in Fig. 2.2. They are, in fact, $\frac{1}{2}hf$, $\frac{3}{2}hf$, $\frac{5}{2}hf$,..., and we can sum these up as $E_n = (n+\frac{1}{2})hf$, where n, which is the first example of what is called a *quantum number*, may take one of the values 0, 1, 2, 3, ... Here f is just the usual (classical) frequency of the SHO, while h is the famous new physical constant, which is central through the whole of quantum theory—it is known as Planck's constant, and it has the value 6.63×10^{-34} Js.

Max Planck (1858–1947)

Fig. 2.1 Max Planck in 1906 [courtesy of Max Planck Institute at Berlin-Dahlem]

Max Planck can justifiably be called the founder of the quantum theory. It was somewhat ironic that he was the one to take one of the boldest steps in the history of science, since he was inherently extremely conservative, in science as in life.

For the first twenty years of his career, Planck worked on thermodynamics, the relationship between temperature, heat, and work, and in 1900, while studying theoretically the spectral distribution of the electromagnetic radiation emitted from so-called black bodies (bodies that absorbed all the radiation incident on them), he found that he could only explain the experimental data by assuming that energy was emitted in elements of hf, where f is the frequency of the radiation, and h was a new constant, which would come to be known as Planck's constant, the fundamental constant of quantum theory. Planck was able to obtain a very good value for h.

His formula for the energy levels of the simple harmonic oscillator was $E = nhf$, where $n = 0,1,2\ldots$. It was only with the discovery of modern quantum theory by Heisenberg and Schrödinger in 1925–6 that it was realized that the n should be replaced by $n+\frac{1}{2}$, as given in the text. Planck's argument was unaffected, since it used only *differences* between energy levels.

Planck had a supremely successful academic career, becoming full professor at the University of Berlin as early as 1892, but his personal life was less happy. His first wife, Marie, died in 1909 and his oldest son, Karl, was killed in the First World War. One of his daughters, Emma, died in childbirth in 1917, while her twin sister, Grete, who had married her sister's widower, tragically died in the same way in 1919. Lastly, his second oldest son, Erwin, who had been a prisoner of the French in the First World War, was involved in the plot against Hitler and was brutally executed in 1945.

Planck himself hated Nazism, and continued to support Einstein's work even when this was politically extremely unwise, but, as a conservative, he found it difficult to stand out effectively against the government, and he preferred to protect what elements of science that he could. His compromises are sensitively discussed by John Heilbron in [1].

In 1948, the internationally renowned research institutes known as the Kaiser-Wilhelm-Institutes were renamed the Max-Planck-Institutes.

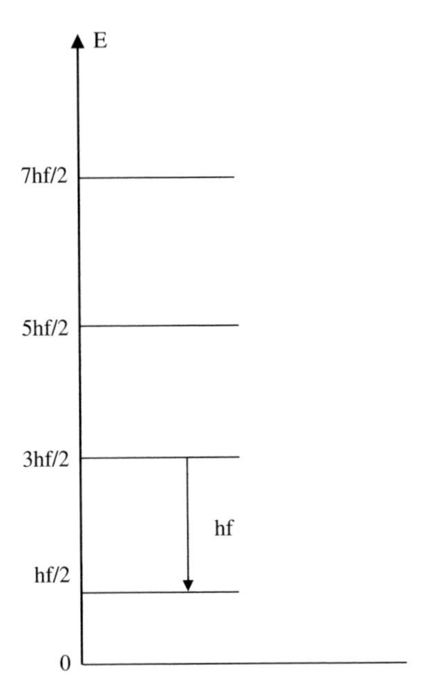

Fig. 2.2 Energy levels of a simple harmonic oscillator. The energies can be written as $(n+\frac{1}{2})hf$, where $n = 0, 1, 2, \ldots$

Planck's constant is clearly extremely small, showing us that quantum theory will generally be important where the systems involved are themselves small—atomic in size. This new constant takes its place with the speed of light, c, and the mass and charge of the electron, m and e, as the fundamental constants characterizing modern physics, which is based very strongly on quantum theory. The unit of h is joule-seconds, making it clear that we can think of it as an energy multiplied by a time; equally though it can be regarded as a momentum multiplied by a distance, and both approaches will be useful later on, particularly when we meet the Heisenberg principle. Actually later in the theory we shall nearly always meet not h itself, but $h/2\pi$, which is always written as $\hbar$ and called h-bar.

The physical quantity of angular momentum, which is extremely important in classical physics, but is perhaps even more important in quantum physics, is itself actually defined as a momentum multiplied by a distance, and on the atomic scale angular momentum will always be expressed as a multiple of $\hbar$.

We see that the lowest energy allowable occurs when $n = 0$, and it is $\frac{1}{2}hf$. This in itself is extremely surprising, as clearly it is greater than 0. This implies that, even at the lowest temperatures, the energy of the SHO cannot be 0. The state of the SHO when $n = 0$ is called the *ground state* of the system, and $\frac{1}{2}hf$ is the *ground state energy* or the *zero-point energy*. The existence of the zero-point energy is another very important difference between classical and quantum theory.

When $n = 1$, the energy is $\frac{3}{2}hf$; when $n = 2$, it is $\frac{5}{2}hf$; and so on, and states with $n > 0$ are called *excited states*. These energies can be plotted on an energy-level diagram as in Fig. 2.2. When the system makes a transition in which the quantum number n drops by 1, energy hf is released. When Planck originally came up with these very novel ideas, as discussed in the historical box, the type of oscillation he was considering was electromagnetic in nature. (We noted above that electromagnetic oscillations are mathematically identical to SHOs.) For such oscillations, this energy hf takes the form of a photon, essentially a particle of light. The idea of the photon will be discussed in detail later.

So the word 'quantum' just means a physical quantity increasing or decreasing by a step of some size, that is to say discontinuously, rather than continuously. The quantum is essentially just the size of the step. The phrase 'quantum leap' or 'quantum jump' is often used by, for example, politicians, to suggest that living standards or national income have increased not gradually over a period of years, but suddenly and dramatically as a result of some new policy. Usually a critic makes fun of the politician by pointing out that in quantum theory a quantum leap is actually exceptionally small because of the tiny size of h. The real point, of course, is that the change is indeed discontinuous rather than continuous, but the poor voter may be left feeling rather bemused.

To get back to quantum theory proper, another extremely important example historically has been the energy of a hydrogen atom. The hydrogen atom consists of a nucleus, which is just a proton in this case, and an electron, and to an extremely good approximation we can picture the nucleus as stationary, and the electron as moving in the electrostatic or Coulomb field of the nucleus. For the moment we will be particularly concerned with situations where the electron does not possess enough energy to escape completely from the nucleus. (We can say that the electron is *bound* to the nucleus.) Since it is usual to define the zero of energy as one where the electron has enough energy *just* to escape to an infinite distance from the nucleus, we will be thinking about situations where the total energy of the electron is negative; we must give the electron energy to get it up to zero energy when it *can* escape. (This is rather like when a bank account has unfortunately been allowed to go into the red or become negative. You must give money to the bank to get up to the zero mark.)

Classically the energy can have *any* negative value in this situation. We can say that the allowed energy values are continuous. The big change in quantum theory is again that the energy we may measure in these circumstances takes only discrete rather than continuous values. The mathematical expression for the lowest allowed energy, or *ground state energy*, of the hydrogen atom is a little complicated. It includes several of the important constants we have already met: m and e, the mass and charge of the electron, and of course h, although as we said, it is more convenient to use $\hbar$ or $h/2\pi$. Then we can write the ground state energy as E_1, which can be written in terms of e, m and $\hbar$, and also the usual constant for any electrostatic interaction, in this case between the electron and the hydrogen nucleus or proton. As we have said, this ground state energy is a negative quantity.

Then the general energy, the energy for higher values of n, is given by $E_n = E_1/n^2$ with quantum number $n = 2, 3, \ldots$, so the energies for the excited states are all less negative, and so actually larger, than E_1. As n gets larger and larger, these energies get closer and closer to 0. The corresponding energy-level diagram is shown in Fig. 2.3.

While the energy of the system is obviously a very important quantity, there are other very important quantities that also exhibit discrete measurement results. One of the most significant from the point of view of this book relates to the spin angular momentum (or just *spin*) of an electron. Spin is a quantum mechanical variable, and the electron spin is usually described as being analogous to the spin of a spinning table-tennis or snooker ball.

We will say that the electron is a spin-$\frac{1}{2}$ particle. This actually refers to a measurement of the z-component of the spin, which we will write as s_z. This quantity is perfectly precise mathematically, and effectively tells us how much of the spin can be regarded as being 'around the z-axis'. It turns out that the result will always be either $\hbar/2$ or $-\hbar/2$. In fact, as we said, the natural unit of angular momentum at the atomic level is just $\hbar$, so we say that, using this unit, or, if you like, taking this unit for granted, the measurement result will be either $\frac{1}{2}$ or $-\frac{1}{2}$.

For s_z, we see that the possible measurement results are again discrete. In this case, in contrast to the two cases considered for energy, where the quantum number, n, can take an infinite number of values, here we see that there are only two possible results, and the appropriate quantum number, which in this case is always called m_s, takes one of only two values. We can say that $s_z = m_s\hbar$, where $m_s = \pm\frac{1}{2}$, and is another

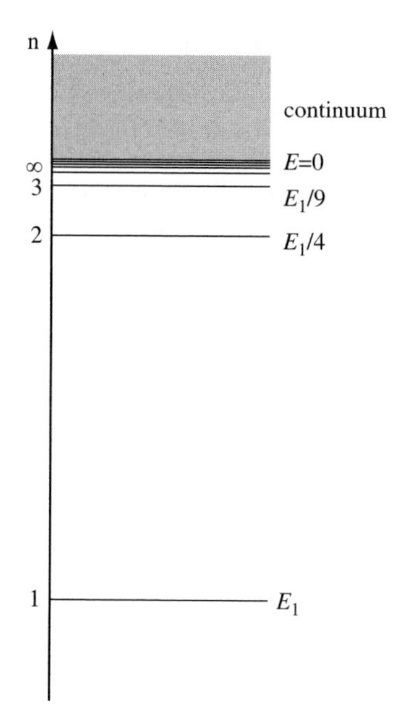

Fig. 2.3 Energy levels of a hydrogen atom. The ground-state energy is E_1 and it is negative since it is for a bound state. There are excited states for $n = 2, 3, 4 \ldots$, and the corresponding energies are E_1/n^2 tending to 0 as n tends to infinity. For positive values of energy, which relate to unbound states, energy levels are continuous.

quantum number. We will very often refer to the two cases as 'spin up' and 'spin down'.

We have looked at some quantities for which the values obtained in a measurement are discrete—this can be said to be the typical quantum case. However, we should also point out that there are other cases for which the values are continuous, just as in the classical case. Very important examples are position and momentum. When we measure the position or the momentum of a particle, we can obtain *any* value, not one of a discrete list as in the examples we have been looking at.

Energy is an interesting case. We have had some discrete examples, but there are also continuous cases. Let us consider again the hydrogen atom case, but now consider positive electron energies, for which the electron can escape from the nucleus altogether, because it has enough energy to reach an infinite distance away from the nucleus where, as explained already, the potential energy is zero. (These are cases where, in our banking analogy, we are 'in the black'.) The electron is still affected by the nucleus, of course, particularly when it is in its vicinity. In this case, the energy levels, though, are continuous. Thus for the hydrogen atom we have discrete energy levels for $E < 0$, and continuous energy levels (or often we say *a continuum*) for $E > 0$. Again this is marked in Fig 2.3.

This is an example of a general result. For energies for which the particle cannot totally escape from the potential, we say that it is *bound*, and the energy levels are discrete. For energies for which the particle *can* escape, we say it is *unbound* and the energy levels are continuous. As we have seen, the hydrogen atom is an example where both types of behaviour are found in different energy regions. For the SHO, because the potential energy tends to infinity at large distances, all energy levels are discrete, as we have already seen.

An example for which we get only a continuum is the free particle, where the potential energy is zero everywhere. Thus our particle has only kinetic energy, so the total energy must be either zero or positive. However, for all positive energies, the particle can travel without any restriction at all, so the energy levels form a continuum, just as for the classical case.

Reference

1. J. Heilbron, *The Dilemmas of an Upright Man: Max Planck and the Fortunes of German Science* (Cambridge, MA: Harvard University Press, 2000).

3
The Schrödinger equation

The time-independent Schrödinger equation, eigenfunctions, and eigenvalues

The property of discreteness explained in the previous chapter is certainly surprising from a classical point of view, but, considered on its own, it would be possible to put together an explanation that is readily understandable. We have consistently stressed that what we have been discussing is the result of *measuring* the particular variable rather than a value it might have independent of any measurement, but it might be suggested that this care is unnecessary; it could be that the variable itself can only possess these values, and that when we measure it, we are just recording a value that the system already possesses.

And indeed quantum theory does allow us to move a considerable way in this direction. To understand this, let us return to the Schrödinger equation; at first we will actually be using the time-independent Schrödinger equation, or TISE, as discussed in Chapter 1.

To study this equation we need to insert the particular form of the potential, that is to say, the simple harmonic oscillator (SHO) potential for that particular case, the electrostatic Coulomb potential for the case of the hydrogen atom, zero for the free particle, and so on. Then by mathematical analysis for the relatively simple cases we have mentioned, but with the use of computers for more complicated cases, we will be able to use the TISE to obtain the various energy levels for each case. (This is exactly the job that the TISE is set up to do!) In the previous chapter we listed the energy levels for the SHO and the hydrogen atom, while, as has been said, all positive energies are allowed for the free particle.

Importantly, the TISE also gives a mathematical function corresponding to each energy. The technical name for this function is an *eigenfunction* and the corresponding energy we have just discussed is called an *eigenvalue* of energy. (These words are traditional, but rather unfortunate and they don't have much meaning in English, or indeed even in German where the '*eigen*' prefix comes from, except for that provided by this use of the words in quantum theory and other areas of mathematical physics.)

Erwin Schrödinger (1887–1961)

Fig. 3.1 Erwin Schrödinger by the River Liffey in Dublin [courtesy of Cambridge University Press]

Schrödinger was the inventor of *wave mechanics*, one of the two seemingly disparate theories produced in 1925–6 that provided a rigorous basis for quantum theory, the other being Heisenberg's *matrix mechanics*. It was soon shown, by Schrödinger and several others, that the two theories could best be regarded as two different ways of representing the same fundamental mathematical structure.

Wave mechanics was based on de Broglie's theory of the wave nature of the electron. During the first half of 1926, Schrödinger produced a magnificent series of papers, not only presenting the basis of the new theory, but demonstrating its analogy to the classical nineteenth-century theory of William Hamilton which related mechanics and optics, and showing how the theory could be applied in the widest range of circumstances.

Schrödinger had hoped that his theory, with its wave-function, could provide a far more attractive approach to the quantum than what he regarded as the sterile and repugnant mathematical approach of Heisenberg, which did not provide any physical model at all. However, it soon became clear that Schrödinger's waves were also most easily regarded as mathematical rather than physical in nature.

Schrödinger, though, continued to oppose the views of Heisenberg and Bohr on quantum theory for the rest of his life; he and Einstein were practically alone in this rearguard action for many years. The famous *Schrödinger's cat* argument was an argument designed to show the inadequacy of the current views of measurement; if the cat remained unobserved, he claimed that these views implied that it could be in a superposition of dead and alive!

Schrödinger was Austrian, and, after wartime service, a series of short-term posts, and six years at Zurich, in 1927, as the result of his discovery of wave mechanics he became Planck's successor at the University of Berlin. However, his hatred of the Nazis led him to resign his post in 1933, and to take up a temporary post in Oxford. In 1936, though, he ill-advisedly moved back to Austria to the University of Graz. When German troops invaded Austria in 1938, he was dismissed and had to flee to Rome and thence to Geneva. Fortunately, Éamon de Valera, Prime Minister of the (then) Free State of Ireland was in Geneva and was keen to recruit Schrödinger to be Director of the School of Theoretical Physics in his new Institute for Advanced Studies in Dublin. Schrödinger stayed in Dublin from 1939 to 1956, when he returned to Vienna for his final years.

During his Dublin years, his greatest achievement was his book based on a course of lectures [1], in which he applied the methods of physics to the gene. Many physicists were stimulated by this book to study biological problems, including Francis Crick, James Watson and Maurice Wilkins, who shared the 1953 Nobel Prize for Medicine for their work on the structure of DNA.

The most complete biography of Schrödinger is by Walter Moore [2].

Because this is so central a feature of quantum theory, let us sketch a few examples. For the SHO, we slot the appropriate potential into the TISE, and when we solve the equation we obtain a list of energies (or eigenvalues) each with its appropriate eigenfunction. The eigenvalues are precisely those from Chapter 2: $E_n = (n + \frac{1}{2})hf$ with $n = 0,1,2,\ldots$, and for each there is a corresponding eigenfunction, which we will write as ϕ_n. We have so far considered, as explained already, the SHO in 1-dimension, which we will call the x-dimension, so our ϕ_n are functions of x.

For the hydrogen atom, we insert the Coulomb potential into the TISE. Again mathematical manipulation will give us a list of eigenvalues, which are precisely those listed in the previous chapter, $E_n = E_1/n^2$, with in this case the allowed values of n being $n = 1,2,3,\ldots$. (Incidentally, Bohr was able to obtain these energies some years before the creation of rigorous quantum theory by use of the so-called *Bohr atom*, a model of the atom that was *semi-classical* in the sense that it imposed a few quantum results on a basically classical description of the atom. This model was extremely influential in the work moving towards the new rigorous quantum theory of 1925–6, but it was totally superseded when the new theory emerged.)

The association of eigenvalues and eigenfunctions is technically a little less straightforward in this case of the hydrogen atom. Because we are naturally in 3 dimensions, the ϕ_n are functions of all 3 coordinates, x, y, and z. Actually, because the situation has spherical symmetry, we usually prefer to use a coordinate system that takes advantage of this symmetry—some readers may know these coordinates as *spherical polars*—rather than the Cartesian coordinates, x, y, and z. This though is just a matter of mathematical detail.

Niels Bohr (1885–1962)

Fig. 3.2 Niels Bohr [courtesy of MRC Laboratory of Medical Biology, Cambridge]

Bohr is universally regarded as one of the very most important and influential physicists of the twentieth century. His 1913 theory of the hydrogen atom was a supremely important step towards the full rigorous quantum theory of Schrödinger and Heisenberg, which was achieved in 1925–6. Bohr's theory was the first work to apply quantum theory to the atom—until then it had been restricted to studies of radiation, and it demonstrated the fundamental significance of energy-level diagrams, transitions between them and the related frequencies of radiation emitted. However, his theory was a mixture of classical and quantum ideas, and it was superseded when the new theories of Heisenberg and Schrödinger described atoms rigorously.

However, Bohr retained his personal significance as the Director of the Institute of Theoretical Physics in Copenhagen, which was the main centre for discussion of the emerging quantum theory through the 1920s and 1930s, nearly all those responsible for the development of the theory spending considerable periods in Copenhagen working with Bohr and with each other.

During the late 1920s Bohr also produced the theoretical framework of *complementarity*, the aim of which was to provide a way of handling the conceptual problems of quantum theory by means of the so-called *Copenhagen interpretation*. For at least 40 years, it was taken for granted that Bohr's solution was complete and final, but in more recent years, it has been subject to increasing challenge [3]. Other important areas of physics illuminated by Bohr include the atomic basis of the Periodic Table, explained in the early 1920s; the liquid drop model of the atomic nucleus, produced in the 1930s; and identification of the mechanism of nuclear fission, achieved around 1939–40.

Bohr was a devoted and celebrated citizen of Denmark, though he spent an important time in England in 1911–12, where, under the influence of Ernest Rutherford, his quantum theory of the atom was developed. In 1943, when Denmark was occupied by the Germans, he fled to Sweden and thence to Los Alamos, where his involvement with the atomic bomb project was mainly restricted to studying the morality of nuclear weapons, and the future of nuclear research for weapons and for the production of power after the war. For the rest of his life, he believed that nuclear secrets should be shared by the scientific community.

Of somewhat more significance is the point that in this case, although, as always, each eigenfunction corresponds to an eigenvalue, except for the lowest energy eigenvalue there are several eigenfunctions for each eigenvalue. (The technical term is that there is *degeneracy*, or that the eigenfunctions corresponding to the same eigenvalue are *degenerate*.)

This does not affect the kind of conceptual matters described in this book, but it does make some technical details a little more complicated. Because of this degeneracy, we cannot just write ϕ_n for the eigenfunction. Rather we must provide a different label for each degenerate eigenfunction, for example ϕ_{n1}, ϕ_{n2},..., each of which will correspond to eigenvalue E_n. (Actually in practice in such situations we will usually choose a label that relates more directly to the essential *differences* between these functions that all correspond to the same energy rather than just labelling them 1, 2, 3,....)

We now discuss the free particle, and actually, just to keep things as simple as possible, we shall think of the particle moving in only one dimension. Here, of course, since the particle is free, the potential energy to be inserted into the TISE is just zero. In this case, as we have stressed, we have a continuum—any (positive) energy is allowable. Instead of E_n, then, which is appropriate for the discrete case, we write $E(k)$, because there is a value of energy for *any* value of the *continuous* variable, k. Again there are eigenfunctions corresponding to each energy eigenvalue; actually there is degeneracy and there are two eigenfunctions for each eigenvalue (except for the special case where the energy is actually zero, for which there is just a single eigenfunction). The two eigenfunctions correspond to the particle moving in opposite directions with the same energy.

Having looked at these cases, which are actually the simplest that can be considered, we briefly interrupt the conceptual discussion to mention that the technique described is at the heart of many, perhaps most, of the applications of quantum theory, both in pure research and in technical development. In principle we insert the appropriate potential into the TISE and obtain the various energy levels for the system.

In practice the procedure is much more difficult, or really just much more time-consuming. To start with atoms, the Schrödinger equation tells us that, perhaps rather surprisingly, the computations for an atom with Z electrons (or we may say an atom with *atomic number* Z) must be carried out in a mathematical space, not of 3 dimensions, but of $3Z$ dimensions, so 6, for example, if Z is 2; this is obviously a massive task if Z is at all large. Tremendous progress has been made, though usually with the help of severe approximations. Molecules and solids can in general terms be studied using the same procedure, though, as would be expected, with still further increased computational needs. In all these cases, the basic interaction is, of course, just the Coulomb interaction.

When one turns to nuclei the situation is rather different. The nuclear forces are not known in any detail, and indeed, while the same type of procedure can be used, the aim is more to obtain information about the interactions from whatever is known experimentally about the energy levels, than to determine the actual energy levels.

We can go further and recognize that supposed 'elementary particles' such as the proton and the neutron act as though they are composed of three particles known as *quarks*, though isolated quarks have never been found. In the subject of *quantum chromodynamics*, the forces between the quarks are investigated by broadly the same kind of technique as described in this section, although the treatment must be relativistic. It is an important aspect of quantum chromodynamics that, in contrast to Coulomb forces, the magnitude of the forces *increases* rather than *decreases* with distance apart, and this is the fundamental reason why isolated quarks do not exist.

To sum up, we have seen the very wide use of the TISE to find energy levels for a tremendous range of systems. Of course when we wish to study *processes* in which systems change, we certainly need the time-dependent Schrödinger equation (TDSE) and we shall get some idea of how this works in the next section.

The time-dependent Schrödinger equation and wave-functions

Everything we said in the previous section relates only to the general type of system we are discussing. For example, we discussed the energies and eigenfunctions of a general SHO at *any* time. In this section, we look at a *particular* system, and note that the subject of the TDSE is a function called the *wave-function*, always referred to as ψ (psi), which depends on both space and time. The wave-function for a particular system can tell us a considerable amount about the properties of the system, or, we should say more explicitly, what result or results we may get if we measure a particular property of the system at that particular time. We shall say much more about this later.

First, though, we discuss the dependence of ψ on time. We should at once note that, if we know the form of ψ at any particular time, the TDSE will tell us its form at any later time, or at least until a measurement is performed. In terms we shall use in the following chapter, we can say that its behaviour is *deterministic* or that ψ *evolves* deterministically.

In fact, if the wave-function at any particular time is equal to an energy eigenfunction, the idea of which was discussed in the previous section, the variation with time is very straightforward. To explain this, we shall first say that it is useful to write the wave-function as a product of two terms. The first gives its magnitude. The second can be called the *phase* of ψ. To explain this latter term, we may think of a circle with a radius vector starting off along the x-axis, and steadily rotating around the circle. The angle that the vector makes with the x-axis can increase from 0° to 360°, and then we are back where we started. The phase is related directly to this angle.

Then we can say that in the variation of ψ with time, for this case where the wave-function is equal to an eigenfunction of energy, its magnitude remains constant, and its phase changes continuously. And this makes things especially simple. The change of phase is actually irrelevant when we consider the result of performing a measurement, which depends only on the magnitude of the function, and so in these cases, the result of a measurement does not vary with time.

We shall now move on to discuss the relationship between the fixed eigenfunctions, and the wave-function of the particular system at a particular time. At the beginning of this chapter, we raised the obvious suggestion that, if we obtained a particular value of energy when we performed a measurement on the system, the reason could be that the system actually had that value of energy even before we performed the measurement.

Translating this suggestion into the terms of the wave-function of the system and the various eigenfunctions, the suggestion is that at all times the wave-function of the system is always equal to one of the eigenfunctions, apart from, as we said in the previous paragraph, a steadily advancing phase factor for ψ. As we also said in the previous paragraph, when we perform a measurement of the energy the phase factor is irrelevant, and we obtain the result corresponding to the particular eigenvalue, and the important point is that this result is just the corresponding eigenvalue. (We shall often ignore this phase factor, since, as we have said, in these circumstances it has no effect.)

We can introduce a new piece of terminology: if the wave-function of the system is equal to a particular eigenfunction, so that when we measure the energy we must get the corresponding energy or eigenvalue, we can say that the system is in the corresponding *eigenstate*.

Taking the SHO, for example, the suggestion is that, before the measurement the wave-function, ψ, might, for example, be equal to the eigenfunction for $n = 3$. If this is the case, the result obtained by the measurement will be E_3, which is equal to $\frac{7}{2}hf$. Clearly it will be assumed that the energy had this value all along. If this argument worked, the conclusion would be that, although the discrete nature of the possible energies obtained in experiments like this was clearly very strange from the classical point of view, there was a relatively straightforward explanation. The wave-function of the system was always equal to one of the eigenfunctions, and so the energy would always take the corresponding allowed value, even before any measurement.

Yet this simple and pleasant explanation is immediately ruled out by a fundamental property of the TDSE—superposition.

References

1. E. Schrödinger, *What is Life?* (Cambridge: Cambridge University Press, 1992).
2. W. Moore, *Schrödinger* (Cambridge: Cambridge University Press, 1989).
3. A. Whitaker, *Einstein, Bohr and the Quantum Dilemma* (Cambridge: Cambridge University Press, 2006).

4
Superposition

Superposition

The suggestion we have been tentatively considering is that the wave-function of a system must always be equal to one of the eigenfunctions, and its energy is thus equal to the corresponding eigenvalue. A measurement of energy thus yields this result in a straightforward way, and the energy eigenvalue and eigenfunction remain unchanged in the measurement process.

But unfortunately this cannot be right. It is a fundamental property of the time-dependent Schrödinger equation, or TDSE, that if a particular wave-function, which we can call ψ_a, is a solution of the TDSE for a particular system, and a second wave-function, ψ_b, is another solution, then the sum of these is also a solution. From the mathematical point of view this is a result of what is called the linearity of the equation, and it is a property shared with many of the most important equations of physics, such as the wave equation, and the equations for thermal conduction and diffusion. Technically it arises because ψ appears on its own in every term of the equation, rather than, for example, as ψ^2.

The TDSE is often called, rather loosely, the 'Schrödinger *wave* equation'. This name is not actually very helpful, as it perhaps suggests that the wave-function in all cases is a simple wave, which is certainly not the case. Nevertheless the terminology *is* helpful in reminding us that ψ in all cases shares the best-known property of waves. This property is that they can add together and interfere, leading to the very well-known phenomena characteristic of waves, for example the colours seen when looking at an oil film on water, or the well-known Young's slits or Newton's rings experiments.

Once we have realized this, the attractive suggestion at the beginning of this section becomes untenable. Let us again take the simple harmonic osciallator (SHO) as an example. If the wave-function is equal to the ground state eigenfunction, which we will call ϕ_0, for example, if we measure the energy we will expect to get the result E_0 or $\frac{1}{2}hf$. It will seem natural to say that the energy was E_0 even before the measurement was performed. Also it seems equally natural to accept that the wave-function will not change at the time of the measurement, so a subsequent measurement of energy will yield the same value E_0.

Similarly there are no conceptual problems if the wave-function before the measurement is ϕ_1, the eigenfunction for the first excited state. The result of a measurement of energy must be E_1 or $\frac{3}{2}hf$, we can assume that the system actually had this value before any measurement and we will further assume that the measurement has not changed the wave-function, so a subsequent measurement of energy will also give the result $\frac{3}{2}hf$.

So far, so good. However, the principle of superposition tells us that we must also consider situations where the initial wave-function is a linear combination or superposition of ϕ_0 and ϕ_1. The simplest mathematical possibility is $\phi_0 + \phi_1$, though for technical reasons that will soon become apparent, we will actually choose $c(\phi_0 + \phi_1)$, where c is a constant that we will quite soon evaluate.

In this case, it is far from clear what result we will get if we measure the energy of the system. We might guess that we should get the value $\frac{1}{2}(E_0 + E_1)$. However, that equals hf, and the very start of our argument in Chapter 2 was the point that we can *never* get this value, since it is not one of the discrete set of values E_n. One might feel that it is very unlikely that we would obtain E_2 or E_3, or indeed E_n for any value of n other than 0 or 1, but equally there seems no possible reason to choose one or the other of E_0 or E_1. We have met a major conceptual difficulty.

Indeed with this appreciation of the importance and significance of superposition of wave-functions, we must face the fact that quantum theory is not only strikingly different from classical physics, as was clear from the fact of *discreteness* as discussed in Chapter 2, but it must present numerous conceptual challenges, which will be analysed in the following sections and chapters.

The Born probability rule or interpretation

What happens in a measurement of energy if the wave-function at the time of the measurement is $c(\phi_0 + \phi_1)$, i.e., a linear combination or superposition of eigenfunctions corresponding to different measurement results?

Actually the answer that corresponds to experiment was given by Max Born very shortly after modern quantum theory was produced by Werner Heisenberg and Erwin Schrödinger in 1925–6. The answer is expressed in probabilistic terms. It is that there are equal probabilities of obtaining the energy eigenvalues corresponding to each of ϕ_0 and ϕ_1, i.e., a probability of $\frac{1}{2}$ of obtaining the result E_0, which is $\frac{1}{2}hf$, and an equal probability of obtaining the result E_1, which is $\frac{3}{2}hf$. To put it another way, if we perform the measurement on a large number of systems, each with wave-function $c(\phi_0 + \phi_1)$, we will obtain the

Max Born (1882–1970)

Fig. 4.1 Max Born [courtesy of Max Planck Institute at Berlin-Dahlem]

Born was a hugely resourceful theoretical physicist who did excellent work in many areas of physics including solids, liquids, and optics. In 1921 he became Professor of Theoretical Physics at the University of Göttingen, where he founded an important school of atomic and quantum physics.

Heisenberg was temporarily working with Born when he made the discovery of his 'arrays of numbers' that could be used to set up the first rigorous quantum theory. Born knew much more mathematics than Heisenberg, and realized that these 'arrays' were well known mathematically as *matrices*. During 1926, Born and his assistant, Pascual Jordan, first on their own, later assisted by Heisenberg, who had by then undertaken a crash course in matrix algebra, produced a full account of the new theory.

Born's most celebrated contribution to quantum theory was his idea that the wave-function could only be used to predict the *probabilities* of different results being obtained in measurements, specifically that the square of the wave-function (more technically, the square of the modulus of the wave-function) represents a *probability density*. This can be called the *statistical interpretation of quantum theory*.

Born was exceptionally disappointed not to share Heisenberg's 1932 Nobel Prize for Physics, and worse still, the following year, as a Jew, he had to leave Göttingen following the rise of Hitler. After 3 years in Cambridge, he became Professor of Natural Philosophy in the University of Edinburgh, where he stayed until 1953. On retirement, he had to return to Germany to obtain a respectable pension until at last, in 1954, he received his own Nobel Prize for the statistical interpretation of quantum theory.

For around 40 years, Born corresponded with Einstein, and these letters were later published as *The Born–Einstein Letters* [1]. The letters give a moving account of disagreement on quantum theory—Born was a great supporter of Bohr and complementarity—but general agreement on the politics, the humanity and the requirement to save all that could be saved from the ravages of Hitler.

result $\frac{1}{2}hf$ for about half the measurements, and $\frac{3}{2}hf$ for the other half.

In retrospect, the answer may appear almost obvious—it is difficult to think of any other solution that works and makes at least general sense, but it must have taken considerable courage for Born to state this the first time, as the suggestion does run very much counter to the whole of physics up to that date, and really exemplifies rather than solves the conceptual difficulties caused by superposition. We shall follow this line in the following section.

Here we just make a few technical points. For the case we have considered so far, ϕ_0 and ϕ_1 have been equally weighted in the wave-function at the time of measurement, and so implicitly we have assumed that the probabilities of obtaining the values of E_0 and E_1 are equal. Since total probability always equals 1, each probability must be $\frac{1}{2}$. Born's rule is actually that the probability of each is c^2, and so c^2 must equal $\frac{1}{2}$, so c must equal $\frac{1}{\sqrt{2}}$. So we write the initial wave-function as $\left(1/\sqrt{2}\right)(\phi_0 + \phi_1)$, as we said before. We can say that the wave-function is now *normalized* in the sense that the sum of the measurement probabilities does indeed equal 1 as it must. Alternatively we can say that $\frac{1}{\sqrt{2}}$ is a *normalizing constant* for the (*unnormalized*) wave-function $(\phi_0 + \phi_1)$. This paragraph is all basically mathematical bookkeeping.

Let us move on to consider the case where ψ is still a superposition of ϕ_0 and ϕ_1 but the two eigenfunctions are not necessarily equally represented, so ψ can be written as $c_0\phi_0 + c_1\phi_1$, where c_0 and c_1 are not necessarily equal. Born's rule, which we just stated, tells us that, in a measurement of energy, the probability of obtaining the value E_0, which corresponds to ϕ_0, must be c_0^2, while the probability of obtaining the value E_1, which corresponds to ϕ_1, must be c_1^2. Since the total probability must, as ever, be 1, if we are to use Born's rule properly, we must have $c_0^2 + c_1^2 = 1$.

So, for example, $\sqrt{3/7}\phi_0 + \sqrt{4/7}\phi_1$ is properly normalized since, $3/7 + 4/7 = 1$ and this tells us that, in a measurement of energy, the probability of getting the result E_0 is $3/7$, while the probability of getting the result E_1 is $4/7$, the probabilities adding, of course, to 1.

Finally, to generalize, the expression for any wave-function ψ at a particular time can be expanded as $c_0\phi_0 + c_1\phi_1 + c_2\phi_2 + c_3\phi_3\ldots$, where the sum can go over all the eigenfunctions. Then, provided

$c_0^2 + c_1^2 + c_2^2 + c_3^2 \ldots = 1$, the probabilities of the various measurement results are $P(E_0) = c_0^2$, $P(E_1) = c_1^2$, $P(E_2) = c_2^2$ and so on.

The conceptual challenges posed by superposition

Having examined the technical issues involved in measurement in the previous section, we now return to the much more interesting conceptual problems.

In this section, we follow the prescription of orthodox quantum theory, as propounded by Bohr in particular; it has often been called the Copenhagen interpretation of quantum theory. The aspect of this interpretation that we are most interested in at the moment was mentioned in Chapter 1. Now that we have introduced the wave-function, we can express it as follows. If the wave-function does not provide a definite value for a given physical observable, that observable simply does not possess an exact value. Similarly, if the wave-function provides only a distribution of values for a physical observable, either discrete (as for the pre-measurement cases we have been looking at in the previous section) or continuous (for some of the cases we looked at in Chapter 2, such as the free particle), then that observable just does not exist to a greater precision. And lastly, if the wave-function gives no information at all about an observable, that observable is totally undetermined; it just has no value. To put things another way, the statement is that no information is to be included in the theory over and above what is available in the wave-function.

If we did allow such added information, it would be termed *hidden variables* or *hidden parameters*. Discussion of what may happen if hidden variables *are* allowed will occur later in Part I, and also in the rest of the book.

However, as explained already, here we *are* following the Copenhagen interpretation and so are not allowing hidden variables. Thus, when we are asked what the value of the energy actually was before the measurement took place, we are in something of a quandary. We must admit that, apart from the exceptional cases where the wave-function is actually equal to a single eigenfunction, the energy of the system just does not possess a definite value before the measurement takes place. Note how important it was that, right back in Chapter 2 when we discussed the question of discreteness, we did not talk of the values energy actually *has*, merely of those that would be obtained in a measurement.

This admission is a statement of one of the big conceptual problems of quantum theory—loss of *realism*. Realism, which tells us that all variables pertaining to a system have values at all times, whether or not they are measured, had been taken for granted in physics up to this time. In quantum theory, or to put it more exactly, in the Copenhagen interpretation of quantum theory that we are following in this section, we must admit that it does not hold.

Another totally accepted belief of classical physics also must be relinquished. This is *determinism*, the belief that the present state of system totally determines its future state. We may say that determinism assures us that, if two systems are identical at a particular time and they are treated identically, they cannot differ at a later time. We noted in the previous chapter that, in the absence of measurement, the Schrödinger equation tells us that the wave-function ψ does indeed evolve deterministically.

But, as we have seen, ψ does not relate directly to measured properties of the system. When we look at measurement results, we see that quantum theory, buttressed by the Copenhagen interpretation, tells us that two identical systems treated identically may give *different* results. Two systems can have identical wave-functions, each say $\left(1/\sqrt{2}\right)(\phi_0 + \phi_1)$ for the SHO case, and the Copenhagen interpretation assures us that they are indeed identical—there are no additional variables for which they might have different values. Yet, when the energy is measured, one system may yield the result $\frac{1}{2}hf$, and a second $\frac{3}{2}hf$.

Certainly the combination of quantum theory and its Copenhagen interpretation is exerting a heavy conceptual price. We shall say that the three big conceptual issues in quantum theory concern determinism, realism, and locality. Determinism and realism we have discussed in the past few paragraphs. We shall come to locality at the beginning of Part II of the book.

We can add to the conceptual price of quantum theory by bringing in the idea of the *collapse of wave-function* or the projection postulate introduced by John von Neumann. This idea attempts to answer the question—what happens if, having obtained a particular value in a measurement, we immediately repeat the measurement?

For the sake of example, let us take the SHO case yet again, and let us assume that the initial wave-function is $\left(1/\sqrt{2}\right)(\phi_0 + \phi_1)$. We know that we expect to obtain the value E_0 or E_1 with equal probabilities. Let us say that in a particular case we get E_1. If we measure the energy again immediately, it would seem totally natural that we should get the same result, or in other words E_1 again. (If we want to be more convinced of this, we might move from measurement of energy, which is actually all we have considered so far, to more general measurement, which we shall consider in the following chapter. In particular let us think of measurement of position. Suppose we measure the position of a particle, find it at a particular place, and then immediately repeat the measurement. If it were *not* found at the same place, we would have to say that it had moved a non-zero distance in zero time, thus travelled at an infinite speed! This would be totally in contradiction to the basic tenets of relativity. For the case of position measurements, therefore, an immediately repeated measurement *must* give the same result as the first measurement, and it is natural to assume that the rule applies to measurements of any variable.)

The implication is, then, that at the time of measurement, and if the result E_1 is obtained, the wave-function must have *collapsed* to the corresponding eigenfunction, ϕ_1. Similarly, if the result is E_0, the wave-function must have collapsed to ϕ_0. In each case, then, an immediately repeated measurement will indeed give the same result as the first. Everything seems to work well, but the supposed collapse is itself a very peculiar phenomenon, and we shall consider this further in the following section.

The measurement problem of quantum theory

As von Neumann pointed out about the collapse idea, it implies that the quantum system develops in two entirely different ways. He called the development when the system evolves freely without measurement Type 2, and the development when measurement takes place as Type 1. In Type 2 development, the system develops purely following the Schrödinger equation. We have already said that if the wave-function at a particular time is a single energy eigenfunction, for example, ϕ_0 in the SHO case, as time moves on it merely gains a phase factor, and in fact this factor is equivalent to a mathematical oscillation at a frequency equal to E_0/h. If the wave-function is equal to ϕ_1, the factor is equivalent to an oscillation at frequency E_1/h. In the more complicated situation where the wave-function at a particular time is a superposition of more than one energy eigenfunction, each term in the superposition oscillates mathematically at the appropriate frequency, so the situation is a little more complicated in detail.

Mathematically, though, it is very straightforward, and indeed very much analogous to the relevant mathematics for such problems as waves on strings. As we have said, Type 2 development is deterministic; if the form of ψ is known at a particular time, it is trivial to work out its form at any later time, provided, of course, no measurement has intervened.

We can also say that Type 2 development is *reversible*. The first thing that this implies is essentially the mirror image of the statement of the previous paragraph. If we know the form of ψ at a particular time, we can work its form at an *earlier* time. It also means that, at least in principle, we can recover the earlier situation from the later one. In simple cases, this can actually be quite easy; in more complicated ones, it will almost certainly be much more difficult, but possible, as we have said, in principle.

Type 1 development, the type of development at a measurement, is very different. We have already pointed out that, except in the simplest

John von Neumann (1903–57)

Fig. 4.2 John von Neumann [courtesy of Marina Whitman, from I. Hargittai, *The Martians of Science* (Oxford: Oxford University Press, 2006)]

John von Neumann was born in Hungary, but studied and lectured in Germany from 1921 to 1930. He taught at Princeton University from 1930 until 1933 when he became one of the foundation mathematics professors at the newly founded Institute for Advanced Study in Princeton, where he stayed for the rest of his life.

He was indisputably a mathematical genius, making major contributions in pure mathematics to set theory, measure theory, mathematical logic, group theory, ergodic theory, and the theory of real variables. His work on economics included the theory of games and the famous minimax theorem. He played an important part in the Manhattan Project developing the atomic bomb, and he also worked out key steps in the nuclear physics involved in thermonuclear reactions and the hydrogen bomb. In this work he became concerned with the development of computers, designing the so-called von Neumann architecture in which data and programme are mapped onto the same address space; this is the system used by nearly all computers today.

His work in applied mathematics and physics included contributions to statistical mechanics and the theory of shock waves. As a mathematician, he put quantum theory on what he considered a more rigorous basis than its founding physicists, analysing it as the mathematics of linear operators on vector spaces (Hilbert spaces) in his famous book of 1932 [2]. However, his *projection postulate*, and his attempted proof that quantum theory could not have an underlying framework of hidden variables, were 'contributions' to quantum theory that have been subjected to close scrutiny and criticism!

cases, it is not deterministic. Two systems may have the same wave-function, but a measurement of energy may yield different answers, and, according to the collapse postulate, the wave-function immediately after the measurement will be different in the two cases. (We would again remind the reader that we are talking within the Copenhagen interpretation in which there can be no additional hidden variables before the measurement which might distinguish between the two systems.)

Also a Type 1 process is certainly not reversible. Again in the SHO case, we can obtain a result of, say, E_0 in a measurement of energy, or we can know that the wave-function after the measurement is equivalent to ϕ_0. Neither statement allows us to say very much about the wave-function *before* the measurement. It *may* have been ϕ_0. All we can really say, though, is that, if the wave-function before the measurement was equal to $c_0\phi_0 + c_1\phi_1 + c_2\phi_2 + c_3\phi_3 \ldots$, the value of c_0 was not zero. Provided that was the case, there was a non-zero probability of the measurement result being E_0, and the wave-function collapsing to ϕ_0.

What we have been discussing in the previous section and this one can be called the *measurement problem* (or sometimes the *measurement paradox*) of quantum theory. The Schrödinger equation dictates that superpositions must always remain as superpositions. Yet it seems that at a measurement we actually obtain one particular result; a superposition breaks down to a single term. In other words, at a measurement it seems that the basic law of quantum theory is decisively broken. As John Bell, whom we shall meet much more in the course of the book, has pithily put it, at a measurement we get a change from *and* to *or*, and this is against the fundamental prescriptions of quantum theory.

And, as Bell also pointed out most clearly, there is a further conceptual complication. Up till now in this book, and this is very much in the style of the usual presentation of quantum theory, we have taken the term *measurement* to be a primitive defining term. In other words, it has been assumed that, from first principles, one can distinguish a Type 2 process, in which interactions between particles and forces due to electromagnetic fields can determine how the system develops, from a Type 1 'measurement'.

Yet a 'measurement' must actually be constructed from exactly the kind of interactions and forces involved in Type 2 processes. A 'measurement' may be distinguished from a 'non-measurement' in terms of the aims of the experimenter, but it is very difficult to see how, at the level of the fundamental behaviour of the system and its wave-function, there can be such a total difference between the two types of behaviour. Of course, as Bell says, in practice, and with sufficient experience and knowledge, we are able to make quantum theory work in a totally pragmatic way, or 'for all practical purposes—FAPP' as he says. We know in practice which manipulations of the system are to be given the status of a measurement, and which are not. But what is lacking is a rigorous axiomatic statement of the foundations of the theory, one that does not require any discretion at all from the quantum theorist.

It can be mentioned that this argument has a relevance right back to the discussion of Born's probabilities. Textbooks on quantum theory often imply that Born's ideas are additional to, or a completion of, the standard laws of quantum theory. The idea may be that Schrödinger's equation (the TDSE) tells us what happens when there is no measurement, and Born's postulate completes our understanding of the theory by telling us what happens when there *is* a measurement. But we now see that this distinction is illusory. Since we cannot determine rigorously and fundamentally what constitutes a measurement, we must admit that Born's ideas are in no sense a rigorous completion of quantum theory. If we cannot unambiguously describe how a measurement result is actualized, we certainly cannot regard talk about the probability of obtaining different outcomes as having a rigorous basis, highly useful as Born's rule is, of course, in a practical way.

In these past two sections, we have seen many of the most troubling problems involved with quantum theory (again, one should say, if it is coupled with the Copenhagen interpretation). First, in the absence of a measurement, in most cases a variable just does not have a value—this is loss of realism. (We have discussed energy as the variable, but we shall shortly see that it also applies to other variables.) Coupled with this is loss of determinism. Secondly we see all the problems associated with measurement; quantum theory appears to rely fundamentally on the idea of a measurement, but the behaviour of the system at measurement seems very difficult to describe or understand, and lastly it is actually not even possible to determine exactly what is and what is not a measurement!

Hidden variables

When discussing the conceptual problems raised by quantum theory in the past two sections in particular, we stressed that we were using one aspect of the Copenhagen interpretation, which is that there are no parameters or hidden variables in addition to the wave-function that could characterize each system. This implies that two systems with the same wave-function are identical.

It may well have occurred to the reader that many of the problems could be removed if we were prepared to lift this prohibition. Let us say we look yet again at the case of the SHO, and let us imagine two systems, each with its wave-function at a particular moment given by $\left(1/\sqrt{2}\right)(\phi_0 + \phi_1)$. However, let us suppose that, in addition to the wave-function, there is a hidden variable for each system that has the effect that for the first system the energy actually *is* E_0, while for the second it *is* E_1.

It may seem clear that, at a stroke, and indeed a trivially simple stroke, we seem to have removed virtually all our difficulties. First, we have restored realism; each system clearly does have a value of energy. Secondly, we have restored determinism. The argument that two identical systems have led to different results no longer holds water as the systems are *not* identical. Each system follows its pre-ordained course to its individual but deterministic conclusion. We need no collapse of wave-function since it is the hidden variable rather than the wave-function that guarantees that the second measurement will provide the same result as the first. Equally neither wave-function nor hidden variable need have changed at the point of measurement, so the situation is practically trivially reversible.

To establish Born's rule, we will generalize slightly and say that, if the wave-function of a collection or *ensemble* of systems is $c_0\phi_0 + c_1\phi_1$, which we may assume is normalized so that $c_0^2 + c_1^2 = 1$, the value of the hidden variable for a fraction c_0^2 of the ensemble will be E_0 and that for the remaining fraction c_1^2 will be E_1. This assures that the correct distribution of results will be obtained when the measurement is performed on each member of the ensemble.

The reader must surely be bemused at this stage. Why have we spent so much time building up the difficulties if they can be disposed of so simply? And why were the 'founding fathers' of quantum theory so adamantly opposed to hidden variables when they seem so useful? (When we talk of 'founding fathers', a term that Bell, in particular, used a lot, we include, as well as Bohr, two younger physicists, Heisenberg and Wolfgang Pauli, who not only made important contributions to quantum theory as it developed, but helped to lay down the rules for how it should be understood in the Copenhagen interpretation.)

First, it should be freely admitted that we have considered a particularly simple case where hidden variables work almost trivially. In less simple cases where we may, for example, need to consider possible measurements of different variables, working with hidden variables is, at the very least, much trickier. Two of those who did not wish to accept the Copenhagen synthesis, Einstein and Louis de Broglie, each constructed a hidden variable model soon after the creation of modern quantum theory, but each soon abandoned it, Einstein because he realized that his theory was flawed, de Broglie because his presentation was savaged by Pauli at the 1927 Solvay Conference.

Yet the refusal of the founding fathers even to contemplate the use of hidden variables still seems totally unjustified. In fact, opinions became even more hard-line when, in his famous book of 1932, von Neumann provided a 'proof' that quantum theory could not be supplemented with hidden variables. This proof was fine as a piece of mathematics, but the significance of the assumptions that von Neumann had made was not immediately obvious, and 30 years later Bell was able to show that one of the assumptions was quite inappropriate; the outcome was that hidden variable theories *were*, in fact, possible. This was far from the end of the matter, however; the hidden variable question will remain important through much of the rest of this book.

We will now pick up a topic closely related to that of hidden variables, that of the relation between the use of probability in physics theories

Louis de Broglie (1892–1987)

Fig. 4.3 Louis de Broglie [courtesy of Cambridge University Press]

Prince Louis-Victor de Broglie became intensely interested in physics in the early 1920s, and in 1924 he submitted his PhD thesis in which he trumped Einstein's suggestion that light had a particle-like as well as a wavelike nature, by arguing that electrons, and other entities usually thought of as particles, should also have a wavelike nature; he produced the famous expression for the de Broglie wavelength equal to Planck's constant, h, divided by the momentum of the particle. It was only as a result of the enthusiastic support of Einstein that the rather bemused examiners accepted the thesis, but within two years de Broglie's ideas had been confirmed by experiment, and had become an essential part of the new quantum theory and the direct source of Schrödinger's wave mechanics. The experiments were carried out by Clinton Davisson and Lester Germer in California, and George Thomson in Aberdeen; Davisson and Thomson shared the Nobel Prize for Physics in 1937.

De Broglie himself had been awarded the Nobel Prize for Physics in 1929, but apart from that his reputation scarcely prospered. Initially, like Einstein and Schrödinger, he was an opponent of Bohr's complementarity, and he put forward a hidden variable approach in which one had not wave *or* particle, but wave *and* particle. However, this suggestion was crushed by the supporters of Bohr at the 1927 Solvay Conference, in particular Wolfgang Pauli, a far superior mathematician to de Broglie; de Broglie himself had no option but to accept Bohr's approach until 1952. In that year David Bohm arrived at a similar theory to that of de Broglie 25 years before, and, in contrast to de Broglie, Bohm was able to defend his theory against Pauli.

Wolfgang Pauli (1900–58)

Fig. 4.4 Wolfgang Pauli [courtesy of CERN Pauli archive]

Pauli was an extremely brilliant physicist. With Bohr and Heisenberg, he was one of the three foremost advocates of complementarity, and he was well known as a destructive critic of those with whom he disagreed. His most famous contribution to quantum theory was his famous exclusion principle, which clarified the theory of atoms and our understanding of the Periodic Table by stressing that each of the electrons in the atom must have a different set of quantum numbers. For this contribution he was awarded the Nobel Prize for Physics in 1945.

In this chapter his importance is as the strong critic of all daring to challenge complementarity and to support such ideas as hidden variables—de Broglie, Bohm, even Einstein (though Pauli makes a brief appearance in *The Born–Einstein Letters* [1] as a peacemaker between Einstein and Born).

and the abandonment of determinism. In the nineteenth century, work on the theory of gases or *kinetic theory*, particularly that of Ludwig Boltzmann and James Clerk Maxwell, made much use of probabilities, for example probabilities of different energies for individual gas molecules, and at first sight it might seem that this implied that the theories were not deterministic. The implication seemed to be that there might be a probability of one thing, another probability of another thing and (perhaps) so on.

Yet this was not actually the case. The actual underlying assumption was that there was an atomic (or molecular) level at which everything was deterministic. Probabilities were introduced because it was not necessary, possible nor in any way desirable to study the behaviour of individual atoms or molecules. Rather their individual behaviour was averaged over, this producing probabilistic behaviour at the level of actual observations.

Much the same possibilities exist for quantum theory. We have said that Born's probability statement implies loss of determinism if but only if hidden variables are ruled out. If they are allowed, the probabilities are essentially statistical averages over these hidden variables, and indeed it is helpful to call such theories 'statistical' rather than 'probabilistic'.

In such theories as that of the Bohr atom, transitions between atomic energy levels are said to be random or 'probabilistic', and it is often implied that, when Einstein used this feature in a 1916 reconciliation of Bohr's work with that of Max Planck, he was necessarily sacrificing determinism, which he was later, of course, so keen to uphold. However, in both cases, we can insist on the idea of an underlying level of deterministic hidden variables dictating when the transitions occur, and, in Einstein's work, the direction in which the photon produced in the transition moves off. (For a little more detail see the historical boxes on Planck (Chapter 2), Bohr (Chapter 3), and Einstein (Chapter 1).)

References

1. M. Born (ed.), *The Born–Einstein Letters: Friendship, Politics and Physics in Uncertain Times* (London: Macmillan, 2005).
2. J. von Neumann, *The Mathematical Foundations of Quantum Mechanics* (Princeton, NJ: Princeton University Press, 1955; originally published in German in 1932).

5
Further complications

Measurement of other observables

In Part I of the book, though we introduced the concept of electron spin and its measurement in Chapter 2, and we will return to this important topic in the next chapter, we have concentrated so far on discussing the energy of the system. We have considered both its measurement and the fact that eigenfunctions of energy evolve in a simple way, just oscillating mathematically in time and gaining a largely irrelevant phase factor. Restriction to consideration of energy has enabled us to uncover and discuss most of the conceptual problems of quantum theory, and of course, in any case, energy is probably the most important quantity in physics.

Yet there are many other variables that we may wish to measure. For example, we can measure the position or the momentum of a particular particle. When we consider the hydrogen atom, as well as its energy, other very important variables include the total angular momentum and the components of this angular momentum; by 'components', we mean broadly how much of the angular momentum can be said to be about the various axes: x, y, and z.

Things can get quite complicated! Each observable, as we can call them, has its own set of eigenfunctions and corresponding eigenvalues. In some important cases, the eigenfunctions of this new observable may be the same as those of the energy, but in many other cases they are not. Just as for a measurement of energy, the result of measuring a particular observable will be a particular eigenvalue of that observable, and we are certain to obtain this result if the wave-function at the time of measurement is the corresponding eigenfunction. However, in the general case, the wave-function before the measurement will be a linear combination or superposition of a number of eigenfunctions, and we can obtain different results with various probabilities. In this book fortunately we will need to consider this procedure only for a very few special cases.

Let us start by thinking about measurements of position and energy, and for concreteness let us again think of our system as being in an SHO potential and in fact having the lowest energy, E_0 or just $\frac{1}{2}hf$. We have not discussed the shape of eigenfunctions so far. In this case we will say that ϕ_0, the corresponding ground state eigenfunction (Fig. 5.1), has its maximum at the bottom of the potential well. (For a pendulum, this would be at the place where the bob would be when the pendulum is stationary; we will call it $x = 0$.) As we move away on either side of this position, or as x increases or decreases from 0, ϕ_0 decreases steadily to 0, reaching 0 only in the limit as the distance away tends to ∞.

In this case, then, the wave-function ψ is equal to ϕ_0. We are in an eigenfunction of energy, so, of course, we know what value we will get in a measurement of energy. However, it is not certain what value we will get if we measure the position of the system; our answer will be probabilistic. And in fact the probabilities can be stated simply. Since x varies continuously, we must actually talk about a *probability density, P*, the probability of finding the particle in a given region per given range of x.

And the general answer is that P is equal to the square of ψ. This is actually the application of the Born rule that we met previously to the case of measurement of position. Thus, in the case we are considering, we are most likely to detect the particle at the bottom of the potential well, and the probability of detection at any position decreases steadily as we move away in either direction. We *may* find the particle any distance from the minimum of potential, but it is less and less likely as we move further away. And, of course, the mathematics is arranged so that the total probability of finding the particle anywhere is equal to 1, as it must be. For other energy eigenfunctions of the SHO, ϕ will be different, of course, and for different forms of potential we will get entirely different sets of energy eigenfunctions, but in all cases the probability density, P, will be equal to ϕ^2, if the system is in an eigenstate. And if it is *not* in an eigenstate, so ψ is not equal to one of the ϕ's, still P is equal to ψ^2.

An obvious question is 'what is the form of an eigenfunction of position?' this must be of the form that when we perform a measurement of position we are bound to obtain a particular answer—say x_0?

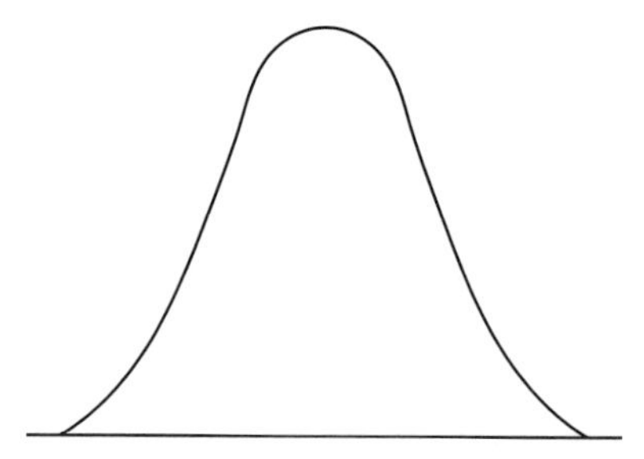

Fig. 5.1 A sketch of the ground state eigenfunction for the simple harmonic oscillator. The maximum value of the eigenfunction occurs at the bottom of the potential well, and the eigenfunction decays on each side. Technically it reaches 0 only at an infinite distance from the maximum.

The answer is also fairly obvious, at least, perhaps, once we are told! It is just a spike at x_0. If *all* the wave-function, so *all* the probability, is at x_0, a measurement of position *must* give us the answer x_0. The reader may suspect that handling this spike mathematically will be tricky, and this is certainly the case, but fortunately we do not need to concern ourselves with that in this book.

The next obvious question may be—what about a measurement of momentum? For the SHO ground state, ϕ_0, the situation is actually rather analogous to our answer concerning a measurement of position; we do not know what answer we will get. The highest probability is that the measurement of momentum gives an answer around 0, and, as the momentum, p, increases or decreases from 0 (where a negative value of momentum just means that the momentum is in the $-x$ direction), the probability of finding the momentum in a given range decreases steadily to 0.

However, while position and momentum behave similarly in this way, in other ways they behave totally differently, one might say in opposite ways. Later in this chapter, we shall call position and momentum *complementary* variables in a sense defined by Bohr. In particular, whereas an eigenfunction of position can be said to be completely localized in space, which, as we have said, makes good physical sense, an eigenfunction of momentum is completely *non-localized*. This means that, if we perform a measurement of position on a particle in an eigenstate of momentum, there is an equal probability of finding the particle in any range of x between $-\infty$ and ∞.

The actual form of an eigenfunction of momentum is a waveform with the wavelength, λ, given by h/p, where h is, as ever, Planck's constant and p is the momentum. This expression is the famous one produced by Louis de Broglie. (See the de Broglie box, Chapter 4.)

At first sight, the content of the two previous paragraphs seem contradictory. A wave has peaks and troughs, and even when we square the waveform to get the probability density, we might expect that the result will still have maxima and minima, the minima actually being zeros. How, then, can the probability density be constant as we said two paragraphs ago?

The answer is mathematical and rests on the fact that the eigenfunctions are mathematically *complex*. They each have two parts or components (one always called 'real' and the other 'imaginary', although the terns should not be taken too literally), and though each part has peaks and troughs, when they are put together mathematically, the sum of the squares is a constant, and this gives us the probability density. (Don't worry about this; we only mention it to *stop* you worrying about the perceived contradiction.)

So if a particle is in an eigenstate of momentum, the value that would be obtained in a measurement of momentum is totally determined, but we can say that the value that would be obtained in a measurement of position is totally *undetermined*. Similarly if it is an eigenstate of position, the value that would be obtained in a measurement of position is, of course, totally determined, but we now add the point that the value that would be obtained in a measurement of momentum is completely *undetermined.*

Wave and particle, and the Heisenberg principle

The preceding discussion brings us directly to some of the most celebrated aspects of quantum theory, particularly from the historical point of view. First let us consider, for simplicity, a free particle (though, as we shall see, we must be careful with the word 'particle') such as an electron. We can choose to observe this in various ways. First we can study a collision or interaction between the particle and another particle or the surface of a solid. This interaction will take place at a point or in a very small region. Essentially we are performing a measurement of position, our particle is localized, and indeed it is behaving just like a classical particle. There is no surprise here.

However, when we do not interact with the particle directly but allow it to propagate, through a slit for example, or through a periodic crystal, we are allowing it to remain non-localized and it propagates as a wave. For example, an electron propagating through a periodic crystalline potential is very much analogous to light travelling through a diffraction grating (a system of very many tiny slits), and we obtain a rather similar diffraction pattern, with maxima and minima.

We usually refer to an electron as a 'particle' because such systems behave as particles classically, but clearly it sometimes behaves as a particle, sometimes as a wave. This type of behaviour is often called *wave–particle duality*. It was de Broglie who first suggested that 'particles' might behave in this fashion, as well as giving the formula for wavelength already mentioned. (See again the de Broglie box, Chapter 4.)

Actually de Broglie's suggestion was an echo of the much earlier suggestion of Einstein that light, which is classically thought of as a wave, should also exhibit particle-like phenomena. (See Einstein box, Chapter 1.) Just like de Broglie's electron, Einstein's light interacts as a particle, which has been given the name of a *photon*, and Einstein particularly discussed its interaction in the *photoelectric effect*. Like de Broglie's particle, light propagates as a wave, and this is, of course, the behaviour with which we are very familiar.

We have met photons especially in connection with the light emitted from atoms and other systems with discrete energy levels, such as the simple harmonic oscillator or SHO. In Chapter 2, we discussed the energy levels in particular for the SHO and the hydrogen atom. An important question is how the system moves from one energy-level to another, and the simple answer is that it drops from a particular level to a lower one by emitting a photon. The energy of the photon is equal to the energy lost by the system, or in other words, the difference between the two energy levels, and the energy of the photon is just equal to its frequency multiplied by Planck's constant, h. Similarly, for the system to jump from one energy level to a higher one, it must absorb a photon, and, of course, the difference in the energy levels in this case will be equal to the energy of the photon absorbed.

For the SHO, the detailed rules of quantum theory tell us that the system can only make transitions between energy levels for which the quantum numbers n differ by 1. Because of the form of the energy levels given in Chapter 2, the energy of the photon emitted or absorbed will always be exactly hf, and the frequency of the photon is exactly f, the classical frequency. This is a very interesting result.

For atomic hydrogen, on the other hand, there are no restrictions on the values of n between which transitions can be made. This leads to a complicated structure in the spectrum of the photons emitted or absorbed, and this structure was very much the stimulus for Bohr's model of the hydrogen atom mentioned earlier.

Werner Heisenberg (1901–76)

Fig. 5.2 Werner Heisenberg [courtesy of Max Planck Institute at Berlin-Dahlem]

By the age of 22, Heisenberg had been recognized as the most brilliant protégé of Niels Bohr, and it was only a few years later in 1925 that he invented the first rigorous quantum theory, which became known as matrix mechanics. His approach in the development of the theory was to base the physics on observable quantities rather than models of physical systems. In the full working out of the theory he received considerable help from Max Born and Pascual Jordan, and rather strangely the development of Heisenberg's crucial ideas by these three men was also carried out practically simultaneously by a young unknown mathematician from Cambridge called Paul Dirac. (See Dirac box.)

Heisenberg became equally famous as the inventor in 1927 of what is most often called the Heisenberg uncertainty principle, though it is more properly called the *indeterminacy principle,* as explained in the text. Heisenberg was much influenced by Bohr in his final version of the principle.

His Nobel Prize for Physics was awarded in 1932 for the creation of quantum mechanics, but the citation stressed the discovery that molecular hydrogen exists in two different forms, one in which the spins of the nuclei are aligned, one in which they are opposed. Throughout his career Heisenberg also performed important work on nuclear and elementary particle physics, and in addition he pioneered the application of quantum theory to ferromagnetism.

Heisenberg rose rapidly through the ranks of German physicists, becoming Professor of Theoretical Physics at the University of Leipzig as early as 1927. However, like many of those whose biographies are told in the boxes in this chapter, his life and reputation were greatly affected by the coming of Hitler.

Heisenberg was a nationalist but he definitely never became a Nazi, and indeed his continued support for Einstein's physics into the age of Hitler left him vulnerable at the hands of the advocates of 'German physics' (in effect, 'Nazi physics'). Indeed he was severely attacked at one time as a 'White Jew', and it was only his family connection to Himmler that may have saved him from extreme peril.

Nevertheless he subsequently did his post-war reputation considerable harm by his ambiguous involvement in Hitler's atomic bomb project, and particularly for his infamous visit of 1941 to occupied Copenhagen, when it appeared, at least to Bohr, that he wanted to discuss the progress each side in the war had made towards the construction of the atomic bomb. The story has been dramatized by Michael Frayn in his celebrated play *Copenhagen* [1]. The standard biography of Heisenberg is by David Cassidy [2].

Pascual Jordan (1902–80)

Jordan was Max Born's assistant when Heisenberg produced his first account of modern quantum mechanics, and Jordan and Born constructed the full theory, Heisenberg also taking part once he had assimilated matrix algebra. Jordan later made major contributions to quantum field theory.

Although he had close contacts with many Jewish physicists, Jordan joined the Nazi Party in 1933, and also became a storm trooper, and during the war he worked on the development of rockets for Hitler. As a result, after the war he had to be declared 'rehabilitated'. Once he had achieved this, he regained his academic status and indeed he entered Parliament, where his advocacy of nuclear weapons caused fresh friction with his former physicist friends.

Fig. 5.3 Pascual Jordan [courtesy of Professor Jürgen Ehlers]

Paul Adrien Maurice Dirac (1902–84)

Fig. 5.4 Paul Dirac (right) pictured with Pauli in 1938 [courtesy of CERN Pauli archive]

Dirac was a research student in mathematics at Cambridge University when he came to know in 1925 of Heisenberg's early work in quantum theory. In an extremely short time he became an expert in the theory, developing the central ideas at the same time as Heisenberg, Born, and Jordan. He developed his own formalism and notation for quantum theory, which was an abstract generalization of the wave ideas of Schrödinger and the matrix ones of Heisenberg, and for the rest of his life he would be regarded as one of the outstanding physicists of the world.

His greatest achievement was his relativistic quantum theory of the hydrogen atom. (Schrödinger's theory had been non-relativistic.) There seemed for a few years to be a major blemish in Dirac's theory—it predicted negative-energy electron states as well as the expected states with positive energy. However, this was turned into a great triumph by Dirac's later idea that all the negative-energy states were normally occupied and unobservable, and that it was *holes* in this band of states that would be observable, and indeed were soon to be observed by Carl Anderson (Fig. 5.5). They are now known as positrons, particles with exactly the same nature and properties as the electron, except that they have a positive charge. This was the beginning of the idea of particle and anti-particle, matter and anti-matter, so central across the whole field of particle physics.

Dirac shared the Nobel Prize for Physics with Schrödinger in 1933, Heisenberg having been awarded the whole prize the year earlier. (Anderson was to share the 1936 prize.) Dirac became Lucasian Professor of Mathematics at Cambridge in 1932, a post he held until 1969. He married the sister of Eugene Wigner, who plays quite a large part in this book. There are many stories about Dirac's naivety or literal-mindedness, and his recent biography is called just *The Strangest Man* [3].

Fig. 5.5 Carl Anderson ready to observe cosmic rays at the mountain top [courtesy of World Scientific]

The properties of light are actually rather trickier to discuss than those of particles, which is why we discussed de Broglie's ideas before those of Einstein, and so we return to free particles to discuss the so-called Heisenberg Principle. So far we have looked at two extreme cases. In the first the position is totally determined before any measurement, the momentum totally undetermined; in the second case it is the momentum that is totally determined, the position totally undetermined.

It is natural to ask whether we might be able to have a compromise, in which both are partially determined, or determined within a given range. This would give us something that is closer to our idea of a classical particle, and it is quite possible to achieve this. We can construct such functions mathematically by adding together a suitable combination of functions with definite but different positions, or alternatively by adding together a suitable combination of functions with definite but different momenta.

For the function we produce, we do not *know* what value we will obtain when we measure the position. However, the highest probability is for a particular value of position, which we can call x_0, the probability decreases steadily as we move away from x_0, and it is highly likely that the value will be within a range we can call δx of x_0.

Similarly we do *know* what value we will obtain when we measure the momentum. However, the highest probability is for a particular value of momentum, which we can call p_0, the probability decreases steadily as we move away from p_0, and it is highly likely that the value will be within a range we can call δp of p_0.

The more accurately we wish to have x determined, or in other words the smaller we wish to have δx, the larger we are forced to make δp, or in other words the less accurately p is determined. Similarly the smaller we wish to have δp, the larger we are forced to make δx. The best we can achieve in any compromise is to have the product of δx and δp around $\hbar$. (To be more specific, we must have $\delta x \delta p \geq \hbar/2$.) As we have said, this tells us that if either one of δx or δp is zero, the other one must be infinite.

The rule given in the previous paragraph is the famous Heisenberg principle, produced by Heisenberg in 1928. Unfortunately it is very often called the Heisenberg *uncertainty* principle, and it is often said that the principle concerns *uncertainties* in position and momentum, the implication being that both position and momentum have exact values, but we can only *know* them within certain ranges. An equally common way of stating the principle is that we can only *measure* the quantities to certain precisions. Yet another variant is to say that 'the measurement disturbs the system', and indeed this is often thought of as the central characteristic of quantum theory; again, though, it clearly suggests that both position and momentum exist with exact values (or, in the general case, all observables have exact values), and it is measurement that causes any problems.

But, as we have tried to stress, and again it should be made clear that we are talking within the standard Copenhagen interpretation of quantum theory, all of these statements miss the point, which is that the quantities just do not have values to greater precision than the Heisenberg principle allows, and it is much better called the Heisenberg *indeterminacy principle*.

States of a spin-$\frac{1}{2}$ particle; photon polarization states

We have talked about measurement of different observables for a particle in a potential well—energy, position, momentum, and we mentioned the angular momentum and its various components. Many particles actually have two types of angular momentum with direct classical analogics. In its path around the sun, the earth obviously has angular momentum due to its motion in the orbit; we will call this *orbital angular momentum*. However, it also spins on its own axis, giving us, of course, night and day as it does so; the angular momentum associated with this motion is very naturally called *spin angular momentum*. In quantum mechanics, when one considers the motion of an electron round the nucleus, both types of angular momentum are exhibited. In this book, we do not need to refer to the quantum mechanical orbital angular momentum, but the quantum mechanical spin angular momentum is extremely important.

In Chapter 2, we said that the electron is a spin-$\frac{1}{2}$ particle; a measurement of the z-component of spin must yield a result of either $\frac{1}{2}\hbar$ or $-\frac{1}{2}\hbar$. Here we make the point that, unlike orbital angular momentum, which, like energy, position, and momentum, is related to the wave-function, the spin angular momentum is associated with an entirely different quantity, which we call the *state-vector*. Thus the most complete description of the state of an electron allowable in conventional quantum theory (which is to say excluding the possibility of hidden variables) constitutes a wave-function, *and* a state-vector for the spin.

The state-vector is actually rather simpler than the wave-function. It simply consists of two numbers. If it is $\begin{vmatrix}1\\0\end{vmatrix}$, then a measurement of the z-component of spin angular momentum, or more simply, the z-component of spin, or more simply still, just s_z, is certain to yield the result $\frac{1}{2}\hbar$. We can say that the state $\begin{vmatrix}1\\0\end{vmatrix}$ is just $|s_z=\frac{1}{2}\rangle$ or $|+_z\rangle$ or just $|+\rangle$, where the last notation assumes that the z-axis is being treated as fundamental, as it often, though not always, is in quantum theory. If, on the other hand, the state-vector is $\begin{vmatrix}0\\1\end{vmatrix}$, then a measurement of s_z is certain to yield the result $-\frac{1}{2}\hbar$, so we can refer to $\begin{vmatrix}0\\1\end{vmatrix}$ as $|s_z=-\frac{1}{2}\rangle$ or $|-_z\rangle$ or $|-\rangle$.

Those are the two simplest situations, where we can say that the spin is in an eigenstate of s_z, or that s_z has a precise value. But, exactly as for observables associated with the wave-function, we must also consider states where we have a linear combination of $|+\rangle$ and $|-\rangle$ or, to put it another way, of $\begin{vmatrix}1\\0\end{vmatrix}$ and $\begin{vmatrix}0\\1\end{vmatrix}$. Such a linear combination can be written as $c_+|+\rangle + c_-|-\rangle$ or as $\begin{vmatrix}c_+\\c_-\end{vmatrix}$, where, as before, $c_+^2 + c_-^2 = 1$. As before, we can then say that, if a measurement of s_z is performed on a spin in this state, then the probability of obtaining the result $\frac{1}{2}\hbar$ is c_+^2, and the probability of obtaining the result $-\frac{1}{2}\hbar$ is c_-^2, the probabilities adding to 1, as, of course, they must.

This is the point where things get a little bit more complicated! For, of course, we may wish to measure, not s_z but s_x. (Of course, to be more general we might wish to measure s_y or the component of spin in some other direction entirely.) We can say at the outset that if the spin is either in state $\begin{vmatrix}1\\0\end{vmatrix}$ or in state $\begin{vmatrix}0\\1\end{vmatrix}$, states for which the result of a measurement of s_z is certain, the result of a measurement of s_x is completely uncertain; there is a probability of $\frac{1}{2}$ of getting the result $\frac{1}{2}\hbar$, and also a probability of $\frac{1}{2}$ of getting the result $-\frac{1}{2}\hbar$. (Note, though, that the range of possible values obtained by a measurement of s_z or s_x or the measurement of any other component of spin is always the same—$+\frac{1}{2}\hbar$ or $-\frac{1}{2}\hbar$.

We can, in a sense, invert the question, and ask whether we can find states for which the result of a measurement of s_x is certain. We can, and we call these states $|s_x = \frac{1}{2}\rangle$ or just $|+_x\rangle$, and $|s_x = -\frac{1}{2}\rangle$ or just $|-_x\rangle$. It is easy to express such states as examples of the general linear combinations we have just been writing. In fact $|+_x\rangle$ is actually $\begin{vmatrix}1/\sqrt{2}\\1/\sqrt{2}\end{vmatrix}$, while $|-_x\rangle$ is $\begin{vmatrix}1/\sqrt{2}\\-1/\sqrt{2}\end{vmatrix}$. We can deduce that in either case we are sure what result we will get if we measure s_x, but there are equal probabilities of $\frac{1}{2}$ of obtaining either value of s_z.

The content of this paragraph is really analogous to that of the previous paragraph, with s_x and s_z interchanged, but overall we must realize the intricate nature of quantum theory as applied to the components of spin angular momentum. Many of the conceptual puzzles of the theory can be related to kinds of issue outlined in the past few paragraphs, but on the brighter side, so can many of the startling applications of quantum theory that have come to the fore in the past few years.

Actually, although the original theoretical discussions and suggestions were mostly based on the spin-$\frac{1}{2}$ system, probably most of the experimental work has used instead another system that has precisely the same mathematical basis. This is the polarization of the photon. It is well known classically that light has a direction of polarization. Light is an electromagnetic wave, and consists of time-dependent electric and magnetic fields that are perpendicular to each other, and that are both perpendicular to the direction of propagation of the wave itself. The direction of polarization of the light wave is defined to be that of the electric field.

We can say that the photon is the particle related to the light wave, so it is not surprising that, like the wave, it too has a polarization. Just as the two basic states of the electron spin are $|+\rangle$ and $|-\rangle$ relative to a particular axis, usually the z-axis, so the two basis states of the polarization of the photon are defined as vertical or $|v\rangle$ and horizontal or $|h\rangle$, again with respect to a particular axis. Then the mathematics of the two types of system is exactly the same.

References

1. M. Frayn, *Copenhagen* (London: Methuen, 1988).
2. D. Cassidy, *Uncertainty: The Life and Science of Werner Heisenberg* (New York: Freeman, 1992).
3. G. Farmelo, *The Strangest Man: The Hidden Life of Paul Dirac, Quantum Genius* (London: Faber, 2009).

6
Orthodox and non-orthodox interpretations of quantum theory

As well as studying the mathematical basis of quantum theory, we also need to 'interpret' it—to relate the mathematics of the theory to our own physical experience, to struggle to understand how a world seemingly made up of Newtonian objects behaving classically can be created from the probabilities of quantum theory. How, for example, can one understand the theory telling us that, if one or other of momentum and position has a precise value, the other is totally indeterminate?

Historically by far the most significant interpretation has been the Copenhagen interpretation of Bohr. This was arrived at in 1928, soon after the coming of modern quantum theory. It can be said that the Copenhagen interpretation was the specific application to physics of the more general set of ideas known as complementarity. For 40 years or so, it was so much taken for granted that Bohr had achieved a complete and final solution of the conceptual complexities of quantum theory, that even to question this was to render oneself unemployable as a physicist. Fortunately over the past 40 years, the climate has become much more tolerant, mainly as a result of the excellent science but also of the excellent strategy of John Bell.

We have stressed that one of the main components of the Copenhagen interpretation was that quantum theory was complete—there could be no hidden variables. An equally important component was that we should not talk or think of the particle at the quantum level. Rather we should restrict ourselves to talking of the results of experiments. Thus we might measure the momentum of an electron and obtain a particular result. Alternatively we might measure its position and obtain a result. But it is impossible to set up an apparatus that might measure both momentum and position simultaneously, and so we shall never be put in the position of discussing both, and wondering why they cannot both have exact values at the same time. Thus it is fairly clear how the Copenhagen interpretation avoids the conceptual problems of quantum theory; whether it goes further than avoiding them and actually solves them is another matter.

Another aspect of complementarity is its rule that we can use *either* a particle-like description of physical phenomena *or* a wavelike description. We cannot include elements of both descriptions in any particular account of a physics experiment.

From the perspective of 80 years or so, we may be inclined to suggest that these rules are a very good way of making sure we cannot discuss the awkward problems that quantum theory throws up, indeed that we should not even worry about them, but it is less clear that they actually explain *why* we are justified in avoiding them.

In the remainder of this chapter, we briefly discuss two of the interpretations that *were* suggested in the 1950s in opposition to Copenhagen. It goes without saying that both were subject to intense criticism by supporters of complementarity.

David Bohm (1901–76)

Fig. 6.1 David Bohm as a young man [from Ref. [2]; courtesy of Addison-Wesley]

Bohm was a man of tremendous ability, whose intellectual interests totally transcended physics. During the early 1940s, he worked for his PhD under Robert Oppenheimer at the University of California at Berkeley, and also became involved in left-wing politics. During the war he performed extremely important work for the Manhattan Project, which was making the atomic bomb; Bohm made major contributions to the physics of plasmas, materials where all the atoms have been ionized because of high temperature or low pressure. This would have been of immense use to the United States in their quest for the hydrogen bomb in the 1950s.

However, in 1949, while he was working in Princeton University, Bohm was called to testify before the House Un-American Activities Committee under Joseph McCarthy, which had the task of denouncing communists. Bohm refused to testify, pleading the Fifth Amendment. Princeton refused to re-instate him, and he went first to São Paulo in Brazil, then to Haifa in Israel, then to Bristol; finally from 1961 until his death he was Professor of Theoretical Physics at Birkbeck College London.

It was in Brazil that he produced his hidden variable interpretation of quantum theory. Einstein, with whom he had been close in Princeton, had encouraged him to explore alternative approaches to the Copenhagen interpretation, but he was dismissive of Bohm's actual results; by this stage Einstein favoured far grander mathematical approaches.

In later life, Bohm's interests and publications were in diverse areas that would usually be classified as philosophy, neuropsychology, the problems of society, biology, linguistics, and the theory of art. His most famous book is *Wholeness and the Implicate Order* [1], while his biography has been written by F. David Peat [2].

The first was produced by David Bohm in 1952. He had unwittingly rediscovered de Broglie's theory of 1927, though Bohm was able to present the ideas much more effectively. The theory was one of hidden variables, in that the particle had a position at all times. While in the Copenhagen interpretation we consider, for example, an electron as behaving *either* as a particle *or* as a wave at any particular moment, in the Bohm theory it always has an explicit particle-like nature *and* an explicit wavelike nature. The theory is often called a *pilot-wave theory*, in that the wave dictates the motion of the particle. Bohm's theory was not only a realist one, it was also deterministic, so in several ways it challenged Copenhagen. More than half a century later, Bohm's theory has by no means become accepted by a majority of physicists, but it is extremely popular with some, and remains the subject of considerable further research.

The other interpretation to be mentioned here is what is most often called the *many worlds* or *many universes* interpretation. The usual way of explaining this is to say that, while, if we accept the von Neumann projection postulate, we believe that, following a measurement, just one of the possible results is realized, *many worlds* would say that all results are achieved, each in its own world; in other words, world splitting takes place at every measurement. Clearly the conceptual implications are staggering! Neither is it clear that the interpretation, at least in its simplest terms, achieves what it hopes to. For just as one of the main problems of von Neumann's interpretation is that, because a 'measurement' cannot be defined in fundamental terms, it is not clear when collapse should occur, exactly the same applies to world splitting.

Hugh Everett (1930–82)

Fig. 6.2 Hugh Everett [from Ref. [4] courtesy of Peter Byrne]

Everett was a polymath; even before he encountered quantum theory, he had graduated in chemical engineering and worked on game theory. His invention of the 'relative state' formulation of quantum theory was performed for a dissertation, which was written in 1955 under the direction of the famous physicist John Wheeler.

Discouraged by the negative response of other physicists to his idea, Everett left physics and became a specialist in operations research. He started in weapon evaluation, particularly studying the effects of nuclear weapons and nuclear fallout, and to this end he invented the mathematical technique of generalizing the use of the Lagrange multiplier for maximization. Subsequently he left military employment and applied the same methods to civilian research, founding several companies and became a multi-millionaire.

Only in the 1970s did be show a renewed interest in his interpretation of quantum theory, when his ideas were taken up by Bryce DeWitt.

Everett's thesis and a number of related papers by other physicists appear in a collection edited by DeWitt and R. Neill Graham [3], while Peter Byrne has recently published a biography [4].

In fact, other examples of the same general approach are much less bold philosophically. The person who began it all was Hugh Everett in 1957, with his *relative state* formulation of quantum theory, which he also called his theory of the *universal wave-function*. The central point of all examples of the approach is that the whole wave-function survives a measurement, rather than just one component of it, and that, in each branch of the wave-function, however the word 'branch' may be interpreted, there is a correlation between the particular state of the system and the corresponding state of the observer. These can be called 'relative states'. In recent years, many attempts have been made to provide clearer understanding of such interpretations.

References

1. D. Bohm, *Wholeness and the Implicate Order* (London: Routledge, 1983).
2. F. D. Peat: *Infinite Potential: The Life and Times of David Bohm* (Reading, MA: Addison Wesley, 1997).
3. B. DeWitt and R. Neill Graham (eds), *The Many-Worlds Interpretation of Quantum Mechanics* (Princeton: Princeton University Press, 1973).
4. P. Byrne, *The Many Worlds of High Everett III: Multiple Universes, Mutually Assured Destruction and the Meltdown of a Nuclear Family* (Oxford: Oxford University Press, 2010).

PART II
THE FOUNDATIONS OF QUANTUM THEORY

7
Entanglement

Bohr, Einstein, and complementarity

In Part I we met most of the conceptual problems associated with quantum theory (at least, as we have said many times, provided the theory is considered to be complete, in the sense that there is nothing in the physical universe that is not present in the wave-function—in particular no hidden variables). The central problems we met there were violation of determinism and also loss of realism. Violation of determinism means that we may know all there is to be known about a system at a certain time, but we still cannot predict its future behaviour; in particular in most cases we will not know what result we will get when we perform a measurement of a particular physical quantity. Loss of realism means that we must accept that physical variables often will not have specific values unless and until we measure them.

We have also met the idea of complementarity, a fairly general set of ideas, which, when applied to quantum theory, is called the Copenhagen interpretation of the theory. It says that quantum theory is indeed complete—it does not allow hidden variables. Complementarity avoids many of the seemingly paradoxical aspects of quantum theory by restricting discussion of any particular observable to a situation where an apparatus is actually present to measure it; in effect we can only discuss the results of measurements, not the values of observables that microscopic objects can have in the absence of measurement. Since it is impossible to measure momentum and position simultaneously, complementarity tells us that we must not discuss the two quantities simultaneously. Similarly it tells us that we can consider light or an electron *either* as a wave *or* as a particle, but not both together at the same time.

Of the characters we met in Part I, the main proponents of complementarity were Niels Bohr, who launched it in 1928, Werner Heisenberg, Wolfgang Pauli, and Max Born. We saw that in the 1950s David Bohm and Hugh Everett set up alternative interpretations in opposition to Copenhagen, while the most important opponents of Copenhagen for its first thirty-five years were Erwin Schrödinger, and, of course, in particular Albert Einstein.

We will meet Schrödinger's views later in this chapter, but for now we concentrate on Einstein. He is probably most famous for two things; his announcement of $E=mc^2$, which is one of the central equations of special relativity, and his statement that 'God does not play dice'. The latter is nearly always taken to imply that Einstein's strongest reason for disliking the Copenhagen interpretation was its rejection of determinism, or in other words its introduction of intrinsic probabilities. Actually this was his initial reaction, but it is certain that by the 1930s his views had altered, and his greatest objection to the Copenhagen interpretation by then was that it ruled out realism rather than determinism.

In fact, Einstein did not believe that quantum theory was complete. He did not accept that, for example, just because the Schrödinger equation did not provide exact values for both the position and momentum of a particle, one was forced to believe that they did not actually have exact values. It should be mentioned that, as we have implied, rejection of the completeness of quantum theory would *usually* imply a belief in the existence of hidden variables; there would be extra variables over and above those provided by quantum theory, and these might provide, for example, values for both position and momentum.

However, we also saw that very soon after the coming of quantum theory in 1925–6, Einstein had suggested a model for hidden variables, but he soon became convinced that it led to unacceptable consequences, and subsequently gave up such simple ideas. Indeed although Einstein had encouraged David Bohm to work out alternatives to the Copenhagen interpretation, when Bohm produced his hidden variable interpretation, which was discussed in Chapter 6, Einstein dismissed it, in a letter to Born, published in the Born-Einstein letters, as 'too cheap'.

His own ideas were much more grandiose. Einstein's most celebrated achievement in physics had been his work on relativity, his special theory of 1905 in which gravitation was not included, and his mathematically much more complex general theory of 1916 in which he triumphantly succeeded in including gravitation. His self-appointed task for most of the remaining thirty-five years of his life was to use the general theory of relativity as the basis of work that would be much more mathematically elaborate still. His intention was to include electromagnetism as well, so that the resulting theory would be a so-called unified field theory, in that it would unify the fields of gravitation and electromagnetism. Einstein's belief—or at least sincere hope—was that once this theory was produced, it would give, as an exceptionally good approximation, all the results we obtain at the moment from quantum theory, but he also hoped that the theory would somehow circumvent the conceptual problems of quantum theory that we outlined at the beginning of this chapter.

However, though Einstein produced some interesting mathematics as a by-product of this quest, he never made any genuine progress towards obtaining the kind of theory he hoped for. Today, in fact, we would recognize that there are actually four fundamental fields in physics, not just the two that Einstein was attempting to unify. The others are the strong and weak nuclear interactions.

Until the 1960s, the strong nuclear interaction would have been described as the force between nucleons (protons and neutrons) that binds them together to form nuclei, this force being provided by the exchange of π-mesons between the nucleons.

In the 1960s, however, it became accepted that protons, neutrons, and π-mesons themselves were not elementary particles, but were built up of so-called quarks and anti-quarks. (To be exact, a proton and a neutron consist of three quarks, while a π-meson consists of a quark and an anti-quark.) In this new regime, we must change the definition of the strong interaction, and we now define it as the force between quarks or between a quark and an anti-quark that binds them together to form the nucleons and the π-mesons. This force is provided by the exchange of particles called gluons between the quarks.

Our idea of the weak nuclear interaction has remained unchanged; it is too weak to bind particles of any type together, and in fact it is the interaction that causes radioactive β-decay.

The necessity for assuming the existence of both nuclear interactions was clear from at least the 1930s, so it is rather surprising that Einstein restricted his efforts to the two classical forces. Also it can be remarked that in recent years a measure of unification of interactions has been achieved. The electromagnetic and weak nuclear interactions have been very satisfactorily unified as the electroweak interaction by Abdus Salam and Steven Weinberg in 1967, though by methods totally different from those of Einstein, and there have been many attempts to include the strong nuclear interaction with the electroweak interaction, although these have been less successful. Even to attempt to include gravitation has been much more difficult. Thus it seems that Einstein tackled a very difficult problem, and moreover tackled it from the wrong end! It is scarcely surprising that he failed.

However, if we call this work of Einstein his *positive* efforts at solving the problem, what can be called his *negative* attempt to disprove complementarity were very much simpler. They actually took place in discussions in the late 1920s and early 1930s with Bohr, who could be termed the High Priest of complementarity, at the Solvay Conferences, which were regular meetings of the leading physicists from all over the world, particularly those involved in the creation and discussion of quantum theory. We shall sketch two important attempts by Einstein to show that complementarity must be wrong.

It should be said that the standard account of these discussions was written by Bohr and published in a book containing comments on Einstein's theories by a considerable number of eminent scientists and philosophers [1]. Bohr's contribution is an account of his discussions with

Einstein on quantum theory held over several decades. While there is not the slightest doubt that Bohr presented this account perfectly honestly, it must still be possible that what he presents does not always give an accurate account of Einstein's views, and for his account of the second of Einstein's attempts it seems that this has happened, as we shall explain.

In Einstein's first attempt, made at the 1927 Solvay Conference, he tackled the claim of complementarity that an object could not simultaneously have exact values of momentum, p, and of position, x. The product of the uncertainties, or better indeterminacies, of each, $\delta p \delta x$, it says, must be at least around h, Planck's constant. Einstein aimed to prove this wrong by demonstrating that an apparatus could be devised that could actually measure both quantities simultaneously to better than the degree of accuracy that complementarity allowed. We shall also discuss a different form of the Heisenberg principle, often called the time-energy uncertainty principle. This states that $\delta E \delta t$, like $\delta p \delta x$, must be at least about h. Here δE is broadly analogous to δx or δp, being an indeterminacy in the energy of the system. However, δt cannot be described as an indeterminacy in t; it is better to think of it as the length of time available for a particular process to take place.

Einstein's idea is shown in Figure 7.1. A stream of particles, electrons or photons say, is directed at the slit in A, and, acting as a wave, diffracts into a beam passing towards screen B. Also present is a shutter C over the slit, so that the slit is only open for a limited period. So in Figure 7.1, δx is the width of the slit, and δt is the time for which the slit is open. The different forms of the Heisenberg principle then tell us that there are non-zero values of δp and δE.

So far there is no controversy. Einstein, though, now recognized that the uncertainty in the momentum, δp, of the particle results from an exchange of momentum between the particle and the screen, and the uncertainty in its energy, δE, results from an exchange of energy with the shutter. He now had what seemed a great idea: why not measure the

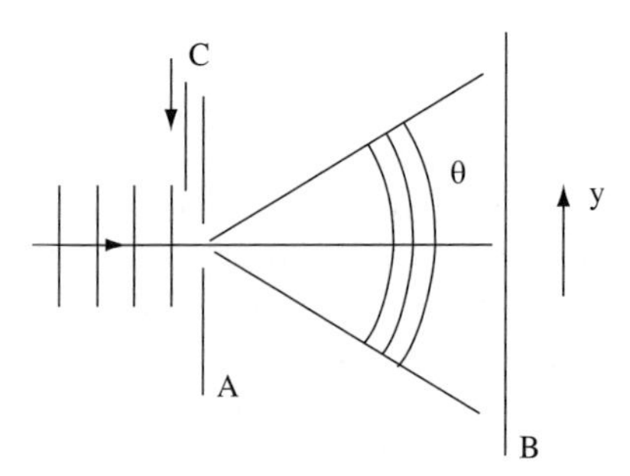

Fig. 7.1 Einstein's slit experiment. A beam of electrons or photons is incident on a slit. Following diffraction the beam is mostly restricted to an angular width θ as it travels to plate B. A shutter C is used to open the slit for a limited period T, so the train of waves passing towards plate B is thus of limited extent.

momentum of the screen and the energy of the shutter by studying the subsequent behaviour of the screen and the shutter themselves? In this way, he suggested, one could calculate exact values for the momentum and energy of the particle, so that δp and δE are zero. Since δx and δt are finite (i.e., not infinite), $\delta p \delta x$ and $\delta E \delta t$ are also zero, in flagrant violation of Heisenberg. (Note that this was a thought-experiment or gedanken-experiment of Einstein in which the particles effectively enter the system one-by-one; a gedanken-experiment is one that cannot be performed, at least with the experimental techniques available at the time, but consideration of which can help our understanding of the fundamental issues involved.)

Bohr, however, was able to show that Einstein's analysis ignored the fact that the shutter and screen themselves must be subject to the Heisenberg principle. Once that is taken into account, elementary calculations make it clear that monitoring their motion cannot provide any useful information about their collisions with the particle itself. Einstein was forced to admit that Bohr had certainly won this particular argument.

At the next Solvay Conference, Einstein came up with a yet more cunning scheme, sketched in Figure 7.2. In this scheme, a box B has a hole in its side that can be held open for a certain period of time, measured by clock C, and then closed by shutter S, during which it can be ensured that a single photon enters it. The box can be weighed before and after this period. The photon has a particular energy, and so, from the equation $E = mc^2$, it has a certain effective mass, which contributes to the weight of the box. So, from the difference between the initial and final mass of the box, the exact effective mass of the photon, and thus its exact energy, can be calculated.

This is the point where we should distinguish between two accounts of the situation. The first is that provided by Bohr. In this account, Einstein says that δE is zero, and δt is just the period for which the hole is left open and so is finite, and so $\delta E \delta t$ is zero. Again it seems that the Heisenberg principle has been violated. There is no doubt that this was what Bohr took as Einstein's argument and it caused him considerable anguish.

However, after much thought Bohr was again able to answer the argument. Figure 7.2 is along the lines of the diagram he himself drew some years later, and in this diagram he made it quite clear that 'weighing' should be regarded as a physical process rather than just an abstract term. Thus he considered the weighing being performed by a spring balance, with the results being given by the position of pointer P on scale A. This position will have its own uncertainty, related by the Heisenberg principle to the uncertainty in the measurement of the momentum of the box.

The key and most surprising element in Bohr's argument was that he used the ideas of general relativity to help to confound Einstein (or at least what he took to be Einstein's argument). According to general

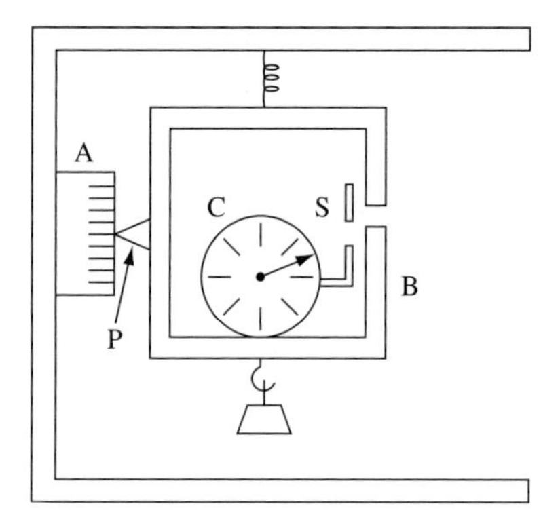

Fig. 7.2 Bohr's analysis of Einstein's photon-box experiment. A clock is used to open the shutter S for a period T. The weighing by the spring balance of box B and its contents is examined explicitly in terms of the position of pointer P on scale A. [Based on a figure in Ref. [1]]

relativity, a clock that moves in the direction of a gravitational field changes its rate. Thus the uncertainty in the position of the pointer gives rise to an uncertainty in the time interval for which the slit is open. When the sums are all done, it turns out that the product of $\delta E \delta t$ is at least of order h, as complementarity would demand.

The standard account says that Einstein was clearly shaken by these rebuttals by Bohr of his arguments. Clearly Bohr had demonstrated the truth of what was a part of the Copenhagen synthesis. It was impossible to measure accurately and simultaneously momentum and position, or energy and time. Completely different and mutually distinct experimental arrangements were required for each measurement—Einstein's suggestions can be seen to be unsuccessful attempts to perform what was impossible.

At the time, and indeed for several decades afterwards, the encounter was seen by nearly all as practically a direct proof of the correctness of Bohr's ideas on complementarity. Bohr said you should not discuss momentum and position simultaneously; the Einstein-Bohr dialogues that we have just discussed showed that momentum and position could not be measured exactly because the two measurements would require completely different and incompatible experimental arrangements. At first sight there seemed a perfect conceptual match.

This was actually illusory. Just as adequate an explanation would be that momentum and position *could* have simultaneous values, but the measurement of position disturbed the system so that it no longer had an exact value of momentum, and vice versa. While quite possible, it would seem extremely difficult to confirm this idea, as it seemed obvious that one could measure either momentum or position but not both.

And it is this idea that the alternative history of the photon-box thought-experiment claims to have been Einstein's actual intention. This account is centred on a letter from Paul Ehrenfest to Bohr in which Ehrenfest tried to explain Einstein's views. Ehrenfest (Figure 7.3) was an excellent physicist who was very close to both Bohr and Einstein, and he often acted effectively as an intermediary between them. This alternative account has been discussed in particular by Don Howard [2], and Louisa Gilder has described the idea very nicely [3].

In this account the photon travels a very long distance and is then reflected, returning after a considerable time to the experimenter at the photon box. The crucial point is that, while the photon is a very long way from the experimenter, we can take a reading *either* of the time at which the photon had been emitted *or* of the difference in energy of the box before and after the emission of the photon (not of course both, as we explained above). If we choose to measure the time, that will enable us to predict exactly the time at which the photon will return, but we will not know its energy. On the other hand if we choose to measure the difference in energy of the box, we will know exactly the energy of the photon when it returns, but will not be able to predict the time when it will return. In other words, depending on which measurement we choose to perform at the photon box, either the energy of the photon or the time of return becomes predictable; we can say that the particular value exists.

But this seems a very strange situation. A measurement at the photon box itself appears to cause an instantaneous change in the quantum state of the photon, despite the fact that, at the time of the measurement, the

Fig. 7.3 Paul Ehrenfest [from N. Blaedel, *Harmony and* Unity (Madison, WI: Science Tech Publishers, 1988)]

photon is an enormous distance from the box. Surely, Einstein suggested, this cannot be the case. If a measurement of the time of emission means that the time of return is determined, while a measurement of the energy difference means that its energy is determined, but neither can actually affect the photon, then surely both must have been determined all the time, before any measurement, and even if no measurement is actually performed. This is the type of argument at the heart of the famous EPR paper, which was not published until 1935. It is indeed interesting to speculate that it was at the heart of the photon box idea five years earlier.

The EPR paper itself was written by Einstein with his younger colleagues, Boris Podolsky and Nathan Rosen [4]. It was this paper that brought in the concept of *entanglement*, though the name would itself would not be suggested by Erwin Schrödinger until later in the same year. Entanglement would be at the heart of the New Quantum Age, though its significance would not be accepted for at least a quarter of a century after 1935. The concept of entanglement itself required an understanding of the idea of *locality*, which we discuss in the following section.

Locality

In Part I, we met two requirements of classical physics that are thrown into question by quantum theory—determinism and realism. Discussion of entanglement brings up a third requirement, which again will be questioned a good deal in this book—*locality*.

The principle of locality says that an *event* at a particular point, which we can call a *cause*, can only have an immediate *effect* at the same point in space. For example, if I hit a beaker with a hammer, we will not be in the least surprised if the beaker immediately shatters. Cause, which is the strike of the hammer, and effect, which is the shattering of the beaker, are at the same point and they are simultaneous. We would, though, be very surprised if striking a hammer had an *immediate* effect across the room or across the universe.

Where cause and effect are at different places, the requirement of locality says that they cannot be simultaneous. Rather some kind of signal must pass from cause to effect, and of course it will take a certain time for it to travel this distance. For example, we may throw a tennis ball, which may travel to a window and break it. The cause is our act of throwing and the effect is the breaking of the window. The signal is the passage of the tennis ball. Indeed we may think the word 'signal' is a little too elevated, since it rather conveys the idea of a passage of information. We may feel inclined to say that the ball does not carry information that causes the window to break; rather it just breaks the window!

But we now consider a couple of more scientific examples. A radioactive nucleus can give off an alpha-particle that travels to a Geiger counter where it produces a click. A heated atom can radiate a photon at a particular frequency; the photon can travel to a detector where not only its presence, but, in principle at least, its frequency can be detected.

In each case the travelling particle—tennis ball, alpha-particle, photon—clearly takes a certain time to travel from cause to effect, and this time can easily be calculated as the distance apart of cause and effect, divided by the speed of the travelling particle. Now the tennis ball will travel quite slowly, at least in comparison with the speed of light; the alpha-particle will travel much faster, but still slower than light, while the photon, of course, being light, travels at the actual speed of light, c, which is, as special relativity tells us, an absolute maximum for the passage of any signal.

Thus we generalize the idea of locality as follows: if locality is obeyed, a pair of events can only be causally connected, that is to say they can only be cause and effect, if the distance between them is less than or equal to the product of c, the speed of light, and the time between the two events. Then it is possible for a signal to go from one event to the other.

We have already mentioned special relativity, and this theory also enables us to discuss locality in another way. The basis of special relativity is to examine events as they appear to observers travelling with respect to each other, for example one observer may be stationary on a railway platform, the other a passenger on board a train moving through the station at high speed. In this circumstance it is quite obvious that the distance between two events may be different for the two observers. For example, the passenger may sneeze, and then a little later sneeze again. She will naturally describe these two sneezes at taking place at the same position, and say that the distance between the two events is zero. However, for the observer on the platform, one sneeze may take place when the train is directly opposite him as he stands on the platform, but the second may take place yards or even miles up the track; he will naturally say that there is a large distance between the two events.

That is obvious, and we certainly do not need relativity to understand it. However, special relativity also tells us—and this is much more difficult to understand—that the time that elapses between the two sneezes is also different for the two observers. The watch of the observer on the train who is actually doing the sneezing will show a certain time between the two sneezes; the watch of the observer on the platform will show a slightly higher time difference. Of course, since the speed of the train must be much less than c, the difference between the elapsed times for the two observers is extremely small, much too small to be observed. In cases where the relative speed is much higher, though, in fact where it is a substantial fraction of c, the difference between the two times would be much larger.

To explain the next bit of the argument, I'll need to introduce the term 'frame of reference'. In terms of our previous example, we may say that the passenger measures distances and times with respect to one frame of reference, essentially an origin and a set of axes for measuring distances, which are attached to the train, and her own clock. The man on the platform similarly has his own frame of reference—his own set of axes, which are attached to the station, and his own clock. And we can consider other frames of reference, all moving with respect to the platform with a particular speed, or, we could just as easily say, moving with respect to the passenger in the train with a related speed.

Now let us consider a pair of events. It is possible that they occur at the same time and the same place, like the hammer breaking the cup earlier in this section, and this type requires no further discussion. All other pairs come in one of two categories. For the first category, it is possible to find a frame of reference in which things are simple: the events happen at the same *time* but in different *places*. In other frames the events occur at different places *and* different times, but concentrating on this special frame, we say that this pair of events is separated by a *spacelike interval*. Incidentally, in different frames the time difference between the two events is itself different, and what is perhaps more worrying, in some frames what we can call event A may occur before event B, while in other frames, the order is reversed: event B comes before event A. We shall return to this seeming difficulty shortly.

First, though, we look at the second category of pairs of events. For this category, we can find a frame of reference in which the events occur at the same *place* but at different *times*, one after the other; let us say that in this frame event A comes before event B. In other frames, the events occur at different times and at different places. In these other frames, the time difference between the two events will itself differ, but an extremely important point is that in all these frames event A comes before event B. Because there is one frame in which only time is different for the two events, we say that this pair of events is separated by a *timelike interval*.

Now let us come back to study of cause and effect. The crucial point is that two events can only be related as cause and effect if they constitute a timelike interval. There is one frame in which they take place one after the other at the same place; clearly it is quite reasonable that one may cause the other; we can say that event A is the cause, event B the effect. In other frames, as we have said, they occur at different places, but event A, which we have called the cause, must always occur before event B, the effect. For example, the cause may be a person setting an alarm-clock, and the effect may be it ringing. Whatever frame of reference we may observe this in, the setting of the clock always occurs before it rings. It would certainly seem strange if the clock rings and then a little later we saw the owner setting it to go off at that (earlier) time!

However, two events separated by spacelike intervals cannot be a cause and an effect. Indeed there is one frame in which the events take place separated in space but at the same time; clearly one cannot be the cause of the other. In other frames they do not occur at the same time, and, as we have seen, the order in time of the two events will be different in different frames. This would be difficult to come to terms with if the events *were* cause and effect. In one frame, for example, two people might raise their hands at exactly the same time. There will be other frames in which person A raises the hand before person B, and still other frames in which person B is first, but, because there is no suggestion that one person raising the hand is the cause of the other person doing the same, there is no conceptual problem in the fact that that the order differs for different observers. Such intervals cannot be cause and effect because a signal at speed c does not have time to travel from one event to the other.

So we now come back to our primary task at the moment: to define locality. In terms of our recent remarks this is straightforward: if events are causally connected the interval between them must be timelike. This means that, in any frame of reference, the distance between the places where the two events occur must be less than or equal to the product of c and the time between the events. In turn this means that there is time for a signal to pass from one event to the other. If the distance is less than c times the time, the signal may travel at less than c, but if the two are equal, it must travel actually at c, and, of course, it naturally will do so if it *is* light.

So far our discussion has been mostly classical. Indeed we can say that it has been more appropriate to a pre-Newtonian situation than to a Newtonian or relativistic one, in that we have been talking about individual causes and effects. In Newtonian physics, of course, the whole of the Universe evolves deterministically, and, when we correct for relativity, the laws themselves imply the restrictions due to locality that we have already described.

Actually, even in deterministic accounts of physics, there is one aspect that that is not treated deterministically—this is the behaviour of the experimenter. It is naturally assumed that the actions of the experimenter, unlike other events, are not determined by what has gone on before. Rather they are always treated in physics as results of *free will*. Thus any action of an experimenter must be regarded as a cause, and, to obey locality, any resultant effect must be separated from that action by a timelike interval; in other words a ray of light must have time to get from the experimenter's action to the effect.

This aspect of locality goes through to quantum theory. An experimenter may, perhaps, change the direction of a magnetic field at a particular point, and this may affect the direction of the spins of electrons at another point. But locality tells us that the effect cannot occur until light has had time to move from the magnetic field to the electrons. Abner Shimony [5] called this aspect of locality *parameter independence*. It is clear that breakdown of parameter independence, by which one means that the spin directions would change earlier, would constitute a breakdown of relativity. Suppose, for instance, that the change took place immediately. Then the experimenter would be sending a signal at an infinite speed, quite contrary to relativity, which decrees that a signal cannot travel faster than c.

However, in quantum theory, there is an additional aspect of locality. We have seen that, in most cases, the result of an experiment is probabilistic. We may, perhaps, get one of two possible results. Now let us imagine that two related measurements of this type are performed some distance apart. If the results are correlated—positively or negatively—that may be a sign of a breakdown of locality. (We shall see that entanglement will give rise to such situations.) Shimony gave the name of *outcome independence* to this aspect of locality. It is important to recognize that outcome independence does not allow a signal to be sent from the position of one seat of measurement to that of the other, because the result of neither of the measurements is in the control of the experimenter. To that extent, violation of outcome independence does not violate relativity. It is obvious, though, that many people, particularly perhaps Einstein, would feel that it is against the spirit of relativity, since, although we are not sending a signal faster than the speed of light, it rather seems that nature itself is doing so.

In the previous paragraph we were cautious. We said that such correlation *may* be a sign of a breakdown of locality. But there is another explanation, for which we can give a couple of classical examples. First we can imagine a coin or a series of coins split in half by a cut parallel to the face of each coin. One half of each coin may be kept in say London, the other half sent to Delhi. When an observer in London and one in Delhi examine their portions of each coin, there will, of course, be perfect correlation. Each time the observer in London sees heads, the one in Delhi will see tails, and vice versa. But of course we do not jump to the conclusion that a signal has passed between the two observers. (If the time between each observation and the corresponding one in Delhi were zero, or sufficiently small, such a signal would constitute a violation of locality.) Rather we recognize that each head and each tail were formed at the process of cutting the coin, before the halves of the coin were separated. Neither of the observations in London or Delhi created the results they observe; rather they just record the correlated information that is already there. We can regard this as a hidden variables situation, the 'heads' and 'tails' corresponding to the hidden variables.

A similar and rather amusing example was given by John Bell, who we will meet a great deal in the rest of the book. Bell's friend Reinhold Bertlmann always wore socks of a different colour, so, as Bell said, if you observed one sock to be red, you knew that a subsequent sight of the other sock would reveal that it is *not* red. But, of course, one does not believe that a signal of any nature was propagated from one sock to the other. The colour of each sock was established as soon as Bertlmann put them on in the morning, so the colours acted as hidden variables, which the glimpses we obtain of Bertlmann's footwear merely record.

In principle it can be quite difficult to determine whether there has been genuine transfer of information between the locations of two experiments, or whether the results of both experiments rely on information already available to both—in quantum terms we can say whether hidden variables are involved.

Entanglement

We now come to the all-important topic of *entanglement*, certainly the most important aspect of the New Quantum Age.

We shall build up the idea gradually. We shall consider two particles each of spin-$\frac{1}{2}$, and it is the spins we shall be considering. From Part I we can say that if it is certain that a measurement of the z-component of spin of particle 1 will give the result $\hbar/2$, we write its state-vector as $|+\rangle_1$; if it will certainly give a result $-\hbar/2$, we write it as $|-\rangle_1$. Similarly for particle 2, we write $|+\rangle_2$ and $|-\rangle_2$. When we look at combined (product) states, a state such as $|+\rangle_1 |-\rangle_2$ tells us that a measurement on particle 1 will give the result $\hbar/2$, while one on particle 2 will give $-\hbar/2$. We can perform either measurement, or we can perform both, in which case it does not matter what order we perform them in; we can say that the measurements are independent, or indeed that the initial states of the two particles are independent. Similarly, we can have $|+\rangle_1 |+\rangle_2$, $|-\rangle_1 |-\rangle_2$ and $|-\rangle_1 |+\rangle_2$. All these are very straightforward.

However, we know from the previous chapter that there are situations where the result that will be obtained in a measurement of s_z is not determined before the measurement. Typical state-vectors for a single spin would include $\left(1/\sqrt{2}\right)\left\{|+\rangle_2 \pm |+\rangle_1\right\}$. For either the + or the – sign in the middle of the expression, the probability of obtaining either $\hbar/2$ or $-\hbar/2$ in a measurement of s_z is $\frac{1}{2}$. (The choice of + or – sign does affect other things, such as the results of measuring other components of spin, but not, as we have said, measurement of s_z.)

Returning to states of two spins, we can easily construct possibilities by multiplying such a state of spin 1 by a similar state of spin 2, for example $\left(1/\sqrt{2}\right)\left\{|+\rangle_1 \pm |+\rangle_1\right\}|+\rangle_2$ or $\frac{1}{2}\left\{|+\rangle_1 + |-\rangle_1\right\}\left\{|+\rangle_2 + |-\rangle_2\right\}$. In both these cases, and in any others constructed using this rule, the results of measurements on the two spins are still entirely independent. We shall call these state-vectors *unentangled*, and they do not present any new conceptual difficulties over those discussed in Part I.

We can make things a little less obvious by rewriting these state-vectors as $\left(1/\sqrt{2}\right)\left\{|+\rangle_1|+\rangle_2 + |-\rangle_1|+\rangle_2\right\}$ and $\frac{1}{2}\left\{|+\rangle_1 + |+\rangle_2 + |+\rangle_1|-\rangle_2 + |-\rangle_1\right.$

$|+\rangle_2 + |-\rangle_1 |-\rangle_2 \}$. Particularly in the second case it is not immediately clear that the state-vector *can* be written as a product of a state-vector for spin 1 and one for spin 2, but of course we know it can be because that is the way we constructed it; if we didn't know this, it would only take a little investigation to see that it could be done.

However, we now move to a state-vector for the combined system of two spins that has a completely different nature: $(1/\sqrt{2})\{|+\rangle_1 |-\rangle_1 \pm |-\rangle_1 |+\rangle_2\}$. In this case, try as one might, the state-vector cannot be written as a product as a state-vector for each spin. The implication is that measurements on the two particles are not independent. If we measure s_z for the first spin, or we can say if we measure s_{z1}, and if we get the value $\hbar/2$, then a measurement of s_{z2} must yield $-\hbar/2$. However, it is equally probable that the measurement of s_{z1} may yield $-\hbar/2$ in which case the measurement of s_{z2} must yield $\hbar/2$.

It may be useful to spell out what we have just said in terms of the collapse idea. If we first measure s_{z1} and get the result $\hbar/2$, we can say that the state-vector will collapse to $|+\rangle_1 |-\rangle_2$, and a measurement of s_{z2} must give the result $-\hbar/2$. However, if we get the result $-\hbar/2$ for s_{z1}, the state-vector must collapse to $|-\rangle_1 |+\rangle_2$, and a subsequent measurement of s_{z2} must give the result $\hbar/2$. Of course if we choose to measure s_{z2} first, everything is analogous with spins 1 and 2 interchanged.

The argument is certainly startling, and we shall proceed to spell out why. We must recognize that, to be entangled, the particles will have been close enough to interact at one time; we shall explain an important example of how this may happen shortly. (Actually there is another way of obtaining entangled states—entanglement swapping—which we will discuss in Chapter 19.) Nevertheless once the particles have interacted they will normally separate and move steadily further apart. We can, at least in principle, wait to perform our measurements until the two particles are on opposite sides of the Universe. Yet it seems that they represent what is know as an *entangled* system, in the sense that a measurement of s_z on either of the particles fixes the value of s_z not only for that particle, but also for the other particle, which may, as we have said, be millions or billions of miles away! From this point of view, any subsequent measurement on the second particle only serves to confirm the value that the original experiment *must* give.

Let us now focus on the time interval between the two measurements. Our collapse story relies on one measurement (it does not matter which) occurring before the other one. It does not specifically cover the case where the measurements take place at the same time. Nevertheless each term of the entangled state-vector has a total s_z of 0, and so it is clear that a measurement on one spin, or we will often say a measurement in one wing of the apparatus, must give the result $\hbar/2$, and that in the other wing must give the result $-\hbar/2$ in all cases.

In any case, our previous remarks on relativity will remind us that it may be that in some frames of reference, measurement 1 comes before measurement 2; in some frames of reference, measurement 2 comes before measurement 1; and in one frame, the measurement occur at the same time. When we say this 'may' be the case, what we mean is that this will be the case if the interval between the two measurements is spacelike. This is indeed just the situation in which we are really interested, because it means that no signal travelling at the speed of light or slower can pass from measurement to the other. If locality is obeyed, the result of neither measurement can influence the result of the other one.

(We can remark that this argument suggests that our collapse explanation of the result of entanglement given above is difficult to uphold in detail; this is why we referred to it at that point as a collapse 'story'.)

This argument is important because there are two obvious possible explanations for what is occurring. The first is that both results actually exist before either measurement is made. Thus the measurements merely record results that are already in existence, and have been so since the entangled state was created when the particles initially interacted. This is certainly a good explanation, but, of course, flies right in the face of the Copenhagen doctrine that no extra information about a system exists, over and above what is in the wave-function or state-vector. In this case it is clear that the values for s_{z1} and s_{z2} are not provided by the form of the state-vector, and so Copenhagen would tell us that they do not exist in the physical universe. They must be created at the time of first measurement. To imagine this is to restore realism in a situation where Copenhagen says that it does not exist, or, we can say, adding hidden variables to the bare description of the situation provided by the state-vector, which is again quite contrary to what Copenhagen allows.

The other possible explanation is that somehow the information on the result of the first measurement has reached the site of the second measurement, thus dictating what its result will be. Yet, as we have just discussed, we are interested in cases where for this to happen locality would be violated.

Thus our brief account of this example of entanglement suggests that it certainly demonstrates important problems for the Copenhagen approach to quantum theory. It argues that either realism must be restored or locality abandoned, neither of which could be palatable to Bohr and his followers. How they did respond to EPR who put the argument forward we shall see shortly.

First, though, we make the point that the kind of non-locality discussed in this section does not allow signals to be sent at a speed greater than that of light, which would be in total violation of the laws of relativity. We cannot select the measurement result in either wing of the procedure, so cannot send information to the other wing. Thus the form of locality that it violates is that of outcome independence rather than parameter independence. Thus we are not explicitly breaking the laws of relativity. Shimony, who was always deeply concerned with international relations, suggested that there was 'peaceful coexistence' between quantum theory and relativity.

Einstein, Podolsky, Rosen

We now come back to the historical record and discover how and when entanglement was first discussed. Earlier in the chapter we considered Einstein's early arguments with Bohr. At this point he was based in Berlin, but in 1932, Einstein was driven from Germany, and was invited to the Institute for Advanced Study in Princeton where he was to stay for the rest of his life. Before coming to Princeton, he had already carried out some work in collaboration with a Russian physicist, Boris Podolsky (Figure 7.4), and when Podolsky also came to Princeton, they re-established these links. In 1935 a young American physicist, Nathan Rosen (Figure 7.5), became Einstein's assistant, and that is how Einstein, Podolsky and Rosen or EPR came to be working together.

The central idea of the EPR paper certainly came from Einstein. Earlier in this chapter, we saw how in his earlier discussions with Bohr, he had attempted to demonstrate how simultaneous measurements of momentum and position, or of energy and time could be carried out, in violation of complementarity. (More exactly his methods attempted to show how the simultaneous measurements could be performed to greater combined accuracy than permitted by the Heisenberg principle.)

When these measurements were to be carried out on the same particle, Bohr's arguments showed that Einstein's attempts failed. However, Einstein believed that the as yet unnamed idea of entanglement would allow him to resume his arguments against Bohr. He had been forced to accept Bohr's position that it was impossible to measure both momentum and position simultaneously, but that did not mean that they did not actually have simultaneous exact values.

The attempt to use entanglement to show that they did in fact do so can be put in various different forms. We shall first sketch the argument put forward in the original EPR paper in 1935, and we make the point that our example of entanglement in the previous section using spins was particularly simple. The state-vector was the sum of just two terms. (Essentially it is the example thought up rather later by David Bohm, which we shall meet shortly.) Unfortunately the example EPR came up with was much more awkward.

It should be mentioned that Einstein himself never became an expert on the technical handling of quantum theory. This did not imply that he had little interest in the theory; quite the reverse—he said that he spent much of his life thinking about the conceptual nature of the theory, far more time, he said, than he devoted to the consideration of relativity.

Fig. 7.4 Boris Podolsky [courtesy of Xavier University Archives, Cincinnati, Ohio]

But it was indeed the conceptual implications of the theory that interested him, not the detailed methods of using it.

Nor did it mean that he thought that quantum theory was, in any real sense, a 'bad theory'; rather he acknowledged it as an outstandingly successful theory, which gave extremely important and accurate (though, he

Fig. 7.5 Nathan Rosen as sole survivor of EPR at the Conference held in 1985 at Joensuu in Finland to celebrate the fiftieth anniversary of the EPR paper. Rosen is on the left, Piron third from left and Peierls second from right. On the right is Max Jammer, an important historian of history of quantum theory. [Courtesy of World Scientific]

suggested, not totally accurate) results. However, he did regard it as broadly equivalent in status to Newton's theory of mechanics or Maxwell's theory of electromagnetism before quantum theory was discovered, extremely good and useful theories, but theories that were not final, and theories whose conceptual understanding had been somewhat undermined by the discovery of quantum theory. Similarly Einstein thought that quantum theory itself was not a final theory, and therefore he was not particularly interested in learning how to use it to solve technical problems.

It is believed that it was Rosen who produced the entangled wave-function that EPR worked with. (As we shall see, it *was* a wave-function rather than a state-vector as in the previous section.) And it was Podolsky who put the argument together and wrote the account of the ideas that was published. But unfortunately in doing so, he irritated Einstein very much, because Podolsky was an expert in logic, and wrote the paper rather as an exercise in formal logic, instead of the comparatively straightforward argument that Einstein thought was possible. Indeed in later years Einstein did produce several quite simple versions of his own argument. In fact the account we give here will not follow any of these original accounts directly, but it will follow what seems the easiest way to explain Einstein's point.

It can be added that, although Einstein remained on good terms with Rosen for the rest of his life—Rosen ended up, in fact, at Haifa in Israel—he never interacted with Podolsky again. This was probably not so much because of Podolsky's poor presentation of the argument, as Einstein saw it, but because Podolsky saw fit to gain publicity for the paper in the *New York Times* before its formal publication in the *Physical Review*. For Einstein this was a cardinal sin, since he believed that the only place where scientific matters should be discussed was the scientific journal, not the popular press.

The gist of the EPR argument as originally presented is as follows. One particle at rest decays into two identical particles that move off in opposite directions. From conservation of momentum, we can say that the magnitudes of the momenta of the particles must be equal; let us say they are both p. It follows that the distances travelled by the two particles must be equal and opposite; we can say that $x_1 = -x_2$.

However, the wave-function is set up so that none of p, x_1 or x_2 has a precise value; each has a distribution of possible values. And so, of course, even more fundamentally the wave-function certainly could not provide precise measurement values for the position and the momentum of either particle.

We can immediately spell out how this example is more complicated than the one we considered in the previous section. In that case, our entangled state-vector was the sum of two terms; the two variables involved, s_{z1} and s_{z2}, each take just two values, and they take one or other of these values in each term. But in the EPR wave-function, which is certainly entangled, p, x_1, and x_2 can each take a continuous range of values, and so, rather than a simple sum of two terms, mathematically EPR had an integral over an infinite number of terms, each say with a specific value of x_2. Mathematically it can also be written as an integral over an infinite number of terms each with specific values of x_1 and x_2.

There are actually two ways of spelling out the argument, one a minimal argument that makes the point clearly, the second a slightly longer argument that hammers the point home more emphatically. The second was used by EPR themselves, and the great majority of those describing or discussing their argument. The first is actually somewhat more close to Einstein's later attempts at putting over the ideas.

For the first we can imagine a measurement being made on the first particle. It can be either of x_1 or of p. From our explanation above, it is clear that, if the measurement is of x_1, we can deduce the position of the second particle; it will have travelled a distance x_2, and, as we have said, $x_2 = -x_1$. But if we know the value of x_2 that obviously implies that x_2 has an exact value, and the wave-function for the second particle is just a spike at that particular position.

But this seems very strange. When we initially described the system, we emphasized that the wave-function told us that x_2 did *not* have a precise value. Yet after our measurement on the *first* particle, it now seems that the wave-function of the *second* particle now *is* one that corresponds to a precise value of x_2. Yet of course the cleverness of using the entangled wave-function is that the measurement on the first particle should not have changed the wave-function of the second particle, if, of course, one accepts locality.

The logic of the EPR argument is then that one must abandon either locality or the fact that, if the wave-function does not provide a precise position, the implication is that such a position does not exist. That is to say one must give up either locality or a belief in the completeness of the wave-function. Or we can say that we must either give up locality or accept realism, in the sense that quantities that the Copenhagen interpretation would say have no precise value actually do have one.

Still following our first way of spelling out the argument, one can alternatively measure the momentum of the first particle rather than its position. The argument then proceeds in exactly the same way. One can deduce the momentum of the second particle, but again the measurement on the first particle should not have affected the second particle if one accepts locality, and so, according to the Copenhagen interpretation, such a value should not exist. Exactly the same conclusions can be drawn.

Now we tackle our second way of spelling out the argument. In this method, we do not need to discuss the wave-function of the system at all. Rather, as before, we argue that a measurement of the position of the first particle will give us a value also of the position of the second particle. Yet, if we retain the principle of locality, it seems that our measurement cannot have disturbed the second particle; therefore it must have had the value of position at times even before the measurement.

We can now repeat exactly the same argument, if, instead of measuring the position of the first particle, we had measured its momentum. We would be able to deduce the momentum of the second particle, again without disturbing the second particle, provided we respect the principle of locality. Therefore the second particle must have had its value of momentum before any measurement was made.

Now of course we cannot measure both the position and the momentum of the first particle; this was what Bohr had demonstrated to Einstein in their earlier arguments. But in the EPR case we do not need to! In a sense, both measurements are hypothetical. The argument is that the mere fact that we *could* measure the momentum of the first particle tells us that the momentum of the second particle has a value in the absence of any measurement at all. And the mere fact that we *could* measure the position of the first particle tells us that the position of the second particle must have a value even in the absence of any measurement. Thus both the momentum and the position of the second particle have exact values. (And of course we should not think that this particular particle is in any way special; the same must presumably apply to the first particle.)

Yet almost the very heart of complementarity is that in no case can the position and the momentum of a particle simultaneously have exact values. To the extent that we consider it successful, we must say that the EPR argument tells us that we must *either* relinquish the principle of locality *or* acknowledge that complementarity is violated in the sense that quantum theory is not complete; hidden variables, values for quantities that may *not* be obtained from the wave-function, *do* exist. We can say we must accept that *either* locality must be abandoned *or* realism must be restored.

Einstein and the EPR argument

We have studied the EPR argument, and now ask—what did Einstein deduce from it? Would he agree to abandon locality, or would he embrace realism? In fact, of course, there was no question about which possibility he would follow. He strongly supported locality; he strongly supported realism. Quite obviously he would retain locality and call for realism to be restored. In other words he strongly argued that complementarity was wrong in denying realism, in other words that there were elements of physical reality that were not provided by the wave-function; in yet other words that quantum theory was not complete. Indeed the title of the EPR paper was: 'Can Quantum-Mechanical Description of Physical Reality Be Considered Complete?', to which EPR answered resoundingly that it could not.

They fully recognized, of course, that logically an alternative solution to their problem was that the assumption of locality might have to be abandoned. However, they scarcely stopped to consider whether this possibility should be taken seriously. To them, or perhaps one should really say, to Einstein, such a possibility seemed ludicrous. It might be mentioned that, in our discussion of locality, we stressed the question of whether a signal could pass from the site of one measurement to the site of the other. Einstein's approach was perhaps a little simpler, or one might say more down-to-earth. He did not believe that two normal systems could 'communicate' with each other, even if there did happen to be time for a signal to pass between them. It would, of course, be different if one of the systems was clearly propagating an actual signal that could be received by the other. But in the absence of this, Einstein considered locality to be obviously true. Indeed he thought that it would be hardly possible to carry out sensible physical analysis if, as he saw it, separated physical systems did not behave independently.

So it is important to recognize that when one speaks or writes of the claims of EPR, one may mean one of two rather different things. One may mean what they actually claimed to have proved: (a) either locality must be violated or realism restored. Alternatively, though, one may mean what they actually believed: (b) locality and realism are both upheld.

The distinction is rather important because even today the EPR analysis is very often criticized, or even ridiculed. It is extremely common to read that EPR has been proved to be wrong by the theoretical analysis of John Bell, backed up by the experiments of, in particular, Alain Aspect. It should be made clear that, provided one restricts oneself to (a) above, that is entirely wrong. The actual analysis of EPR is certainly not obviously wrong or misguided, and is in no way affected by the work of Bell, Aspect or others. Indeed, while there were technical difficulties in the particular example of entanglement that Rosen came up with, I would argue that the EPR paper was both extremely clever and exceptionally important in stimulating the great majority of the work and ideas that will be described in the rest of this book.

Unfortunately it must be admitted that what seemed to EPR the further obvious step of moving to (b) was ill-conceived. It is (b) that, bar the closing of a few loopholes, as will be explained later in this book, has been disproved by the combined efforts of Bell, Aspect and others. But the distinction between (a) and (b) shows that one must be exceptionally careful in making remarks such as 'EPR were wrong' or 'Einstein was misguided'.

Bohr and EPR

However, we now return to 1935, and consider the reactions of physicists at the time, and it was clear that the question virtually all would want to ask was—What does Bohr make of it? In fact he was immensely disconcerted by the paper, at least initially feeling that EPR had built up a convincing case.

However, working tirelessly for a period of weeks together with Leon Rosenfeld, who was acting as Bohr's assistant, and was later to become the most trenchant defendant of Bohr's views, before and after Bohr's death, he came up with what he believed was a complete refutation of the EPR argument. The resulting paper was published later in 1935 in the same journal and with the same title as the EPR paper [6].

By 1935, the scientific establishment had been for a long time convinced that Bohr's analysis of the problems of quantum physics was correct and indeed triumphant, and that Einstein was unfortunately trapped in classical modes of thought, unable to take on board the new approach to physics, which had been heralded by Bohr and which was necessary to understand quantum phenomena. Not surprisingly, there was practically complete agreement that yet again Bohr's arguments showed those of EPR to have been not only just incorrect, but also ill-conceived and what we might term muddle-headed.

Yet the reader of this paper today finds it very difficult to appreciate why they thought in this way. As Howard has written [7], the problem one finds today in attempting to understand Bohr's approach to quantum theory is that it is difficult, not just to see why his arguments were felt to be successful, but even to see what they actually mean. Howard's response is not to reject Bohr's ideas but, along with several authors, to set themselves the task of restating and reinterpreting Bohr's arguments.

My own approach [8] to this paper of Bohr is to identify an ambiguity in what Bohr considered his task to be, and a related one in his actual argument. It will be remembered that Bohr's arguments against Einstein's earlier proposals did not mention complementarity explicitly, but merely analysed Einstein's thought-experiments with the use of simple physical arguments. So Bohr did not actually use complementarity, but it was certainly the case that his arguments suggested strongly that at least some aspects of complementarity succeeded in uncovering much of the underlying truth of quantum theory.

In much of his argument against EPR, Bohr appears to be taking on the same task—he is endeavouring to show, on strictly physical grounds and without mention of complementarity, that EPR's account of their experiment is flawed, in that it misses some essential feature of the situation. And indeed Bohr says that the EPR argument, when analysed rigorously, did not bring in any new features not already present in the previous arguments.

Yet in other aspects of his argument he does explicitly mention and use the ideas of complementarity. He claims to show that complementarity can explain what happens in the EPR experiment in its own terms. This would be, of course, reassuring to Bohr himself and his supporters, but it could scarcely be expected to convince Einstein and his co-workers, since they obviously did not accept the principles of complementarity on which this part of the argument was based.

We shall not discuss in detail how complementarity would discuss EPR's argument. As we have seen, there are several different ways of constructing the EPR argument, and the response of complementarity to each would be different. However, we will mention an argument from Bohr's paper in response to the idea that a measurement of *either* momentum *or* position on the first particle would give the value for the same quantity for the second particle without disturbing the second particle, so both quantities for the second particle must exist even in the absence of any measurement.

Now Bohr explicitly admitted that a measurement on the first particle could not cause a 'mechanical disturbance' on the second particle. But he claimed that there could still be what Arthur Fine [9] would later call a 'semantic disturbance'—what we *say* about the first particle affects what we may *say* about the second one. As Bohr put it: the measurement on the first particle can produce 'an influence on the very conditions which define the possible types of predictions regarding the future behaviour of the system', the 'system' including *both* particles. This might be described as a not especially clear call on the precepts of complementarity.

It may be thought that Bohr's so-called rebuttal of EPR should have been only moderately convincing even to those who fully understood and accepted the principles of complementarity, and not convincing in the slightest to anybody else. Yet as we have said that is not what happened. By this period, the community of physicists had fully accepted that Bohr had shown the way the conceptual structure of physics had to develop in order to account for the recent development of the quantum theory. They were also convinced that Einstein had simply failed to rise to the challenge, being wedded to outdated notions, and having unfortunately lost his flexibility of mind that had been so evident twenty years previously. It seemed natural to assume that this latest disagreement was just another example of this sad truth.

The EPR argument itself was to be almost totally disregarded for 30 years when it was to become the centrepiece of John Bell's own first crucial work in 1964. In the subsequent years, initially slowly but with increasing speed and power, many of the ideas it introduced have become absolutely central in the development of the foundations of quantum theory and the creation of quantum information theory. Yet strangely, and for reasons we touched on in earlier in the chapter, Einstein, let alone Podolsky and Rosen, have very often failed to attract the credit that they undoubtedly deserve for this important work.

Schrödinger's response to EPR

The one physicist who gave strong support to Einstein in his battle against complementarity and the Copenhagen interpretation of quantum theory was Erwin Schrödinger. It can be said that the two men agreed almost completely in their attitude to the reigning orthodoxy, though not necessarily on how it should be replaced. (As an example of the latter point, Schrödinger was by no means convinced that determinism was necessary in any future theory, while Einstein would have been extremely reluctant to lose it.)

After the publication of the EPR paper, between June and October 1935, Einstein and Schrödinger exchanged a considerable number of letters. Schrödinger built on the content of the EPR paper to produce a substantial account of his own ideas. In his own words, 'The appearance of [EPR] motivated the present—shall I say lecture or general confession?' Subsequently the account was published [10].

Schrödinger's whole argument is interesting and readable, but we shall concentrate initially on two points. The first is that he was able to extract from the central feature of the EPR argument and give it the name we have already been using—'entanglement'. He said that 'I would call this not *one* but *the* characteristic trait of quantum mechanics, the one that enforces the entire departure from classical thought'. This comment of Schrödinger was particularly prescient since, as we have already implied, entanglement stands at the centre of the great majority of the surprises and opportunities of quantum theory that we shall highlight in this book.

The other point that we shall stress is his introduction of the famous or perhaps notorious *Schrödinger's cat*. Schrödinger remarked that one could set up what he called 'quite ridiculous cases'. His particular case involved a cat shut up in a steel container along with a small amount of radioactive material. This incarceration takes place for a time equal to the half-life of the radioactive species, so at the end of this period, and in the absence of any 'measurement', by which we mean if we do not detect whether the atom has decayed, we must write the state-vector of the system as $\left(1/\sqrt{2}\right)\left(|decayed\rangle + |surviving\rangle\right)$.

This implies that the state of the atom is a linear combination of the decayed state and the surviving state. If, but only if, we perform a measurement, then we will observe that the atom is either decayed or surviving, each with an equal probability of $\frac{1}{2}$. The linear combination of different states of the atom before any measurement is, of course, strange from a classical perspective, but it is just what we have become used to in quantum theory, and we would be prepared to accept that atoms behave in this way, even though it is difficult to understand.

However, Schrödinger went much further. If his atom does decay, he says, the decay particle produced serves to release a hammer that shatters a small flask of hydrochloric acid that then kills the cat. However, if the atom does not decay, of course the flask does not break and the cat survives. Let us again assume that there is no 'measurement', or in other words the whole setup is unobserved; Schrödinger says that everything is in a strong steel vessel so that there is no possibility of seeing what is happening inside.

So by the same rules as before we must write the total state-vector at the end of the half-life of the radioactive atom as $\left(1/\sqrt{2}\right)\left(|decayed\rangle|broken\rangle|dead\rangle + |surviving\rangle|unbroken\rangle|alive\rangle\right)$, where the three parts of each term in the linear combination refer to the radioactive atom, the flask and the cat, respectively. Clearly, though this is mathematically of the same form as the state-vector of the atom alone that we had previously, it is much more difficult to accept—in fact we will almost certainly refuse to believe that it makes sense even in the rather rarefied atmosphere of quantum theory. While we can accept that an atom somehow straddles the states of being decayed or surviving, it is much more difficult to imagine that a flask straddles the states of being broken or unbroken. Our reluctance is presumably because the atom is microscopic as we have previously defined it, while the flask is macroscopic; we imagine we know how macroscopic objects will behave! And again we are even less willing, much much less willing to accept that the cat straddles the states of being alive or dead. Yet Schrödinger is pointing out that this is the clear application of quantum theory as interpreted by Copenhagen.

Perhaps even more surprising is what happens if, even perhaps inadvertently, we open the steel vessel and thus observe the situation or, we can say, measure the various quantities involved—the state of the atom, flask and cat. Any measurement will collapse the total system down to one of the two terms in the linear combination above. Thus the system may collapse to a state in which the atom survives, the flask is unbroken and the cat is alive. Less happily, though, it may be to a state in which the atom has decayed, the flask is broken and the cat is dead. In that case, straightforward application of the Copenhagen interpretation tells us that it is the opener of the vessel who is actually responsible for killing the cat.

Strange as quantum theory undoubtedly is, it is very difficult to believe that is quite *that* strange! But if not, it seems clear that we must add something to our description of the situation on top of the wavefunction, something to tell us whether the cat is alive or dead. In other words, much like Einstein and EPR, Schrödinger's ideas lead us to the belief that quantum theory is not complete.

References

1. P. A. Schilpp (ed.), *Albert Einstein: Philosopher-Scientist* (Evanston, IL: Library of the Living Philosophers, 1949).
2. D. Howard, 'Nicht sin kann was nicht sein darf, or the Prehistory of EPR, 1909–1935: Einstein's Early Worries about the Quantum Mechanics of

Composite Systems', in: A. J. Miller (ed.), *Sixty-Two Years of Uncertainty* (New York: Plenum, 1990), pp. 61–106.

3. L. Gilder, *The Age of Entanglement: When Quantum Physics was Reborn* (New York: Knopf, 2008).
4. A. Einstein, B. Podolsky, and N. Rosen, 'Can Quantum-Mechanical Description of Physical Reality Be Considered Complete?', *Physical Review* **47** (1935), 777–80.
5. A. Shimony, 'Events and Processes in the Quantum World', in: R. Penrose and C. J. Isham (eds), *Quantum Concepts in Space and Time* (Oxford: Oxford University Press, 1986), pp. 182–203.
6. N. Bohr, 'Can Quantum-Mechanical Description of Physical Reality Be Considered Complete?', *Physical Review* **48** (1935), 696–702.
7. D. Howard, 'What Makes a Classical Concept Classical?', in: J. Faye and H. J. Folse (eds), *Niels Bohr and Contemporary Philosophy* (Dordrecht: Kluwer, 1994), pp. 3–18.
8. M. A. B. Whitaker, 'The EPR Paper and Bohr's Response: A Re-assessment', *Foundations of Physics* **34** (2004), 1305–40.
9. A. Fine, *The Shaky Game* (Chicago: University of Chicago Press, 1986).
10. E. Schrödinger, 'Die gegenwärtige Situation in der Quantenmechanik', *Naturwissenschaften* **23** (1935), 807–12, 823–8, 844–9; translated by J. D. Trimmer as 'The Present Situation in Quantum Mechanics', in: J. A. Wheeler and W. H. Zurek (eds), *Quantum Theory and Measurement* (Princeton: Princeton University Press, 1983), pp. 152–67.

8
The achievement of John Bell

John Bell

John Bell (Figure 8.1) is undoubtedly the main hero of this book. He succeeded in clarifying the totally confused situation left from the 1930s by the work of Bohr and von Neumann, and the scarcely understood criticism of their views by Einstein and Schrödinger. In doing so he achieved a far deeper understanding of the quantum theory and its implications. It may be said that his success was threefold. First, the support of the physics community for the views of Bohr and for complementarity was so strong, that even daring to question these ideas was practically sufficient to prevent oneself from being taken in any way seriously as a physicist. We shall see that the way in which Bell overcame this problem was extremely clever. Second, even to the very limited extent that any credence was given to the fact that Bohr and Einstein had had an interesting and worthwhile debate, it was taken for granted that the arguments were totally esoteric, and could be of no conceivable practical significance. It was 'armchair philosophy'. That Bell was able to demonstrate direct experimental significance to his analysis—the term 'experimental philosophy' became common—was an enormous surprise. And third, and rather amazingly even from the perspective of as late as the 1980s, his work would eventually be used as the basis of an entirely new technology, that of quantum information theory.

Bell [1–3] was born—John Stewart Bell and just Stewart to his family—in Belfast in 1928. Though it was clear from an early age that he was highly gifted, his family was not well-off, and his secondary education was at the Belfast Technical College rather than at one of the more prestigious but more expensive local grammar schools. Even this was lucky—none of his three siblings were able to attend school much after the age of 14. John was able to move on to the Physics Department of the local Queen's University, but for the first year just as a technician; only at that stage was he able to put the funds together to enroll as a student. He proceeded to perform exceptionally well, gaining first-class honours first in physics and then in mathematical physics.

Fig. 8.1 John Bell receiving an honorary degree from Queen's University Belfast in 1988 [courtesy of Queen's University Belfast]

As we shall see shortly, he was already exceptionally interested in quantum theory and, in fact already highly dubious about the Copenhagen interpretation. However, his first priority was to get a job, and in 1949 he was happy to be appointed to the UK's Atomic Energy Research Establishment at Harwell, his task being to assist in the design of the early accelerators. Then in 1953, he was given leave for a year to study at Birmingham University with Rudolf Peierls, one of the leading physicists working in the UK.

During this period Bell independently discovered the so-called and highly important CPT theorem. This theorem says that if you start with a proper physical event, and apply three operators to it, the C or charge conjugation operator, which replaces particles with anti-particles; the P or parity operator, which reflects the event in a mirror; and the T or time-reversal operator, which replaces t in the governing equation by $-t$, what one obtains is another proper physical event. (This was important because it is now known that any one of these three operators applied individually to a physical event may produce an event that *cannot* occur in practice.) The CPT theorem is one of the most important in quantum field theory, but unfortunately for Bell, the theorem was also discovered at about the same time by Pauli and Gerhard Lüders, and it is always now known as the Pauli-Lüders theorem.

However, Peierls was extremely impressed by Bell's work, and he encouraged his transfer to a new group that had recently been formed at Harwell, the Theoretical Physics Division, working on quantum field theory and the theory of elementary particles. We shall see more of Peierls' work and views, and his continuing relations with Bell in Chapter 15. Bell produced excellent work at Harwell, and also took time to marry Mary, a member of the accelerator design group.

However, both husband and wife steadily became concerned that Harwell was moving from pure research to more applied work, and in 1960 they both moved to CERN, the Centre for European Nuclear Research in Geneva. Bell was to work here right up to his death in 1990.

He published many important papers on elementary particle physics and quantum field theory. Perhaps his most important work was carried out with Roman Jackiw (Figure 8.2) in 1969, and it announced and explained the ABJ or Adler-Bell-Jackiw anomaly of quantum theory. (Stephen Adler achieved the same result independently.) The ABJ anomaly solved an outstanding problem in particle physics—the fact that a neutral pion appeared to defy standard theory by decaying into two photons. Rather amusingly, it had been forgotten that Jack Steinberger (Figure 8.3), a close friend of Bell, but an experimental rather than a theoretical physicist, had obtained the correct answer by a simple but seemingly naïve method as early as 1947. When Bell was informed of this by Steinberger, it stimulated his own work, which combined Steinberger's insight with Bell's knowledge and rigour.

Fig. 8.2 Roman Jackiw [© Zentralbibliothek für Physik, Renate Bertlmann]

Fig. 8.3 Jack Steinberger [© Zentralbibliothek für Physik, Renate Bertlmann]

In 1988, Steinberger was to receive the Nobel Prize for physics for his discovery that there were two types of neutrino. In the so-called first generation of particles, there are the electron and the electronic neutrino, but there is a second generation of particles that contains the muon, which has exactly the same properties of the electron apart from a much higher mass, and Steinberger's discovery, the muonic neutrino, distinct from the electronic neutrino. In fact we now know that there is a third generation, which contains the so-called tau particle and the tau neutrino.

The ABJ discovery was just the start of the process of study of such anomalies in many areas of particle physics over the next forty years. Anomalies occur when symmetries of classical physics are lost when we move to quantum physics. Reinhold Bertlmann, who we met in the previous chapter, wrote a substantial monograph on the topic thirty years later [4]. The ABJ anomaly also provided confirmatory evidence of the fact that each type of quark must come in three so-called 'colours'—'red', 'yellow' and 'blue'. (It should be stated that these 'colours' have no connection to the colours of our everyday life!)

Another important contribution was Bell's suggestion of 1967 that weak nuclear reaction should be described by a gauge theory. This was prescient since such theories are now used for *all* the fundamental interactions. Bell's suggestion was taken up by his collaborator, Martinus Veltman, who constructed a suitable theory. It was Veltman's research student, Gerard t'Hooft (Figure 8.4), who showed that the theory worked in practice—unwanted infinities could be removed by so-called 'renormalization', and Veltman and t'Hooft shared the Nobel Prize for Physics in 1999.

Fig. 8.4 Gerard t'Hooft [© Zentralbibliothek für Physik, Renate Bertlmann]

Bell was an honourable man who worked hard at the job he was paid for, as well as providing support and advice to very many younger physicists. However, quantum theory was his passion—he called it his 'hobby', and it was this that was to make him famous.

John Bell and quantum theory—the early years and the Bohm interpretation

While he was still a student, Bell was fascinated by quantum theory, but he was very suspicious of complementarity. Indeed he had a major clash with Robert Sloane, the second-in-command of the Department. Sloane announced that if one knew δx, the uncertainty or indeterminacy in x for a particular particle, then it was easy to work out δp, the uncertainty or indeterminacy in p, using the Heisenberg principle. But when Bell asked him how one obtained a value for δx in the first place, his answer was unconvincing, and regrettably Bell accused him of (intellectual) dishonesty. Sloane would not be the last to be accused by Bell of woolly thinking on these matters! In fact the answer should have been that the value of δx was a result of whatever measurement had last been performed on the particle, but even when he realized this point, Bell felt that it only added to his discomfort over the seemingly central role of measurement in quantum theory.

Bell remained all his life a realist. He considered himself a 'follower of Einstein' in these matters, and, to him, this meant being a great advocate of hidden variables. (From early in the chapter, we will realize that this was a slight misunderstanding of Einstein's actual position.) However, by the time Bell went to Harwell he was aware of von Neumann's so-called proof, which had been almost unchallenged for twenty years, that hidden variable theories of quantum theory were impossible. Von Neumann's book in which the argument had appeared had been published in 1932, but only in German, and Bell could not read German; it would not be translated into English until 1955. At Harwell, a German-speaking physicist friend, Fritz Mandl, tried to explain the gist of von Neumann's argument, but in practice Bell was forced to accept the word of Max Born in a popular book of the day [5] that hidden variable theories were impossible.

But then, in Bell's words [6], in 1952 he 'saw the impossible done' in the work of David Bohm (Figure 8.5), which was discussed in Chapter 6. Bohm provided an explicit hidden variable interpretation of quantum theory with wave *and* particle rather than wave *or* particle as in the Copenhagen interpretation. The wave-function is defined in the usual way, but in addition the particle has a precise position and momentum at all times, and these exist independently of any measurement.

Fig. 8.5 Bohm was among the greatest names of American physics who participated in the famous Shelter Island Conference of 1947 at which many of the problems of pre-war physics, including quantum electrodynamics, but also elementary particle physics, made enormous progress. Seated, from left: J. R. Oppenheimer, A. Pais, R. P. Feynman, H. Feshbach; standing: W. Lamb, K. K. Darrow, V. Weisskopf, G. E. Uhlenbeck, R. E. Marshak, J. Schwinger, D. Bohm. [Courtesy of National Academy of Sciences Archives]

The square of the wave-function gives the distributions of position and momentum across a large number of particles—we may say across an *ensemble* of particles. This part of Bohm's ideas is the same as in the Copenhagen interpretation, but there, of course, the square of the wave-function or probability density gives us the distribution of results *only* if and when we perform measurements. For Bohm we have, as we have said, a distribution of actual positions and momenta completely independently of whether a measurement is performed.

In practice, we very often do use probabilities in the Bohm theory, because we are not actually interested in the individual coordinates of all the particles in the ensemble. This is the same as in the discussion of kinetic theory in the nineteenth century, when the scientists involved did not wish to work with the individual positions and momenta of all the gas molecules; it was the various probability distributions that were useful. But for Bohm, as for the creators of kinetic theory, we could *in principle* work with individual particles, though it would be time-consuming and the results obtained would not actually be very interesting. In the Copenhagen interpretation of quantum theory, prior to measurement we could not work with individual particles even if we wanted to, because they don't exist; we *have* to use probability densities of results that will be produced only if a measurement is performed.

We mentioned in Part I that Bohm had independently come across the same general approach as de Broglie had used a quarter of a century earlier, until he was unfortunately shot down by Pauli. De Broglie's expression for the interpretation was that of the *pilot wave*, and that expresses nicely the fact that the wave may be said to determine the motion of the particle.

What is exciting about the Bohm interpretation is that, totally contrary to the claims of Bohr and von Neumann, the fairly simple set of ideas reproduces the standard results of quantum theory—it must do this, of course, because it is an interpretation of quantum theory, not a new theory—but it remains entirely realist. In fact, Bohm's interpretation is also deterministic, though for Bell in particular that was rather unimportant—it was the realism that was crucial. Incidentally it was highly *non*-local, which would also be extremely important for Bell as we shall see.

Let us now study the various reactions to Bohm's work. We start with Einstein. Since he was adamantly opposed to complementarity, because Bohm's work clearly struck a blow against complementarity, and indeed because Einstein had actually stimulated Bohm to think about these matters, one might have thought that Einstein would have been excited by Bohm's interpretation. Yet we must remember from the previous chapter that Einstein did not believe in simple hidden variable theories, and was searching for something far more grandiose. In a letter to Max Born he dismissed the Bohm theory as 'too cheap'. This was really a massive own goal by Einstein, on the grounds that 'my enemy's enemy is my friend'! Bohm was very hurt and disappointed by Einstein's reaction. (Incidentally Bohm thought that Einstein's dismissal of his work was because, as was just said, it was non-local. This would actually have been a better reason for Einstein's reaction, but it does not appear that it was the case.)

The reaction of the High Priests of Copenhagen—Heisenberg, Pauli, and also Léon Rosenfeld (Figure 8.6), who by this time was probably the main spokesman for Bohr's views—was much less surprising. They dismissed it out of hand. Heisenberg said that the particle trajectories were 'superfluous "ideological superstructure"'—superfluous because the interpretation gave exactly the same results as the orthodox interpretation where they did not appear. Bohm did, in fact, suggest that in the region of very high potentials and exceptionally short distances, his theory might be adapted to produce differences from standard quantum theory, but Heisenberg dismissed this as 'akin to the strange hope that sometimes $2 \times 2 = 5$'.

Pauli also described Bohm's work as 'artificial metaphysics'. As we just mentioned, a quarter of a century earlier, Pauli had criticized the ideas of de Broglie, which could be regarded as forerunners of those of Bohm, so severely that de Broglie had given them up and had become, for this quarter of a century at least, an acceptor if not an admirer of the Copenhagen position. However, Bohm, a more capable theoretical physicist than de Broglie, was easily able to show that Pauli's arguments

Fig. 8.6 Léon Rosenfeld [from N. Blaedel, *Harmony and* Unity (Madison, WI: Science Tech Publishers)]

were simplistic and misleading. Fittingly de Broglie returned to an advocacy of his earlier ideas.

But it was exceptionally difficult for Bohm to convince a typical physicist to be interested in his ideas. To the extent that his interpretation agreed with current ideas, it did seem pointless—'armchair philosophy'. But as soon as he suggested areas where there might be disagreement, areas where in fact the current theory was untested, such as interactions at very small distances, and where there was at least a possibility that Bohm's equations might be adjusted to provide agreement with experiment, the suggestion could not help but seem unconvincing. It appeared unlikely that he could restrict such disagreement to these areas where current ideas were not yet totally established. It will be interesting to see how Bell avoided this problem in his own work.

In fact, Bell was just about the only reader to understand the real importance of Bohm's work, and indeed to be delighted and entranced by it. It was Bell who recognized the close resemblance to de Broglie's earlier work, and he would usually call the interpretation that of de Broglie-Bohm. He realized that it made a very big dent in the *raison d'être* of the Copenhagen interpretation. Its proponents had insisted that loss of realism and determinism was the only way. Bohm had shown that they were wrong.

In a paper published in 1982 [6], Bell was to write:

> Why then had Born not told me of this 'pilot wave' [in Ref. 5, which was published in 1949, long after de Broglie's work, where Born had quoted von Neumann on the impossibility of hidden variables]? If only to point out what was wrong with it? Why did von Neumann not consider it? More extraordinarily, why did people go on producing 'impossibility' proofs after 1952, and as recently as 1978? When even Pauli, Rosenfeld and Heisenberg could produce no more devastating criticism of Bohm's version than to brand it 'metaphysical' and 'ideological'? Why is the pilot wave picture ignored in text books? Surely it should be taught, not as the only way, but as an antidote to the prevailing complacency? To show that vagueness, subjectivity, and indeterminism are not forced on us by experimental facts, but by deliberate theoretical choice?

And of course it is really verging on the dishonest to brand the Bohm interpretation as 'artificial' and, as such, a completely unworthy and unnecessary competitor to Copenhagen. Bohm [7] argued that

> if de Broglie's ideas had won the day at the Solvay congress of 1927 [which could conceivably have happened if he had been supported by a few people more mathematically astute than de Broglie himself], they might have become the accepted interpretation; then, if someone had come along to propose the current interpretation, one could have said that since, after all, it gave no new experimental results, there would be no point in considering it seriously.

In other words I felt that the adoption of the current interpretation was a somewhat fortuitous affair.

James Cushing [8] has written an interesting book exploring exactly this theme.

Bell attempted to create interest in the Bohm interpretation for the rest of his life, but this is not to say that he thought it was the final answer. He wrote that neither de Broglie nor Bohm actually liked it very much. He was presumably arguing from the fact that de Broglie did give up the idea rather easily, while Bohm was discouraged by the generally negative response, and also admitted that his interpretation was not totally satisfactory.

Bohm (Figure 8.7) himself became Professor of Physics at Birkbeck College in the University of London in 1961, and for the rest of his life devoted his interests to a wide-ranging study of the nature of science and its relation to human knowledge and learning. (See David Bohm box in Chapter 6.) In 1987 a collection of essays [7] was published in his honour. The breadth of his interests is indicated by the fact that they include discussions of science, orthodox and unorthodox, but also work that would most naturally be classified as philosophy, psychology, neuroscience, biology, linguistics and the theory of art. However, it must be

Fig. 8.7 David Bohm in later life [courtesy of Mark Edwards/Still Pictures]

admitted that unfortunately he took little part in further study of the topic of this book!

Others did, though it must be said that only a minority of physicists have ever shown any interest in Bohm's interpretation of quantum theory. Those who have done so, though, have been *very* interested. Perhaps the most dramatic application of Bohm's work came in 1979, when, with the ready availability of computers, Chris Philippidis, Chris Dewdney and Basil Hiley produced trajectories for the two-slit or Young's slits experiment. The interference of the waves from two slits is one of the most basic experiments in wave theory; it was influential in gaining acceptance of the wave nature of light in the nineteenth century, and was equally important in discussions of the wave nature of the electron in the twentieth. With the Bohm theory we have, of course, as shown in Figure 8.8, maxima and minima being formed at the right-hand-side of the diagram as in any wave approach, but also, absolutely startlingly for anybody brought up on conventional quantum dogma, actual particle trajectories.

Much detailed work has been performed on the Bohm interpretation and two substantial books have been published, one [9] by Bohm himself and Basil Hiley, who worked closely with Bohm for many years, and another [10] by Peter Holland, another important follower of Bohm.

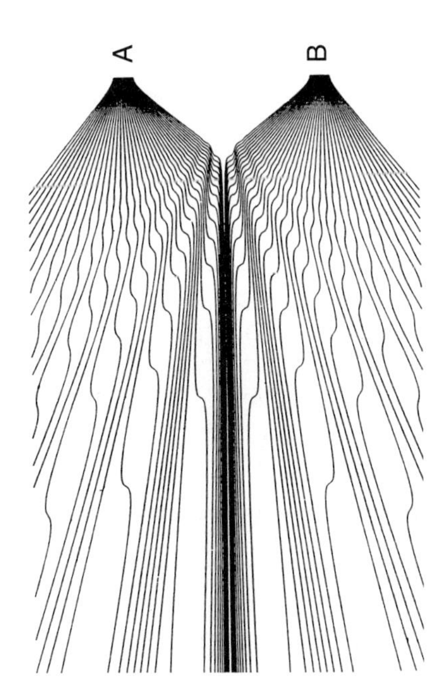

Fig. 8.8 Two-slit interference in the Bohm theory. Particles emerge from slits A and B and move along trajectories. They bunch around the maxima of the usual interference pattern. [From C. Phillippidis, C. Dewdney and B.J. Hiley, *Nuovo Cimento* **52B** (1979), 15–28]

In the latter work, the attempt to drive a wedge between the results of Cope nhagen and that of Bohm has been abandoned, but what is attempted is more interesting. There are certain aspects of physics where the Copenhagen interpretation does not give a clear answer. A good example is that it does not give a straightforward result for the seemingly simple question of the time taken for a particle to travel from one point to another. Since the Bohm theory has its explicit trajectories, it is able to make a particular contribution here, not contradicting the Copenhagen interpretation but supplementing it. Quite a lot of work has been performed studying time in quantum theory using the Bohm interpretation.

But let us return to Bell. He was not particularly interested in the further development of the Bohm theory, but much more so in working out what it told us about the fundamental nature of quantum theory. It was clear, and from Bell's point of view, absolutely excellent, that it showed that hidden variable theories were *not* ruled out, in stark contradiction to the arguments of Bohr and his comrades, and in particular, the theorem of von Neumann. But that meant that it was essential to show exactly where that admittedly great mathematician had erred.

Bohm himself had come up with an explanation; he said that the theorem could be circumvented by associating hidden variables with the apparatus as well as the system being observed. But Bell was to say that this analysis 'seems to lack clarity or else accuracy'. He was able to show that including the apparatus did not always lead to hidden variables, and also that there were situations where you could certainly have hidden variables without explicitly considering the apparatus. So we might say that Bohm's suggestion was not especially helpful.

Bell was also struck by the wholly unwelcome feature that, as mentioned before, Bohm's theory was non-local. Indeed Bell calls it 'grossly' non-local in the sense that, for example, if there are two particles in our system, the trajectory of each particle depends in a complicated way on the trajectory and wave-function of *both* particles. Thus it would seem natural to search for alternative approaches that also rebutted von Neumann by using hidden variables, but, unlike the Bohm theory, were local. Was this possible?

John Bell—von Neumann and the first great paper

Bell thought about these matters fairly consistently for more than a decade, though of course his main efforts went into elementary particle theory, the work for which he was paid. However, a particularly important encounter occurred in 1963 with Josef-Maria Jauch (Figure 8.9), who was a highly respected physicist and Director of the Institute of Theoretical Physics at the nearby University of Geneva. Jauch was perhaps best known for his publication with Fritz Rohrlich eight years before of an extremely important book, *The Theory of Photons and Electrons*. His

Fig. 8.9 Josef-Maria Jauch [courtesy of Professor Charles Enz and ETH-Swiss Federal Institute of Technology Zurich]

visit to CERN was to give a seminar on what he called the strengthening of von Neumann's theorem. Since Bell quite rightly regarded the Bohm papers as a proof that the von Neumann theory required scrapping rather than strengthening, this seminar was something like a red flag to a bull!

Bell had fairly intense arguments with Jauch as he had with Fritz Mandl. In both cases he learned a lot about the arguments for the *status quo*, which was very useful for the time when he would seek to demolish it. Also in both cases his opponent, buttressed by the warm glow of Copenhagen righteousness, probably made little real effort to appreciate Bell's arguments. In 1973, Jauch was to publish a book *Are Quanta Real?: A Galilean Dialogue* [11], which, interesting and amusing as it was, showed no understanding of any position opposed to Copenhagen. Jauch actually died the following year, but it is unfortunate that both Mandl and Jauch have gone down in history as reactionaries opposed to progress in the form of John Bell, when, as we have said, their discussions with Bell were actually very important for him.

Anyway, his encounter with Jauch made Bell determined to think his ideas through once and for all, and to get them known. The opportunity came later in 1963 when he and Mary were granted leave from CERN, and visited the United States. They spent most of the time at the Stanford Linear Accelerator Centre (SLAC) in California, where Mary was fully absorbed in accelerator physics, but they also made shorter visits to Brandeis University at Massachusetts, and the University of Wisconsin at Madison It was during this period that Bell had the time to work out his ideas in detail, to get them onto paper and to seek to get them published.

He actually had a threefold task—first to make definitive statements of his ideas, secondly to avoid being categorized as a crank by powerful physicists, which, as we have seen, was an ever-real danger, and thirdly nevertheless to attract the attention of those who might be interested and would like to take the ideas further. It is fascinating to study how he achieved all these three goals.

One aspect of his success was that his arguments were particularly clear and simple. While Bohm's papers showed immense erudition, they were also lengthy and highly technical. They would take a great deal of time to study, and it was easier for those who were, in any case, totally unsympathetic to the whole idea of hidden variables, and this would be the great majority of physicists, to dismiss the papers as irrelevant, rather than to take time trying to show what might actually be wrong with them.

Bell's arguments were, of course, immensely subtle in their meaning and implications, but the examples he used to put them across were so straightforward, and his logic so simple, that they practically demanded attention. And even more importantly he was able to suggest a crucial experiment. While Bohm had only been able to make the vaguest suggestions of areas of physics where there might just conceivably be divergences from the results of standard quantum theory, Bell was able to pinpoint specific types of experiments where it seemed quite plausible that quantum theory might be violated. Whereas Bohm's ideas, like the Bohr-Einstein debate, could be dismissed as armchair philosophy, with not the remotest relevance to practical physics, Bell's were effectively a call for experimental physicists to get to work!

The first of an extremely important pair of papers to be written (though *not* the first to be published, for reasons to be explained) was titled 'On the Problem of Hidden Variables in Quantum Mechanics'. It was a short paper, published from Stanford, and Bell acknowledged support from the US Atomic Energy Commission; it was submitted for publication in 1964. The centrepiece was an analysis of the von Neumann argument against hidden variables and also two other prominent arguments to the same conclusion, the one Jauch presented in his seminar at CERN, and another one we shall meet shortly. In each case Bell was able to show that assumptions were made in the argument which rendered it invalid.

At the beginning of the paper Bell had in a sense out-trumped Bohm by producing an extremely simple hidden variable model of his own, while at the end he discussed the non-locality of Bohm's model and suggested a generalization.

All this we shall discuss in detail, but we sketch it here to make the point that, at first sight, it seems strange that Bell published this paper in a review journal, in fact the most well-known review journal, the American *Reviews of Modern Physics*. In general, original research is published in regular journals—*Physical Review, Journal of Physics* and so on. Then every so often somebody takes on the task of reviewing all the literature in a particular field, and the resulting review is published in a review journal.

Sometimes what is achieved in a review is little more than a listing or summary of all the papers, which is obviously useful enough for all those working in the field or possibly hoping to enter it, but little more. More often though the author of the review will attempt a much more critical study of the journal literature, evaluating the experimental and theoretical contributions of the original writers, and possibly coming up with new ideas to generalize or explain the various individual results. Indeed there may be a fair amount of substantially new material in a review article. A really thorough review may be cited very heavily in future papers reporting original work, and the author of an excellent review may obtain considerable credit for writing it, though not, of course, as much as for a really novel original piece of research.

Despite that point, though, it is very unusual to publish original research on its own in a review journal. It would be less likely to be seen and noticed by those interested in the field of research. Yet this is effectively what Bell did in this paper. In the text he uses, to an extent, the language of a review. He remarks that, since Bohm's work of 1952, 'the realization that von Neumann's proof is of limited relevance has been gaining ground…However, it is far from universal.' The first point would seem to be an exaggeration, the second a definite understatement!

Bell remarks that he had not found any adequate account of what is wrong with the von Neumann argument, and adds that: 'Like all authors of noncommissioned reviews, he thinks that he can restate the position with such clarity and simplicity that all previous discussions will be eclipsed.' He also speaks of von Neumann, and the other producers of theorems stating the impossibility of hidden variables, as 'the authors of the demonstrations to be reviewed', but fairness would have to say that he is not 'reviewing' but strongly criticizing, if not the actual work itself, at least the far-reaching conclusions drawn from it.

Bell knew his arguments were extremely important, and yet he deliberately, in a sense, undersold them. It is true, of course, that different scientific authors tackle these matters very differently. Some like to present their results in a very matter-of-fact way, almost implying that they are already part of the scientific consensus, and thus should not be open to question or criticism. Others—almost certainly the majority—do the opposite; they stress and often over-stress, wherever possible, the originality of their work and how much it changes the scientific landscape.

It should also be said that Bell was definitely one of the former class. It would by no means be the case that he did not hope and expect to be valued for his achievements. There are extremely few scientists of who that could genuinely be said, and Bell had come a long way in his life, and naturally wanted to be respected for his important work. Yet he did not find it natural to draw attention to himself and his work.

An interesting example is his important paper on CPT invariance mentioned earlier. The paper he wrote reads exactly as though it were already in a textbook. Reading it, one gets no impression of the fact that it is a major advance, and indeed, one obtains no feeling of what would actually be quite helpful, any distinction between the points that are already well known and those where there are new arguments.

Yet, even having said that, the hidden variables paper is still presented in an unusual way. One feels that Bell was stating his points clearly for the record, and was certainly keen to attract the attention of those who might be interested, but still did not want to be seen to make a head-on challenge to the immensely powerful advocates of the Copenhagen position.

With that preamble out of the way, we now study the paper in greater detail. As we have said, Bell presents a very straightforward argument that demonstrates the possibility of a hidden variables approach giving the usual quantum expressions for measurements on a spin-$\frac{1}{2}$ system. Bell uses a hidden variable λ, which takes continuous values in the range $-\frac{1}{2} \leq \lambda \leq \frac{1}{2}$, and he is easily able to produce the required expression for a measurement of the spin component along *any* direction.

Actually we can look ahead to the second great paper where he produces a similar argument that is actually a little more instructive as it is geometrical rather than just using an algebraic expression. In this form of the argument, actually slightly adapted from that of Bell, for the case $s_z = \frac{1}{2}$, or 'spin up', the hidden variables are represented by unit vectors with uniform probability density over the upper hemisphere, that is to say all the vectors have positive components along the z-axis. Different members of the ensemble all have $s_z = \frac{1}{2}$, but each has a hidden variable represented by a vector pointing in a different direction.

Let us now suppose a measurement is taken of the spin along the z-axis. We look at the fraction of the ensemble for which the component of the vector for the hidden variable along the z-axis is positive. Of course this is just 1; as we have said, in this case *all* the vectors have positive components along this axis. We translate this to say that the probability of obtaining the result $s_z = \frac{1}{2}$ in a measurement is again just 1, exactly as we require. Similarly let us imagine a measurement of the spin along the $-z$-axis. For this case we look at the fraction of the ensemble for which the component of the vector along the $-z$-axis is positive and of course this is zero. This corresponds to getting the result $s_z = -\frac{1}{2}$ when we know in fact that $s_z = \frac{1}{2}$; it must indeed be zero. These results correspond exactly with the quantum results and have been obtained extremely easily.

However, we now need to consider a measurement of spin along a general direction. What we do is to find the fraction of the ensemble for which the vector has a positive component along this particular direction. For a general vector some vectors will have a component in the direction of the measurement and some against. If we were particularly lucky, it might turn out that we could obtain agreement with theory by decreeing that those with a positive component would yield a result with the spin in the direction of the measurement, and those with a negative component would yield a result with the spin in the opposite direction. That degree of success would be a little too much to hope for, but Bell was able to perform a little transformation of angles—geometrically we can describe it as a suitable squashing of the hemisphere—to obtain the result given by quantum theory.

Bell was ready to admit that both his models were highly artificial; there was no claim of actual physical significance or indeed generality. As he said, in any case a more complete theory would require an explanation of how the hidden variables behave during the measurement. However, this was completely irrelevant. What Bell had shown, in support of Bohm's work of a decade earlier, was that von Neumann's so-called 'proof' that quantum theory could not be reinterpreted as a theory of hidden variables must be incorrect. Bell's hidden variable model was much less sophisticated than that of Bohm, but in some ways that was an advantage rather than a disadvantage, as it was much harder for opponents to suggest that there must be an error in the argument. With these two working models at his hand, Bell would now be able to see exactly where the various impossibility proofs had gone wrong, and this is exactly what he proceeded to do.

First naturally he looked at von Neumann's argument. It should be said that von Neumann was indeed a very great mathematician. One would not expect to find any mathematical mistakes in his work, and Bell certainly did not do so. But though his great book *Mathematical Foundations of Quantum Mechanics* certainly contributed to improving vastly the level of mathematical rigour in quantum theory, one might be less certain that the mathematics mirrored in a satisfactory way the underlying physical assumptions. This was exactly the kind of mismatch that Bell searched for, and which he found.

To explain this we need to define a quantum mechanical term *expectation value*. To give an example, suppose in a measurement of energy, there is a probability p_1 of getting the answer E_1 and a probability p_2 of getting the answer E_2. Then the expectation value for the measurement is $p_1E_1 + p_2E_2$. In other words the expectation value is a suitably weighted average of the possible results of a measurement. (It may immediately occur to the reader that the term *expectation value* is a very confusing one. For we *know* that we will get one of the values E_1 or E_2. In other words not only do we not *expect* to get the value $p_1E_1 + p_2E_2$, we know we won't get it! But this is in any case the definition handed down to us for good or ill!)

There is now a very interesting relationship in quantum theory. For example, we can consider two components of spin, say s_x and s_y. We can then take a linear combination of these, say $s_x + s_y$, and proceed to calculate the expectation values of s_x, s_y and $s_x + s_y$, which are usually written $\langle s_x \rangle$, $\langle s_y \rangle$ and $\langle s_x + s_y \rangle$. We find that $\langle s_x + s_y \rangle = \langle s_x \rangle + \langle s_y \rangle$. This is a general result applying to all similar cases.

It is very natural to take this for granted as it seems practically a straightforward mathematical result. Yet Bell pointed out that it is anything but straightforward. It must be remembered that we are not talking about physical quantities; we are talking about the results of measurements. The first expectation value is to do with the results of a measurement of spin along the x-axis, the second with the results of a measurement along the y-axis and the third actually deals with the results of a measurement along an axis bisecting the x- and y-axes. It is far from an obvious fact that the third expectation value should be the sum of the first two. The fact is that, as we have said, quantum theory tells us that the result is true, but actually this is little more than a coincidence!

Now let us turn to the hypothetical hidden variables that von Neumann considered in his argument. He assumed that the same relationship concerning expectation values applied to these hidden variables, and this assumption was built into his 'proof' that hidden variables could not exist. It is easy to understand how this assumption could have been made. It was perhaps natural to apply the same rule that applies to the observable quantities that we know about to these additional ones that may or may not exist. But more than that it is perhaps easy to forget that the expectation values indeed refer to results of measurements. As we have said, entirely different measurement schemes are required to measure s_x, s_y and $s_x + s_y$, and it is completely inappropriate to assume that the expectation values can be added as though they were simple physical quantities.

In any case Bell was able to point out that this additional assumption was completely unwarranted. It was great to have the working scheme of Bohm and his own scheme as he was able to show clearly that the assumption was *not* correct in these cases. With this assumption removed, von Neumann's whole argument collapsed. In Louisa Gilder's charming book [12], where the participants in the various events are presented rather as in a drama, she presents Mary Bell reading her husband's paper in draft and commenting: 'Von Neumann's assumption does appear rather foolish when you phrase it like that.' Whether Mary actually used these words we do not know, but if she did it must be said that she had a good point! And yet the theorem based on this assumption had held up the development of arguments opposing or questioning Copenhagen for nearly thirty years.

In his paper, Bell now turned to the argument of Jauch that he had discussed with Jauch himself in Geneva a few months before. The argument had actually been published a few months earlier in a joint paper of Jauch and Constantin Piron. The paper was an exercise in quantum logic, which may be described as a set of rules for reasoning about propositions taking the laws of quantum theory into account. The reader may feel that this definition makes quantum logic sound extremely abstract, and this was almost certainly Bell's point of view as well. However lucid the reasoning might be in purely mathematical terms, he would have felt that the translation from mathematical theorem to physical application would be exceptionally hazardous.

Jauch and Piron's argument was based on projection operators. A projection operator projects any vector into its component along a given direction. For example, in our usual three-dimensional space, we might have projection operators projecting a general vector along the *x*-, *y*- and *z*-axes. In their work observable quantities are represented by projection operators and we discuss their expectation values. If a and b are two projection operators projecting onto particular subspaces, then we define $a \cap b$ as the projection onto the intersection of the two subspaces.

The important part of the argument of Jauch and Piron considers the expectation values of these projection operators. They assume that if, for some state, $\langle a \rangle = \langle b \rangle = 1$, then it follows that for that state $\langle a \cap b \rangle = 1$. Bell asks why they regard this as a natural assumption, and suggests that they are making an analogy with the behaviour of propositions in normal logic. Here we can say that the value 1 corresponds to the truth of a proposition, while the value 0 corresponds to its falsehood, and $(a \cap b)$ means (a and b). So $\langle a \rangle = \langle b \rangle = 1$ means that a and b are both true, and if this is the case, so also is $(a \cap b)$. Once Jauch and Piron have made this assumption, it is fairly easy to show that no hidden variables are allowed in quantum theory.

However, as Bell points out, they have made exactly the same kind of mistake as von Neumann. The projectors a and b are not logical propositions; rather they relate to the results of measurements, and in general different measurements require entirely different setups. In no way can two projectors be added in the way Jauch and Piron suggest. As we have said, it is in fact the case that the statement is true for the observables of normal quantum theory, but there is absolutely no reason to believe it should be true for any hypothetical hidden variable states, and it is not true for the hidden variable models of Bohm and Bell himself. Thus the argument of Jauch and Piron also failed.

Bell now turned to the third of what he called the impossibility proofs. He had been directed by Jauch to the important work of Andrew Gleason published in 1957. Like the paper of Jauch and Piron, Gleason's work is an exercise in quantum logic, and it is a very general study of the axiomatic basis of quantum theory. Thus it was not explicitly a contribution to the hidden variable question, but Jauch pointed out to Bell that its conclusion could easily be used to produce an impossibility (of hidden variables) theorem. Later Bell called this the Gleason-Jauch argument. In fact, unwilling to study the full and extremely complicated Gleason argument, Bell was able to produce this important application directly himself, and so the result is sometimes called Bell's Second

Fig. 8.10 Simon Kochen [© Zentralbibliothek für Physik, Renate Bertlmann]

Theorem, the 'first' being the content of his local realism paper discussed later in this chapter. Actually the result is most often called the Kochen-Specker theorem since it was independently discovered by Simon Kochen (Figure 8.10) and Ernst Specker in 1967.

The Kochen-Specker theorem refers only to systems with at least three distinct observables, and it aims to avoid the problems of the von Neumann argument by restricting itself to observables that may be measured simultaneously with the same apparatus. Essentially the theorem starts by assuming that the hidden variables have specific values before any measurement, very much the assumption of EPR. It then succeeds in using the quantum formalism to demonstrate contradictions, and thus it seems that the initial assumption must be wrong. Thus this argument is a much bigger challenge to Bell than that of von Neumann, or that of Jauch and Piron. Yet he was able to circumvent it.

It is true, he said, that we may measure the two observables, say A and B, with which we are directly concerned, with the same apparatus—that is precisely the point made earlier, the point where the Kochen-Specker theorem is more powerful than that of von Neumann. Yet we must measure at least one other observable together with these two because, as we have said, there must be at least three distinct observables. We may choose, for example C_1 or C_2 as our third observable, and we must admit that A, B and C_1 may be measured simultaneously with the same apparatus; similarly, A, B and C_2. However, Bell pointed out that we may not assume that C_1 and C_2 themselves may be measured simultaneously. Thus the total apparatus setup used for measuring

A, B and C_1 will in general be different from that for measuring A, B and C_2. Thus, Bell concludes, there is no reason to assume that the results for A and B will be the same in the two cases. This renders the argument against hidden variables ineffective.

We have said that Bell thus circumvented the Kochen-Specker theorem, but in a sense he had taken a hit. The EPR assumption, much the simplest idea of a hidden variable, which says that the hidden variable has a value before any measurement, and this value will be the result of the measurement, has had to be been rejected. Instead we must recognize that the result of the measurement depends on the 'complete disposition of the apparatus'. Bell actually quotes Bohr on this point concerning 'the impossibility of any sharp distinction between the behaviour of atomic objects and the interaction with the measuring instruments which serve to define the conditions under which the phenomena appear'. It must have been rather irritating for him to have to acknowledge Bohr in this way, but he did not shirk his duty!

The modern expression for this type of behaviour is that hidden variable theories with three or more distinct observables, or as we would say, of dimension at least three, must be *contextual*, in the sense that any measurement result depends on the full context of the measurement.

Does the Bohm theory obey this rule? It certainly does. Any account of the theory reports that the hidden variable has a value at all times, but that this is not necessarily the value obtained in a measurement. Rather the measurement process consists of an interaction between measured system and measuring system in which the value of the observable may change. This process is rather straightforward, though, rather than being of an esoteric quantum nature. Incidentally the Kochen-Specker argument does not apply to Bell's own scheme because the scheme does not contain three distinct observables.

For the last section of the paper, Bell turned to the Bohm theory itself and demonstrated clearly what Bohm himself had, of course, already stated, its non-locality. Bell considers a system of two spin-$\frac{1}{2}$ particles with magnetic moments so that they may be detected using magnetic detectors. He shows that the trajectory of each particle depends on the trajectory and wave-function of the other one, and so the measurement results for each particle will depend on the disposition of the apparatus measuring the other, even though the two particles and the two measurement apparatuses may be very far separated. Bell actually comments here that what he calls the EPR paradox is thus resolved in the way which Einstein would have liked least—hidden variables or realism restored but only at the expense of locality.

By the time he wrote this paper, Bell had already struggled valiantly to produce other hidden variable schemes that would not suffer from this defect and would be local, but he had failed. He therefore suspected that there must be a general requirement that *any* hidden variable scheme must be non-local, and this is how he finishes the paper; he suggests searching for an 'impossibility proof' that shows the impossibility of hidden variable theories that respect 'some condition of locality, or of separability of distant systems'.

The section is actually titled: 'Locality and separability', and it is interesting to compare the two words. The idea of separability is that two systems separated in space should evolve separately. It was, in fact, very much Einstein's demand in, for example, EPR, because he did not believe that the pursuit of physics was possible unless this requirement was maintained. It was essentially separability that was tested together with realism in the earlier tests of Bell's work (which we will meet in the next chapter).

Locality allows an event in one sub-system to respond to an event in another sub-system, but only if the time between the events is less than that taken by light to travel from one event to the other. It is locality that was tested, again together with realism, in later tests of Bell's work.

In his acknowledgements at the end of this paper, Bell gave special thanks to Mandl, thanking him in particular for intensive discussion back in 1952 when, Bell says, the first ideas of the paper were conceived; he also thanked Jauch. It is to be hoped that this was some compensation for dismissing the ideas that they had presented with such passion to Bell, those of von Neumann, and of Jauch and Piron respectively, in roughly a page each!

This first great paper of Bell had something of a chequered career before seeing the light of day and being published. First it had luck in that the referee was sympathetic. This was certainly not a foregone conclusion, as so many of those physicists best known in quantum theory were totally hostile to the very idea of hidden variables. Louisa Gilder suggests that the referee was Bohm and that certainly makes sense, both because of his generally positive view of the subject, but also because he asked for greater attention to be paid to the measurement process. It will be remembered that Bohm thought that full consideration of the measuring apparatus was absolutely central to any discussion of hidden variables. Bell, though of course recognizing that discussion of measurement was important, particularly in the light of contextuality, felt that many aspects of the hidden variable question could be discussed without giving a complete discussion of measurement.

Bell had to give at least some ground to the referee, but in fact chose to make a minimal response. The eventually published paper contains the words: 'A complete theory would require an account of the behaviour of the hidden variables during the measurement process itself. With or without hidden variables the analysis of the measurement process presents peculiar difficulties, and we enter it no more than is strictly necessary for our very limited purpose.' (He also gives some references that we mention briefly in the following section.) In the paper he had stated that, in his analysis of the Bohm theory, the measurements involved are ultimately ones of position. In a footnote he adds that

'There are clearly enough measurements to be interesting that can be made in this way. We will not consider whether there are others.' He really seems to be saying that he has already addressed the relevant aspects of measurement and does not intend to add any more!

He then sent the paper back to the Editor of *Reviews of Modern Physics*, but that is where any good luck finished. Unfortunately the letter with the revised version of the paper was misfiled at the journal. When the Editor wrote to Bell inquiring why he had not submitted a revised version, he sent the letter to Stanford, the address from which Bell had submitted the original version of the paper. Bell had by this time returned to England, and again unfortunately nobody at Stanford was alert enough to forward the letter to him. It was not until Bell himself at last wrote to the journal asking what was the fate of his revised version that two-and-two were put together and it was realized what had happened. The revised version of the paper was then published but by now it was 1966! (Bell had been very patient, but this is because authors of submitted papers do not like to harry editors; they are likely to get their papers rejected!)

By this time, Bell had actually submitted what I call his second great paper, which is discussed in detail in the following section. This answered the question at the end of the first paper—all hidden variable theories must indeed be non-local, and Bell was able to include this point as a note at the end of the *Reviews of Modern Physics* paper.

Bell and measurement—1966

In this section, I will discuss briefly references added to the paper in connection with the referee's request for more discussion of measurement. They give some clues to Bell's views on measurement at this time, views that were to strengthen over the rest of his life. Included are papers by Jauch, by Eugene Wigner and by Abner Shimony, who had already thought deeply about quantum measurement, and who we shall meet fully in the next chapter.

Wigner (Figure 8.11) deserves to be classified as a genuine hero of our work. He was one of the group of astonishingly able Hungarian mathematicians and physicists who moved to the West between the two world wars; others were von Neumann himself and Leo Szilard, who was perhaps the first to recognize the possibility of constructing nuclear weapons, and who we shall also meet in Part III of the book in connection with his ideas on information.

Wigner won the Nobel Prize for Physics in 1963 for his work on nuclear physics and elementary particles, and he was recognized as an extremely influential physicist for decades before then. Yet, practically uniquely among such highly influential people, he was prepared to admit that there were problems remaining in the interpretation of

Fig. 8.11 Eugene Wigner [courtesy of John C. Polanyi, from I. Hargittai, *The Martians of Science* (Oxford: Oxford University Press, 2006)]

quantum theory and quantum measurement. He came up with some ideas of his own, and, as we shall see at various points in this book, he also supported work of this type by others, even when it was in rather different directions from his own.

Another of Bell's references was to a book by Bernard d'Espagnat (Figure 8.12). D'Espagnat is a (relatively) unsung hero of the study of the foundations of physics. He was Director of the Laboratory of Theoretical Physics and Elementary Particles at the University of Paris XI (Orsay) and a frequent visitor to CERN from the very beginning of the organization. In the early 1960s, both Bell and d'Espagnat had concerns about quantum theory and its current interpretation, but at CERN this was not a thing that you readily admitted. D'Espagnat, though, became particularly suspicious when he saw a book by de Broglie on Bell's bookshelf. (As a proud Frenchman, d'Espagnat may have been particularly pleased that the book was in French!) Eventually discussions did take place, and thereafter d'Espagnat was perhaps Bell's most constant supporter concerning the actual physics.

Fig. 8.12 Bernard d'Espagnat [© Zentralbibliothek für Physik, Renate Bertlmann]

Fig. 8.13 The 1970 Varenna Summer School on the Foundations of Quantum Mechanics: *Scuola Internazionale di Fisica 'E. Fermi'*. In the second row (seated), Bell is 6th, Wigner 8th, Piron 9th from the left; Jauch is 1st and d'Espagnat 3rd from the right. In the front row, Kasday is 4th and Hiley 5th from the left, and Zeh is 2nd from the right. Santos is 3rd from the left in the third row. Shimony is 2nd from the right in the fourth row while Selleri is on his own towards the right in the fifth row. [Courtesy ofAcademic Press]

In 1970, the Italian Physical Society, on the suggestion of Franco Selleri, proposed that d'Espagnat should organize a summer school at Varenna on the Foundations of Quantum Mechanics (Figure 8.13). D'Espagnat received great support from Wigner and Bell, and this school played a crucial role at this important time, bringing together virtually everybody who was involved in the subject. We shall have cause to refer to this school several times later in the book.

D'Espagnat's article 'The Quantum Theory and Reality' [13] was published in the popular journal *Scientific American* in 1979, and was exceptionally important in publicizing Bell's work, which was certainly still necessary even at that comparatively late date, while his book *Conceptual Foundations of Quantum Mechanics* [14], which was published in 1976, provided an extremely useful account of the underlying ideas.

On the philosophical side, their ideas were close but they did differ. While Bell's realism remained untarnished, d'Espagnat came to believe in an unconventional realism. He felt that, while the idea of a reality independent of mind made sense, what science describes is only a set of collective appearances. He believed in an empirical reality or, in the words of the title of one of his books, *Veiled Reality* [15]. On one occasion he sent Bell a preprint that contained the words 'empirical reality'; Bell sent back a note with 'I do not believe in empirical reality' written diagonally across it is large letters! But we shall see that d'Espagnat's contribution to Bell's work was important.

The last of the papers referenced by Bell was by a paper written by himself together with Michael Nauenberg of Stanford University, and published in a book in honour of the distinguished theoretical physicist Victor (or 'Viki') Weisskopf. Both Bell and d'Espagnat were invited to contribute to this volume, and both chose to write on the interpretation of quantum mechanics. Nauenberg at the time was employed at Columbia University and, like Bell, was a visitor at Stanford. In 1966 he was to move to the University Santa Cruz where he had a most successful career for the next thirty years. He was to publish widely in the fields of particle physics, condensed matter physics and non-linear dynamics, but he also developed a substantial interest in the history of physics, studying the works of Hooke, Huygens and, in particular, Newton.

Bell and Nauenberg's paper in the Weisskopf volume, which is included in *Speakable and Unspeakable in Quantum Mechanics,* is called 'The Moral Aspect of Quantum Mechanics', where the word 'moral' is used in a totally unconventional and, it must be said, rather confusing way. They studied quantum measurement processes in some detail, and use 'moral' to mean processes where the state of the measured system is unchanged during the interaction between this system and the measuring device, in contrast to 'immoral' measurements in which the state of the measured system changes.

Actually much more significant from our point of view is what happens subsequent to this interaction between measuring and measured systems, which is, at least according to conventional wisdom of the time, the collapse of the wave-function to a single term, as a result of so-called 'observation'. However, Bell and Nauenberg, while admitting that, of course, this conventional wisdom works extremely well in practice, were able to demonstrate problems with this idea in principle.

First they imagined an experiment in which the measurement interaction occurs instantaneously at three separate times, $t = 0$, τ and 2τ. For an 'immoral' measurement, they were able to show that the probabilities of obtaining various answers depend on whether you apply collapse after each measurement, or just once after the third one.

They extended this idea to wonder whether the collapse could or even should be moved further forward. The measuring instruments, after all, are just large assemblies of atoms, so it seems illogical to discriminate between them and the system being observed, which is also just an assembly of atoms. So perhaps superposition of different measurement states, just like superpositions of different atomic states, could remain uncollapsed.

Indeed Bell and Nauenberg, perhaps in a rather tongue-in-cheek way, considered moving the point of collapse even further up the chain. Theorists, such as Bell and Nauenberg perhaps, are maybe not over-concerned with the actual experimental arrangements, so even the actions of the experimenters may remain as a superposition of different actions, rather than being collapsed down to a single classical-like experiment. Even the carrying out of their duties of the editors of the journal where the work is eventually published may conceivably, they suggest, not require collapse from the point of view of the preoccupied theorists. For theorists, the decisive action or observation that might be said to provide the all-important information about the result of the experiment is just reading the result in the journal. So perhaps it is just this action that should trigger the collapse.

But even this, they suggest, may be to jump the gun. For reading the journal is clearly an act of no more significance than several of those further back in the chain—reading measurement pointers or reading computer output, for example. And, as the authors say, 'at this point we are finally lost'. If we do not collapse at this point we are left with a state vector for the universe containing all possible worlds. Then 'quantum mechanics in its traditional form has nothing to say. It gives no way of, indeed no meaning in, picking out from the wave of possibility the single unique thread of history.'

Thus Bell and Nauenberg reached the conclusion that quantum mechanics is, at best, incomplete, and they looked forward to a new theory in which events may occur without requiring collapse caused by 'observation' by another system. Observation in a sense relates to the consciousness of the being carrying out the observation, and the authors raised the question of whether consideration of consciousness may have to be dragged into physics. However, they expected that physics will have recovered a more objective description of nature long before it begins to understand consciousness.

They admitted, incidentally, that practical people who are uninterested in logical questions did not need to read their paper. They recognized that their own view is a minority one, and that typical physicists imagine that all such questions have been answered long ago, and that they could fully understand how if they could only spare twenty minutes to think about it! Bell and Nauenberg also accepted that the minority view has also been around ever since the beginning of quantum theory, so they agreed that the new theory may well be a long time coming!

For themselves, though, they did believe that quantum mechanics will eventually be superseded. This, of course, it may be said, is the fate of all theories, however successful in their own terms. Newton's mechanics has been superseded by both quantum theory and relativity, and though Maxwell's electromagnetism, the other central constituent of classical physics, triumphantly survived the coming of relativity, it also had to bow to quantum theory. However, Bell and Nauenberg believed that the eventual downfall of quantum theory is somewhat different. They suggested that its 'ultimate fate is apparent in its internal structure. It carries in itself the seeds of its own destruction.'

Bell's views were scarcely to change for the rest of his life, as we shall see, in particular, in Chapter 14, but it is interesting that he made them so clear during this American visit, as well as writing his two great papers. Once he returned to England, as we shall also see, it must be said that he remained rather cagier about promoting his views.

Bell's second great paper—realism and locality

As we said, in his *Reviews of Modern Physics* paper Bell raised the possibility that *all* hidden variable theories might be non-local. In other words, since we are describing hidden variable theories as being *realistic*, the suggestion is that, assuming quantum mechanics is correct, the universe cannot exhibit both locality and realism. In his second great paper, he was able to show that this is, in fact, the case.

The title of the paper was 'On the Einstein-Podolsky-Rosen Paradox' and indeed the argument was totally based on the EPR paper. The fact that Bell was able to draw quantitative conclusions from his analysis of EPR would probably have seemed surprising at the time to those of both camps, the vast majority who had written off the EPR paper as irrelevant, but also the tiny minority who though it made a useful point. Even the latter would have thought of the Bohr-Einstein debate as fundamentally one of conceptual analysis—armchair philosophy if you like. One might say that nobody, not even Bohr or Einstein, would have supposed it could lead to concrete results or tests! Yet amazingly in the hands of Bell it did so.

We shall briefly review the Bohm version of the EPR argument. (Sometimes this is called EPR-Bohm, but mostly, since it is the version that has always been considered since Bohm invented it, it is just called EPR, unfair to Bohm as that may seem.) We here stress a point that was not mentioned previously. The source of the entangled state, the initial state of our system is just a single particle with spin $S = 0$, which breaks up or decays into two spin-$\frac{1}{2}$ particles. Because, of course, the total spin of the system must be conserved, we may say that either spin 1 must be in the $|+\rangle$ state and spin 2 in the $|-\rangle$ state, or spin 1 in the $|-\rangle$ state and spin 2 in the $|+\rangle$ state. In fact though there is no reason for the system to take the first option rather than the second, or vice versa. The system will be in the entangled state $\left(1/\sqrt{2}\right)\left(|+\rangle_1|-\rangle_2 - |-\rangle_1|+\rangle_2\right)$.

Whereas in the previous chapter we had ± between the two products, because it was only entanglement that was being discussed, here we are specifically starting with the $S = 0$ state and quantum theory says that this implies the minus sign. It is also important to recognize that the initial state before decay is spherically symmetric, so the final state must also be symmetric between the x-, y- and z-axes. This is far from obvious in the form in which we have written it, where we have used the usual convention of taking the z-axis as primary in the sense that $|+\rangle$ and $|-\rangle$ refer to the z-axis unless stated otherwise. This means that we could write the state more clearly as $\left(1/\sqrt{2}\right)\left(|+_z\rangle_1|-_z\rangle_2 - |-_z\rangle_1|+_z\rangle_2\right)$. However, as we have said the state is actually symmetric between x, y and z, so we could equally well write it as $\left(1/\sqrt{2}\right)\left(|+_x\rangle_1|-_x\rangle_2 - |-_x\rangle_1|+_x\rangle_2\right)$ or $\left(1/\sqrt{2}\right)\left(|+_y\rangle_1|-_y\rangle_2 - |-_y\rangle_1|+_y\rangle_2\right)$. It is certainly far from obvious that all these ways of writing the state refer to exactly the same state, but the fact is that they do! And indeed we may go further. Rather than just x-, y- and z-axes, the suffix may refer to any direction at all.

The key aspect of Bell's analysis was that, up to that time, the very few people who had considered the EPR argument had nearly always considered measurements of the component of both spins along the z-axis. Just occasionally they had mentioned the component of one spin along the x-axis, but no other components were even considered. Bell's argument centres on analysing measurements carried out in *general* directions. In hindsight and with the benefit of a further nearly fifty years of understanding, this may not seem the greatest of advances, but that is to neglect the fact that, in the thirty years since the publication of the EPR argument and right up to Bell's work, it was taken literally for granted that nothing could be achieved by making this step.

Bell's argument takes place in two stages. In the first, the calculations are practically trivial, being indeed essentially a rerun of the EPR-Bohm argument, but the conclusions Bell draws from it are central for his own argument. He imagines a measurement along *any* axis in one wing of the EPR-Bohm experiment, and the same axis in the other wing, and says that if the measurement on the first spin gives the result $\hbar/2$, then the measurement on the second spin *must* give the result $-\hbar/2$.

He then brings in what he calls a requirement of 'separability or locality', in fact two requirements. First he remarks: 'Now we make the hypothesis, and it seems one at least worth considering, that if the two measurements are made at places remote from one another, the orientation of one magnet [which measures the spin component] does not influence the result obtained with the other.' We note that that this requirement of locality is one of parameter independence; a setting of the apparatus at one point should not affect experimental results at another.

Bell immediately adds another locality requirement. He says that because we can predict the experimental result in one wing by knowledge of the result in the other wing, we must assume that that each result is, in fact, predetermined. This implies the existence of hidden variables or, as Bell puts it, 'a more complete specification of the state'. It is an assumption of locality, but this time of outcome invariance—a measurement result in one wing of the apparatus cannot have an immediate effect on the result in the other wing.

It is interesting to consider why Bell introduced these ideas in the rather limp way—'a hypothesis... at least worth considering'. Surely locality was a very natural concept that would hardly have been challenged—at least explicitly. But here Bell was effectively saying that if, as Copenhagen did, one denied the existence of hidden variables, one was also denying locality. Again he seems to have been following the path of making his own views clear, without going out of his way to offend any more than necessary the highly powerful Copenhagen élite.

It should not be questioned that Bell himself was hugely concerned about the existence or otherwise of locality. For him, realism was the most important requirement for physics, and locality somewhere behind it. This would be as true for Bell as for Einstein, but after that there was a distinct difference between the two men; determinism was also very important for Einstein but scarcely important at all for Bell.

Although this first part of the paper was very simple, indeed, as we have said, very much a rerun of the original EPR-Bohm argument, Bell considered it very important in making clear the assumptions behind his argument. Twenty years later he was to write that in EPR locality is assumed, not determinism, which is inferred as part of the argument. He mentioned 'a widespread and erroneous conviction that for Einstein determinism was always the sacred principle'. In a footnote he adds: 'And his followers. My own first paper on the subject starts with a summary *from locality* to deterministic hidden variables. But the commentators have almost universally reported that it begins with deterministic hidden variables.'

Bell now proceeds to the main mathematical argument, though he first gives some background. For his hidden variable he uses a parameter λ. As he says, it is completely unimportant whether λ denotes a single variable, a number of variables, or a number of functions, or whether the variables are discrete or continuous. Some people, he suggests, might prefer the hidden variables to fall into two sets, one for each wing of the experiment. But that refinement, he says, is just an example of his highly unspecific assumption.

The central point, he says, is that A, the measurement result in the first wing should depend only on **a**, the direction of the field in that wing and λ, *not* on **b**, the field direction in the other wing. Similarly B, the result in the second wing, should depend on **b** and λ, not on **a**. It should also be stressed that A does not depend on B, and B does not depend on A, over and above the fact that both depend on λ.

Bell now shows that a number of simple experimental results are easily predictable on this basis. The first is just the measurement of spin on a single particle and we looked at this earlier in the chapter. We imagine the hidden variable λ being represented by vectors of unit length over a suitable hemisphere, and a suitable though admittedly *ad hoc* rule gives us the correct probabilities of a measurement of a component of spin in any particular direction giving the result $\hbar/2$ or $-\hbar/2$.

The second task easily performed is to reproduce the results of an EPR-Bohm experiment for the cases where the fields in the two wings are either along the same direction or along perpendicular directions. In this case Bell uses a uniform distribution over all directions for his vectors representing λ, and the required result is obtained trivially. We remember that these were the only two cases of the experiment discussed up to this point, so the fact that there are trivial local hidden variable explanations for them makes it absolutely clear why there was no hope of Bell's result being obtained until he analysed more general cases.

Lastly he shows that there is no difficulty in reproducing the result for the general case in EPR-Bohm *if* we were to allow the result of the measurement in one frame to depend on the measurement setting, or in other words the direction of the field, *in the other frame.* We can then apply exactly the same kind of *ad hoc* trick as in the single particle case we just mentioned. However, the trick does require knowledge of **b** when obtaining the measurement results for A, and knowledge of **a** when obtaining the results for B. This is, of course, exactly what is ruled out by our insistence on locality.

So now he turns to his main argument which assumes local realism. This is the only part of the analysis used in any experimental work, so such work tests exactly local realism; the argument from locality to realism and determinism in the earlier section of Bell's argument plays no part in this. And as stressed before, the great advance on any earlier discussion is that the measurements in either wing of the experiment may be along *any* direction.

Bell's argument is straightforward but subtle. He analyses three different experimental arrangements, each of which involves two of the three field directions that he considers, so each field direction occurs twice. He thus produces measurement probabilities $P(\mathbf{a}, \mathbf{b})$, $P(\mathbf{a}, \mathbf{c})$ and $P(\mathbf{b}, \mathbf{c})$. When he relates these probabilities, he is able to show that an inequality, which he has shown is a direct consequence of the assumption of local realism, is not obeyed, either accurately or arbitrarily closely, by the usual quantum mechanical expression. We may call this result 'Bell's theorem'. The inequality is the first of many so-called 'Bell inequalities', though the inequality now used almost universally was not produced by Bell himself; it is discussed in the following chapter.

Before discussing the implications of this exceptionally important result, we look at another proof of the same result, which was produced by Nick Herbert [16]. It is very short and simple in form, but again it is quite subtle.

It actually relies on an interesting property of EPR-Bohm pairs. We know that if the measurement directions are the same, for example both along the z-axis, we will always get opposite results; in one wing we will get the result $\hbar/2$ and in the other the result $-\hbar/2$. Or we can say that the probability of getting the same result is zero. However, we can express the same facts in another way. We can consider the field as being along the z-axis in the first wing, but along the $-z$-axis in the other wing. With this slight change in definition, we can now say that we are bound to get the same result in both wings. If the measurement of spin gives a result along the z-axis in the first wing, the measurement result in the second wing is certain to be along the $-z$-axis. We may say that the probability of obtaining this result is 1.

Now we keep the measurement direction in wing 1 the same, but we rotate the direction in wing 2 by an angle θ. A straightforward use of quantum mechanics tells us that the probability of obtaining the same result is now, as we should expect, less than 1; to be exact is $\{1 + \cos \theta\}/2$. So the decrease in the probability of agreement is just $1 - \{1 + \cos \theta\}/2$, which is $\{1 - \cos \theta\}/2$. Again it is a standard mathematical result that, for small values of θ (and in this argument we shall assume all angles are small), $\cos \theta$ is very nearly equal to $1 - \theta^2/2$, so we may write the probability of disagreement as $\theta^2/4$. The appearance of the *square* of θ is perhaps unexpected; it is actually typical quantum behaviour, and we shall see that it is the cause of the disagreement with a local hidden variable approach. Any classical approach would give a dependence on θ rather than θ^2, and this dependence would *not* lead to disagreement with local realism.

In the local hidden variable model, realism implies that as we swing through θ, a certain number of vectors representing the hidden variables change from contributing to agreement, which they all do when θ is zero, to contributing to disagreement. We can say that, when we have turned through θ, the probability of disagreement is $d(\theta)$. Then when we turn through a further angle θ, we introduce a further probability of disagreement $d(\theta)$. Thus after both rotations the probability of disagreement will be $2d(\theta)$. But we can now imagine performing the full rotation of 2θ all at once and the probability of disagreement after this rotation should be $d(2\theta)$.

Thus we have obtained the equation $d(2\theta)=2d(\theta)$, and we may say that this result is obtained from a hidden variable or realist approach. Yet when we look at the quantum predictions, we see that the probability of disagreement after a rotation of θ is $\theta^2/4$, so $2d(\theta)$ is $\theta^2/2$, but $d(2\theta)$ is $(2\theta)^2/4$ or just θ^2. We immediately see that there is total disagreement between the hidden variables approach and the quantum expressions. (On the other hand, it is easy to see that, for the classical case, where the probability of disagreement is proportional to θ, we get agreement between hidden variables and theory.)

We have tried to keep the argument simple to get the general point across. Yet we must admit that it is, in fact, over-simplified because we have not explicitly mentioned the idea of locality, which is obviously wrong since it is *local* hidden variable models we are discussing.

Where we have actually assumed locality in the above argument is that we have taken it for granted that the decrease in probability of agreement is the same when the angle of rotation increases from 0 to θ as when it increases from θ to 2θ. In each case we have assumed that it is $d(\theta)$. There are two points we now need to consider—first, why we may avoid this assumption if we do not take locality into account, but second, why we must stick with it when we *do* include locality.

The first point is actually straightforward. Provided we have a reference point for our zero, we are by no means obliged to assume that the hidden variables will have the same effect when the angle increases from θ to 2θ as it did when it increased from 0 to θ. Indeed this is actually the freedom Bell exploited in what we called earlier in this section his first prediction of simple experimental results, that for a measurement of component of spin for a single particle.

However, we must now bring in locality. We talked of a reference point for our zero, relative to which we may measure our angle in wing 2, but a little thought tells us that the only possible reference point is the measurement direction *in wing 1*. It is with reference to that direction that we have spoken of our angle of rotation in frame 2, from 0 to θ to 2θ, and this justifies our handling the two parts of the rotation separately.

But the principle of locality rules out consideration of the measurement setting in wing 1 when we are analysing measurements in wing 2. When we take locality into account we cannot discuss the two rotations of the measurement direction in wing 2 as first 0 to θ, and second, θ to 2θ. Rather all we can say is that in each rotation the angle increases by θ. This in turn justifies our initial statement that for each of the rotations, from 0 to θ, and from θ to 2θ, the probability of disagreement increases by the same amount, which we called $d(\theta)$. Thus our original proof did indeed assume locality as well as realism, so it is a proof that a local realistic model cannot reproduce the results of quantum theory.

(Parenthetically we note two points. First, when we examine the situation a little more closely, we recognize that some of the disagreement built up during the first rotation may be cancelled in the second rotation. Thus our equation becomes $d(2\theta) \leq 2d(\theta)$. However, we note that this adjustment makes the conflict between local realism and quantum theory more rather than less pronounced. Second, it is pointed out that our argument connects with what Bell called above the third simple prediction of an experimental result. It shows that there is no difficulty in providing a hidden variable approach for the general EPR case *if* we can make use of the measurement direction in the second frame. This is the argument here.)

At this point, though, we return to Bell's paper. He concludes by referring to possible experiments, commenting that his derivation 'has the advantage that it requires little imagination to envisage the measurements actually being made'. His conclusions were, of course, exceptionally interesting and important in their own right. But they also had the great merit of suggesting types of experiments that could check fundamental aspects of the Universe. This may be compared with the half-hearted suggestions of Bohm that there *might* be areas of physics where his approach *might* give results differing from those of quantum theory; he was broadly dismissed as an armchair philosopher.

In contrast Bell had been able to suggest an area of physics where there was a genuine possibility that quantum theory might give the wrong result. The assumption was presumably that the true theory of microscopic objects, unlike quantum theory, respected local realism. Its results happened to be experimentally indistinguishable from quantum theory in nearly all cases, but obviously where quantum theory differed from the requirements of local realism, quantum theory must be wrong.

Was this likely? Bell, at least, *hoped* it might be true. He was to write much later that 'First, and those of us who are inspired by Einstein would like this best, quantum mechanics may be *wrong* in sufficiently critical situations. Perhaps, Nature is not as queer as quantum mechanics.' If it were found to be true, there would be two very good reasons for Bell to rejoice. First, it would show that nature did respect realism and locality, and this would have given intellectual and even aesthetical satisfaction to Bell, as it would have done to Einstein. But it would also have massively increased his scientific standing, as the person who demonstrated the limitations of quantum theory, and, at least to an extent, where and perhaps how one might look for its successor. (Bell, it might be remarked, was not a conceited or demanding man, but all scientists thrive on interest in and support for their work.)

He admitted that the likelihood of quantum theory being overthrown was probably slight. In a letter to John Clauser, who is discussed further in the next chapter, he wrote:

In view of the general success of quantum mechanics, it is very hard for me to doubt the outcome of such experiments. However, I would prefer these experiments, in which the crucial concepts are very directly tested, to have been done and the results on record. Moreover, there is always the slim chance of an unexpected result, which would shake the world!

It must be remembered that there are in principle two types of experiment, the first in which the settings of the measurement devices are unchanged during the period of the experiment, and the second in which the settings are changed during the flight of the particles. The first allows there to be an exchange of information between the two wings of the experiment, while in the second the constant changes of settings are specifically designed to rule out such information exchange. The first type of experiment rigorously could only rule out or allow the combination of realism with separability, and it is only the second that actually rules out or allows local realism, though this distinction is not always made clear. In fact, the early experiments on Bell's inequalities that tested only local separability were usually referred to as test of local realism, and the move to testing local realism was usually described as 'closing the locality loophole'. It would be taken for granted that the second type of experiment would almost certainly be enormously more complicated than the first.

In Bell's paper, it is frankly not made clear if he knew whether experiments of the first type had already been carried out. If any such experiments had been performed, it is clear that they must have agreed with quantum theory, since any disagreement would have attracted considerable publicity. Several experiments at least broadly along the lines of that suggested by Bell had been performed, and he probably did not wish to spend time ascertaining whether they had done exactly what was required to check his work. It would indeed be a tricky task when it was carried out by Clauser and others, as we shall see in the following chapter.

Clauser considered that at this point Bell 'hedged his bets'; he says that Bell appears to imply that there was already experimental evidence that quantum theory was correct for the case of the static measurement directions. Bell goes on to suggest [6] that 'The situation is different if the quantum mechanical predictions are of limited validity. Conceivably they might apply only to experiments where the settings of the instruments are made sufficiently in advance to allow them to reach mutual rapport by exchange of signals at speeds less than or equal to that of light.' He thus proposes what we have called the second type of experiment.

Clauser, though massively enthusiastic of Bell's ideas in general, believed that he was 'clearly bluffing' on this point. He believed that for Bell to promote what Clauser calls the 'second-generation' experiment, one with changing measurement directions is 'clearly silly' unless one is sure that the 'first-generation' experiment with static measuring devices has already been carried out and has given results agreeing with quantum theory. Clauser comments that 'I personally will claim much of the credit for calling his bluff.' We shall follow this story in the next chapter.

So Bell finished the paper with a clear if not particularly strident call for experimental work to decide between local realism and quantum theory. In his two papers, he had at last made giant steps in a process that had begun for him back in 1952, if not, in fact, in a more general way, even earlier in Belfast. Much later he was asked by Paul Davies [17] how long it had taken him to derive his main result. He replied that the general question had been continually in his mind for a very long time, that he had been thinking intensely about the subject for a number of weeks, presumably since he had arrived in Stanford, and finally he obtained the all-important result over just one weekend.

Bell completed a first draft of this paper in Brandeis University, and the final version in the University of Wisconsin, giving seminars on the topic in each of these universities. Interestingly he thanked two colleagues for useful discussions. One was Myron Bander, at that time a postdoc at Stanford, but from 1966 a Professor at the University of California at Irvine. He is an elementary particle physicist, but has maintained an interest in the foundations of quantum theory. The other was John Perring, who was based at Harwell, where he was still working in Bell's old group. Just as Bell was on leave from CERN in Stanford, Perring had also come to Stanford on leave from Harwell.

During their stay in Stanford, Bell and Perring also produced a rather remarkable paper that was published in the top physics journal, *Physical Review Letters*. It was only a very short time since Jim Cronin and Val Fitch had obtained a rather sensational result. It will be remembered that Bell had been one of those who had produced the CPT theorem a decade earlier, which said that physics was invariant under the combined operation of the C, P and T operators, which were defined towards at the beginning of this chapter. In fact until 1957 it was taken practically for granted that physics was actually invariant under any one of the three operators. In that year it was found that it was not invariant under C or P individually, but it still seemed practically certain that it was invariant under CP or under T.

However, in 1964 Cronin and Fitch showed that, in the decay of the K-meson to two pions, CP conservation was broken at the 1% level. Since CPT *must* be conserved—this is what Bell had shown to be a rigorous requirement of quantum field theory—it follows that neither is T conserved. This in turn implies that even at the level of elementary particles, there is—at this very low level—an arrow of time. A process takes place in a particular direction; the exact reverse process would not take place. This may be called loss of time-reversal invariance. For this contribution, Cronin and Fitch were awarded the Nobel Prize for Physics in 1980.

Naturally there were many attempts to explain this result, of which John Polkinghorne [18] comments that the most interesting and responsible was due to Bell and Perring. They pointed out that the experiment was not carried out in a CP neutral environment because the laboratory, like the rest of the galaxy, was made of matter rather than anti-matter. They suggested that there might be a weak long-range force that discriminated between particles and anti-particles. Thus the effect discovered by Cronin and Fitch might not be intrinsic, but merely apply in regions where matter dominated.

As Polkinghorne says, the Bell-Perring idea fits the facts and made a clear prediction; if the energy at which the effect was observed was doubled, the effect would be seen four times as strongly. The experiment was performed, but there was no change in the strength of the effect. Had things been otherwise, as Polkinghorne says, Bell and Perring might well have received a Nobel Prize of their own, and 1964 would have been an even better year for Bell!

However, let us return to the local realism paper. To attract attention, it would have seemed sensible to submit this to the *Physical Review*. This was by far the best-known physics journal for papers, as distinct from short letters announcing brief ideas, theories or experiments, for which, as we have said, the most prestigious journal would be *Physical Review Letters*. Yet Bell chose to submit the paper to a totally new journal called simply *Physics*, and his paper appeared in the very first volume of this journal.

This journal actually has an interesting story [19]. It was the brainchild of Philip Anderson, who won the Nobel Prize for Physics in 1977 for his theory of disordered materials, and Bernd Matthias, very well known for his work on superconductivity. The full name of the journal was *Physics Physique Fizika*, because it accepted papers in all three languages, as its aim was to be a journal of great prestige; its subtitle was: 'An international journal for selected articles which deserve the special attention of physicists in all fields'.

Bell's rationale for this choice of journal was that *Physical Review*, like most other mainstream journals, had a system of page charges. The authors of published papers were asked to pay a certain amount for each page of their paper. This meant that the publisher of the journal could keep the price of the journal down and so make it more accessible to readers. For scientists publishing a paper from their permanent address, the charge would normally be paid by their employer without any problem. Indeed today the system usually works more mechanically; when a group of scientists apply for a grant to carry out a particular experiment, a relatively small amount of money is included to pay the page charges of the papers that will be published during the course of the work.

The awkwardness for Bell was that he was a guest at Stanford and he did not want to have to ask his host to pay up for these page charges, especially for so unorthodox a paper. Fortunately to publish in *Physics* seemed to be the way out, since, as part of its policy, it had no page charges, and actually paid contributors a small amount. Indeed it aimed at being the scientific equivalent of something like the *New Yorker*, accepting only papers of enormous general interest. It seems that Bell's paper was accepted because Anderson thought, obviously totally incorrectly, that it demolished Bohm's interpretation of quantum theory.

Bell had to admit, though, that the scheme had a drawback. As compensation for the page charge, *Physical Review* provides the author of a paper with a number of free reprints of the paper to distribute to other workers in the area. *Physics* has no page charges but, unsurprisingly, no free reprints, so Bell had to buy these himself, and he ended up out-of-pocket after all. And sadly the journal itself lasted only four years.

While Bell's account of the incident makes sense, one cannot help but wonder whether it is the only explanation for the rather bizarre choice of journal. Again it seems to fit the pattern of, first, getting priority for the ideas; second, enabling those who might be interested to study them, and, in this case, perhaps even to perform the experiment; but, third, not to cause too much hostility from the majority of physicists who would have found even the discussion of these ideas to be practically heresy.

The question would be: Would his work in fact be visible enough to attract useful attention from those who might be sympathetic? This is the topic of the next chapter.

Einstein and Bell

However, we first briefly take stock of where Bell's work leaves the position of Einstein. There is a widespread belief—it could be called a myth—that Bell's work somehow undermined Einstein's beliefs, and in particular it totally demolished absolutely everything that was said in the EPR paper. In part this belief is based on the results of the experiments to test Bell's conclusions; these experiments are often said to show that the conclusions of the EPR paper were false. We shall leave this aspect of the matter until we have examined the experiments themselves. Here we concentrate on the theory of the EPR paper compared with that of Bell's two great papers that we have just discussed, mostly the second one.

Just to give a flavour of the disagreement, I mention that John Gribbin famously called Bell 'the man who proved Einstein wrong' while Robert Romer retorted that he was 'the man who proved Einstein was right'. One can understand Gribbin's remark, as we shall see, but I certainly would give more support to that of Romer.

To start with, we may recall that Bell called himself a 'follower' of Einstein. He was totally at one with Einstein over their requirement of realism in physical theories. For both of them this was the central tenet of their approach to scientific work. Bell felt perhaps a very little less strongly about locality than Einstein did. As we shall see, he was actually prepared to sacrifice Einstein's approach to relativity in order to produce what he felt was a more coherent vision of the Universe. And Bell was actually rather indifferent to determinism, in quite strong contrast to Einstein.

But perhaps Bell's greatest identification with the views of Einstein was over their shared very great distaste for what both might have called the obscurity and self-delusion of complementarity. Bell confided to Jeremy Bernstein [20] his feeling that: 'Einstein's intellectual superiority over Bohr, in this instance, was enormous; a vast gulf between the man who saw clearly what was needed, and the obscurantist.'

Also of course the very name and the whole structure of his second great paper centred on the EPR idea. The EPR paper, of course, brought the idea of entanglement into the mainstream of physics, and entanglement was the main tool to be used, not only in Bell's work, but in virtually all further studies of the foundations of quantum theory and of quantum information theory. Bell clearly owed a great deal to Einstein for providing what might be called the tools of his trade.

Of course, though, this does not necessarily imply that Bell's arguments could not end up by contradicting those of EPR, so let us analyse the connection, starting with the EPR paper itself. It is certain that the main argument presented in the paper is clear and convincing, as was discussed in the previous chapter. As explained there, provided one restricts oneself to what EPR actually demonstrated, which we called there (a), where we either have realism (hidden variables) or we do *not* have locality, there is no possible conflict with any further theory or experiment.

There is certainly no conflict with Bell. With (i) non-realism and locality ruled out by EPR, we are still allowed (ii) realism and locality, (iii) non-realism and non-locality or (iv) realism and non-locality. Bell now rules out (ii) but clearly that allows (iii) and (iv), so there is certainly no direct disagreement with the actual argument of EPR.

It is true, of course that, as stressed in the previous chapter, Einstein and EPR opted strongly and naturally for what we called (b) there. Here it is (ii) and it is indeed precisely what Bell rules out. It is amusing to note that we may say that EPR ruled out Bohr's favourite option, (i), while Bell ruled out Einstein's. From this point of view, it is perhaps fair to say that Bell proved Einstein wrong, and indeed that experiments that come out against local realism prove Einstein wrong, but it would make just as much sense to say that Bell proved *Bell* wrong, since Bell was probably just as much in favour of (prejudiced in favour of, if you like) local realism as Einstein! In our quote a few paragraphs back concerning 'Einstein's intellectual superiority over Bohr', Bell had to continue to say that 'Einstein's idea doesn't work. The reasonable think just doesn't work.'

Bell writes [6] that:

It would be wrong to say 'Bohr wins again', the argument was not known to the opponents of Einstein, Podolsky and Rosen. But certainly Einstein could no longer write so easily, speaking of local causality 'I still cannot find the fact anywhere which would make it appear likely that the requirement will have to be abandoned'.

Yet there is undoubtedly still prevailing a belief that EPR was flawed at a much more fundamental level than just hoping and expecting that a perfectly reasonable conjecture would turn out to be true. A typical and amusing example is told by Abner Shimony, whom we shall meet a great deal in the next chapter. Shimony already had a PhD in philosophy, and so, it is to be presumed, a fair amount of self-confidence when his adviser for his PhD in physics, the famous theoretical physicist Arthur Wightman, gave him the task, as an exercise, of going through the EPR paper and reporting back with an account of what was wrong with it. Shimony reported back that he didn't think there was anything wrong with it at all!

It is interesting to wonder why there is still such prejudice against the EPR paper. Part of it perhaps comes from the unfortunate fact that Schrödinger—with the best intention in the world—christened the argument the EPR *paradox*, and the name stuck. There are actually two different meanings to the word 'paradox'. The first, which is certainly most common in mathematical circles, is that a paradox is an argument that seems to be rigorous yet leads to a conclusion that is obviously incorrect. For example a typical mathematical paradox might lead to the conclusion that $1 = 0$, or that all triangles are equilateral.

The second definition of the word is that it is just an argument that is *para docens*, or 'against teaching', that is to say that it appears to conflict with current beliefs. This is undoubtedly the way Schrödinger intended the word to be taken—he used it this way several times in various publications for ideas that were surprising but in no sense self-contradictory. And it certainly fitted the EPR argument perfectly.

Those members of the scientific community that were at all interested, though, certainly thought in terms of the first definition. It must be remembered that EPR used what were essentially hidden variables, which were thought by all to have been ruled out by von Neumann. The view certainly became common that EPR had just *assumed* a combination of determinism and realism, both of which were regarded as illegitimate, and—surprise, surprise!—reached a paradoxical result. It was only natural that given new results—theoretical or experimental—there was a great temptation to remark that these were in conflict with EPR without feeling the necessity to justify the statement at any level higher than the superficial.

What is even more surprising is that, even among those working in quantum information theory, where entanglement is centrally important, and the EPR situation is constantly used, there is sometimes a reluctance to accept that Einstein, Podolsky and Rosen themselves deserve particular credit. Indeed the word 'paradox' often convinces the people involved that all EPR left behind was confusion and misunderstanding, which had to be cleared up by others!

However, let us turn back to the undoubted fact that Einstein *did* support local realism and that Bell *did* rule this out. D'Espagnat [21] has reported very cogently on this point. Towards the end of Bell's life, he reports that Bell was asked to spend a day in discussion with a private organization in Geneva. Since a day seemed a long time, Bell asked d'Espagnat to share the burden, so d'Espagnat was able to report on Bell's approach to this type of question. Surely, Bell was asked, his proof of non-locality had weakened Einstein's position and inevitably strengthened that of Bohr.

In his reply Bell admitted that Einstein's standpoint had been very much based on locality, that his own work meant that this aspect of it could not be maintained, and so the EPR attack on Copenhagen was to this extent obsolete. He agreed that an independent observer might consider this a point for Bohr and Copenhagen. However, he insisted that this did not touch the core of Einstein's and his own chief requirement, which was not for locality at all but for realism. He stressed that theories exist that are non-local but 'ontologically interpretable', such as the de Broglie-Bohm theory, and this condition, in his opinion, was the only crucial one for a physical theory to make sense.

So Bell felt his work was broadly supportive of Einstein, and definitely not the reverse. In addition, of course, it was the Bohm theory that he felt undermined totally the Copenhagen synthesis, so his remarks showed not the faintest hint that he felt his own work gave any support to the position of Bohr, apart perhaps just for contextuality!

Bell and relativity

When Bell published his collected papers on the foundations of quantum theory as *Speakable and Unspeakable in Quantum Mechanics* in 1987, shortly before his death, he included a paper entitled 'How to Teach Special Relativity'. This paper had been published in the very first volume of the rather obscure journal *Progress in Scientific Culture* in 1976. In the paper he says that he had long thought how he would teach special relativity if he had the opportunity to do so. Nevertheless he gives no real explanation of why he decided to publish this paper, and certainly neither the paper itself nor anything in *Speakable and Unspeakable* gives any indication of why the paper is included with all the ones on quantum theory.

Incidentally, in his discussion with Bernstein, Bell confided that he wanted to write half a dozen books, which, he said, presumably meant that he wouldn't write any of them. The one he particularly mentioned was a book on hidden variable models of quantum theory and the reactions of physicists to them, but given his stated interest in special relativity, perhaps this also is the content of a book he might have written.

It is clear, in fact, that the inclusion of the paper in *Speakable and Unspeakable* was because he thought that his approach to relativity removed or at least softened the blow of having to accept non-locality.

To understand what Bell took to be the most pressing problem with non-locality, we review the problem of the ordering of events as discussed in a standard undergraduate course on special relativity, and as mentioned in the previous chapter. The usual Lorentz transformation is used to find the time coordinates of a pair of events in what we may call frame 2, given that in frame 1 they occur at t_A and t_B with $t_B > t_A$, so that in frame 1 event A occurs before event B. If the events occur at times t_A' and t_B' in frame 2, then in some cases it may be that $t_B' < t_A'$ so that in frame 2 the ordering of events is the opposite of the ordering in frame 1.

At first sight this seems a somewhat horrific result because it might seem that event A might be cause and event B effect. In frame 1, cause occurs before effect, which is fine, but in frame 2 it might seem that effect comes before cause. However, standard relativity tells us that this problem does not exist. We must distinguish between timelike intervals and spacelike intervals. For timelike intervals between the two events, we have $c^2 \Delta t^2 > \Delta x^2$, where Δt and Δx are the time and the distance between the two events. For such intervals, the two events may be cause and effect, but this creates no problems because there is a fixed order of events. For spacelike intervals, we have $\Delta x^2 > c^2 \Delta t^2$. For these intervals, the ordering of the events is indeed different in different frames, but this does not cause conceptual problems because the events are sufficiently far apart that a signal cannot pass from one event to the other. This last point is indeed exactly what is meant by locality.

This means then that when, stimulated by Bell's work, we must discuss a non-local universe, we are implying specifically that events separated by a spacelike interval may be cause and effect. This in turn means that, for such pairs of events, cause will occur before effect in some frames, but in other frames the effect may come before the cause. This is the problem Bell is attempting to deal with in his approach to relativity.

His general position is to reverse what would usually be viewed as the enormous conceptual advance taken by Einstein in 1905. Over the previous fifteen years or so before 1905, a number of giants of late Victorian or turn-of-the-century physics—in particular Joseph Larmor, Henrik Lorentz, Henri Poincaré and George Francis FitzGerald—produced a highly functional approach to the subject of physics as experienced by moving objects or moving observers. They were asking questions like: Given that we know the laws of physics in what we may call 'the rest frame', do they vary, and if they do how do they vary, in 'moving frames'? Given that an object has a length ℓ when at rest, what will its length be as measured by an observer moving with speed v with respect to the object? If two events occur at a particular point, and the time between them is τ as measured by an observer stationary at that point, what time will elapse between the events as measured by a second observer moving with speed v? If an event has coordinates x and t in 'the rest frame', what will the corresponding coordinates be in 'the moving frame'?

We have written these questions somewhat tentatively because they are among the most basic, but also the most subtle in science. And indeed for several centuries they were thought to have been answered conclusively. This refers to Newton's Laws of mechanics, but even before that to the writings of Galileo, which could be described as a forerunner for some aspects of Newton's Laws.

Galileo liked to consider people travelling on a boat. Later in the early twentieth century, Einstein would use a train rather than a boat. Today we usually talk of a spaceship, but all these modes of transport perform exactly the same job! Galileo stressed that, provided the boat was travelling at constant velocity, passengers on the boat could carry on their lives in exactly the same way as when on shore—walking about, drinking soup and so on. In no way did they have to keep reminding themselves, consciously or subconsciously, that they were moving. In particular, to take a more modern example, if a badminton game was being played, the player hitting the shuttlecock in the direction of the travel of the boat would have no advantage over the opponent hitting in the opposite direction. (Onc might well suspect there *would* be an advantage without the benefit of a little experience or a little thought.)

None of this applies, of course, if the velocity of the boat, train, spaceship or, to take a more homely example, bus is *not* constant. (Velocity not being constant is another way of saying that the system is being accelerated.) If, for example, we are travelling on a bus and the brakes are jammed on, we will certainly be aware of this and may well be thrown forwards in the bus. Now let us imagine the bus taking a very sharp curve at constant speed. We may remember that a change in velocity, so a non-zero acceleration, may imply a change in speed *or* a change in direction (or, of course, both), so in this case velocity is not constant, or to put it another way, the acceleration is not zero. We will be well aware of the change of direction of the bus and may indeed be thrown to the outside of the bus.

So we are coming to an appreciation of the fact that physics is relatively simple in the earth's frame and in frames travelling at constant velocity with respect to the earth—we call these *inertial frames*. It seems that the laws of physics (though we must recognize that we are talking only of mechanics at the moment) are the same in all these frames. (Thinking of frames is always a little awkward, so we can just say—for observers on the earth and observers travelling at constant velocity relative to the earth.) In frames accelerating with respect to these frames, the laws are less simple. We call these *non-inertial frames*, but we shall not need to refer to them very much.

The alert reader may feel uneasy, recognizing that, in fact, the earth is rotating round the sun and also spinning about its own axis, so strictly speaking it cannot be considered to be an inertial frame. This is true, and very sensitive experiments can detect this effect, but it is indeed such a very small effect that, to an extremely good approximation, the frame of the earth behaves as an inertial frame.

This set of ideas, as we have said, comes from Galileo's broadly empirical observations, backed up with considerable mathematical insight. When Newton produced his famous Laws, Galileo's insights were an integral part of them. So Newton's Laws have the same form in all inertial frames. Indeed Newton's First Law just says that if no force is applied to an object, its velocity is constant—it will have zero acceleration. It is important to recognize that the law applies in *all* inertial frames, not just in that of the earth. On the other hand, it does *not* apply in non-inertial frames; when you are sitting on a bus and the driver jams the brakes on, there is no force on you, but you still may end up rolling on the floor!

This may lead us to the next advance in understanding. We have rather naturally been thinking of the frame of the earth as 'at rest', and other frames, even other inertial frames, as 'moving'. But now we see that this is both unnecessary and restrictive. It is much more elegant to think in terms of *all* inertial frames having exactly the same status. Let us imagine, for example, a set of axes, xyz, defining a frame stationary on earth—we may say the 'rest frame' of earth. Now we imagine a train travelling along the x-axis of this set of axes, defining the rest frame of the earth, at speed v. We may now place a set of axes, x' y' z', to define the rest frame of the train. We may now equally describe the relative motion of earth and spaceship by saying that the earth travels along the $-x'$ -axis of the rest frame of the train at speed v. We see that all motion of inertial frames is *relative*—motion of one inertial frame relative to another, or with respect to another. (This is why, as we said at the time, we introduced these ideas a little tentatively at the outset.)

This is a beautiful set of ideas, which we may call Newtonian Relativity, but we must remember that, as yet, it only applies to Newton's Laws. Things were to look much less clear in the second half of the nineteenth century when electromagnetism came to the foreground of attention.

Before we move onto this, though, we need to consider the way in which coordinates transform between inertial frames. In other words, for example, if an event takes place at coordinates x' y' z' and at time t in the earth's rest frame, what would be the coordinates x' y' z', and time t', of the event in the frame of the train? (Actually because the relative motion is along the x- or x'-axis, it is fairly straightforward to recognize that $y' = y$ and $z' = z$, so it is only x and x', and t and t' that we need to be concerned with.)

There is a seemingly obvious way of performing this transformation known as the Galilean Transformation. (Again we must be cautious in word choice here, because the obviousness of this way of doing things will be strongly questioned shortly.) To derive this transformation, we may imagine the train passing through a station. At a time which we may say is $t = t' = 0$, a clock in the waiting-room at the station has exactly the same x-coordinate as a clock in a compartment on the train. (Remember that the x- or x'-direction is the direction of the relative motion of the station and train, or in other words the direction of the track.) At this moment we say that a passenger on the train sneezes. This is our first event, and the coordinates in the different frames are related by $x_1 = x_1'$, and $t_1 = t_1' = 0$. (x_1 or x_1' is just the distance of the passenger from the position where, rather arbitrarily, we have the origin of the axes for the rest frame of the station.)

Now we imagine the train travelling for a time t, at which point the passenger sneezes again; this is our second event. What is the x' coordinate of the event? Remember that dashed coordinates are in the frame of the train, and clearly relative to the train the passenger has not moved between the two events. So $x_2' = x_1'$.

But now we must think out the value of x_2. It must be remembered that the undashed coordinates are relative to the frame of the station, and clearly, relative to the axes stationary in the station, the passenger has travelled a substantial distance, in fact a distance equal to vt. So its x-coordinate has increased by vt, or we have $x_2 = x_1 + vt$.

What we require is something slightly different. We want to relate x_2', the coordinate of the second event in the rest frame of the train to x_2, its coordinate in the frame of the station. But we have $x_1 = x_1' = x_2'$, so finally we obtain $x_2' = x_2 - vt$.

Now we must consider the relation between t_2' and t_2. But here again it will seem entirely obvious that there can be no difference in the time of the same event (the second sneeze) as measured by the passenger in the train and somebody standing on the platform. In other words we will automatically take totally for granted that $t_2' = t_2$. (Yet again, shortly we shall have to reconsider this argument.)

Our subscript 2 just relates to the second event, but we may now drop it, as the equation is true for any event to give our final equations as

$$x' = x - vt$$
$$t' = t$$

These, as we have said, are the equations of the Galilean Transformation between one inertial frame and another. As well as using them to relate the coordinates of an event in the two frames, we can use them in a different way. We may write down Newton's Laws in, say, the undashed frame, and transform the equations to the dashed frame. What we must find, of course, and fortunately do find, is that Newton's Laws have exactly the same form in the dashed frame as in the undashed frame.

However, as we have said, let us now fast forward to the second half of the nineteenth century. Probably the greatest achievement in nineteenth-century physics, indeed one of the very greatest in the entire history of physics, was James Clerk Maxwell's theory of electromagnetism. Maxwell built into his theory, which was developed between 1855 and 1873, Coulomb's Law, which gives the electrostatic field between two electric charges; Ampère's Law, which gives the magnetic field in terms of the electric currents present; and Faraday's Law of Electromagnetic Induction, which gives the induced electric field in terms of the rate of change of magnetic field.

Maxwell was highly influenced by Michael Faraday's visionary idea of the electric and magnetic field. Faraday had stressed the significance of the medium between the electric charges and currents, and made this medium the centre of the electromagnetic action. He was not adept at formal mathematics, and Maxwell's synthesis may be described as a turning of Faraday's crucial physical insights into respectable mathematics. Maxwell included some crucial insights of his own, and the result was a complete theory of electromagnetism, always called Maxwell's Equations, that has been central to the whole of physics and technology ever since. Maxwell's work now stands with the mechanics of Newton and the thermodynamics of Rudolf Clausius and William Thomson (Lord Kelvin) as the main components of classical physics.

The most interesting part of Maxwell's theory was the prediction of a form of waves, now known as electromagnetic (EM) waves. Maxwell was able to work out the predicted speed of these waves and it turned out to be very close to that of light, which is 3×10^8 ms^{-1}, and is always written as c. It thus seemed clear that light was just an example of Maxwell's EM waves.

I say an example because light, which we can say is the form of EM wave that our eyes are sensitive to, only occupies a small range of wavelength and of frequency. (For any waveform, the product of the wavelength, λ, and the frequency, f, is just equal to the speed of the wave, so for EM waves, we have $\lambda f = c$.) EM waves with other values of wavelength and frequency, or we can say waves in other parts of the EM spectrum, include X-rays, gamma-rays, ultraviolet, infrared, microwaves and radio waves. So there is much interesting physics and a wealth of useful and often lucrative applications.

Practically the only bad news is what we, in this section, are most concerned about. If we assume Maxwell's Equations are true in one particular inertial frame, and we apply a Galilean Transformation to find out what happens in another inertial frame, it turns out that Maxwell's Equations are *not* obeyed. Transforming Maxwell's Equations with the Galilean Transformation does *not* give us a form of relativity analogous to Newtonian Relativity.

This, in fact, implies that there must be a unique rest frame of the Universe in which Maxwell's Equations are obeyed; in all other inertial frames they will not be obeyed (at least as long as we stick to the Galilean Transformation). And it is very easy to see how we must think of this frame. All the wave forms with which we are familiar are waves *in a particular medium.* Water waves are obviously waves in water, waves on a string are equally obviously on a string, sound waves are waves in air, or whatever medium they are passing through, and so on. It thus seems absolutely natural to assume that EM waves must also be waves in a particular medium, and this medium has been called the *ether* (or the *electromagnetic ether*). The frame of the ether is then the one in which Maxwell's Equations are obeyed.

It may be remarked that in the last quarter of the nineteenth century physicists devoted an enormous amount of time to explain the nature of this ether. Its properties must be very strange indeed! On the one hand it must be extremely tenuous, because normal objects can pass through it without any resistance at all, but on the other hand, to be able to support the kind of wave that light is—it is what is called a *transverse* wave—it must be infinitely strong.

Rather more profitable was the work of the scientists mentioned towards the beginning of this section—Larmor, Lorentz, FitzGerald, Poincaré—who established the relationship between the physics of objects at rest in the ether frame and that of moving objects. And here, we must remember, in contrast to the physics of Newton, as we have just seen, and also in contrast to the physics of Einstein, as we shall shortly see, we would expect to have no concerns about using the terms 'moving' and 'at rest'.

Unexpectedly things do not work out quite as simply as that. Given that Maxwell's work encourages us, or even dictates to us, to believe in a rest frame for the Universe, which is just the rest frame of the ether, it is natural to want to find it, or in other words to find its velocity—speed *and* direction—relative to the frame of the earth. A crucial experiment to attempt this task was carried out in the 1880s by Albert Michelson and Edward Morley.

To understand this experiment, we need to consider the speed of a ray of light (or other EM wave) travelling as measured by observers in different frames. Let us start in the rest frame of the ether. Here, by definition as it might be said, the light obviously travels at speed c. But now think of other frames. The easiest way of thinking about this is to think of a scientist in the rest frame of the ether shining a torch. Clearly the observer will see the light travel away at speed c. Now imagine a spaceship travelling away from this observer at speed, v, which we will say is less than c. It seems totally obvious that the speed of the light as measured by an observer will be $c-v$. Similarly if another spaceship is

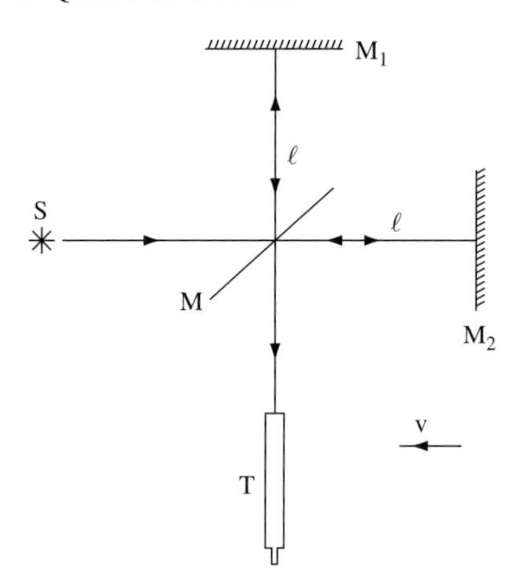

Fig. 8.14 The Michelson-Morley experiment. Light from source S reaches a semi-silvered mirror M, and subsequently reaches telescope T by travelling along path MM_1T or MM_2T. Interference between light that has travelled along these paths is observed. The distances MM_1 and MM_2 are assumed to be exactly equal, and are given by ℓ. The velocity of the ether in this frame is indicated by v.

travelling in the opposite direction again at speed v, it seems equally obvious that, to an observer on board *this* spaceship, the sped of the light will be $c + v$.

Could we test this directly? In other words, can we measure the one-way speed of light? The standard answer would be that, because of the very large value of c, this is impossible. It may be remembered that Galileo did attempt this, flashing light from one hilltop to another, but he soon realized that the reaction time of the observer was far longer than the time the light took to travel. The standard textbook methods for measuring c involve light travelling from A to B, being reflected at B and travelling back to A; the time taken for the light to make this journey is determined by registering an independent event at A that takes the same amount of time.

The Michelson-Morley experiment is a variation on this theme In. Figure 8.14, light from source S reaches the semi-silvered mirror M. A fraction of the light is reflected to M_1 then back to M where a component is transmitted to the telescope T. Light can also reach T by travelling along the path MM_2MT. The velocity of the ether in the frame of the earth is indicated on the diagram. (Of course we do not know this, but our analysis of the experimental results should allow us to determine it.)

From our previous considerations we may see that along MM_2, we would expect the speed of the light in the frame of the earth to be $c - v$, while along M_2M it should be $c + v$. With ℓ marked on the diagram, it is easy to calculate that the time taken to travel along MM_2M is not $2\ell/c$, as we might think at first, but

$$\frac{2\ell / c}{1 - v^2/c^2}.$$

Although it is not so obvious, the time take for the light to travel along MM_1M is also greater than $2\ell/c$, but in this case it is slightly different and is equal to

$$\frac{2\ell / c}{(1 - v^2/c^2)^{1/2}}.$$

To observe the difference between these two expressions, Michelson and Morley had arranged for the whole apparatus to be mounted on a massive stone slab and floated on mercury; at this point they rotated the apparatus through 90°. This meant that the two wings of the apparatus were interchanged, and the interference pattern seen through T should change by an amount dependent on the velocity of the ether relative to the frame of the apparatus.

The details of the analysis do not actually matter much because, although there was a clear prediction of an effect, and although the beautifully designed and constructed apparatus was easily sensitive enough to detect this effect, no effect was observed! The Michelson-Morley experiment, which was one of the very most famous in the history of physics, gave a null result.

Naturally there was much discussion of how this could be. The most famous suggestion is normally known as the Fitzgerald-Lorentz (or FL) contraction. This is that when travelling through the ether, the length of a body contracts. To discuss the amount of contraction, it is helpful to define a quantity γ (gamma) defined as

$$\gamma = (1 - v^2/c^2)^{-1/2}.$$

This quantity is central throughout relativity theory, and it is shown in Figure 8.15. It is equal to 1 for $v = 0$, but rises more and more sharply as v increases, and heads to infinity as v gets closer and closer to c. Then the FL contraction says that, when travelling through the ether, the length of a body is equal to ℓ/γ, where ℓ is its real length.

In terms of γ, we can see that, in our previous analysis that the times for the light to travel along MM_2M and MM_1M were $2\ell\gamma^2/c$ and $2\ell\gamma/c$, respectively so it is easy to see that an FL contraction applied to MM_2M, the path where the apparatus is moving through the ether, will indeed reduce the difference in path lengths to 0, giving us precisely the Michelson-Morley null result.

It is very often stated that the FL contraction was totally *ad hoc*, merely a clever device to avoid the extreme awkwardness of the Michelson-Morley result. Nothing could be further from the truth. Bruce Hunt [22] has described the actual course of events in detail.

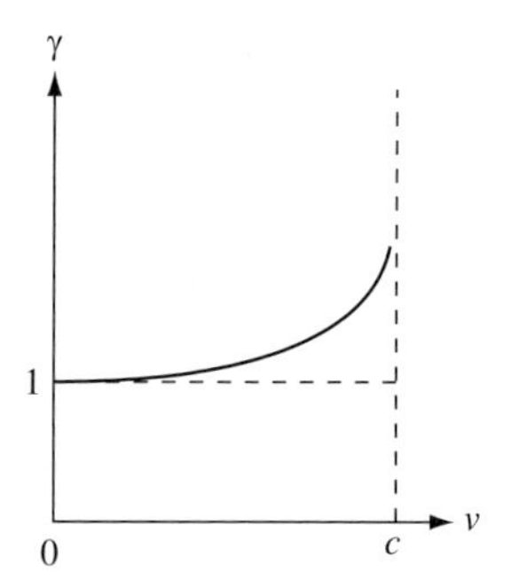

Fig. 8.15 Gamma (γ), the quantity central to any analysis in relativity. It is equal to 1 for small values of speed, v, but increases steadily as v increases and tends to infinity as v tends to c.

FitzGerald was one of an informal group of physicists and mathematicians working in the 1880s and 1890s on the elaboration of Maxwell's theory; Maxwell's death in 1879 had left the theory with enormous potential but with many unanswered questions, and many applications to be followed up. As well as FitzGerald, the group included Oliver Heaviside, telegraphist and mathematician, and Oliver Lodge, a well-known physicist, while from around 1890, Larmor and Heinrich Hertz, the German physicist who first produced the EM radio waves that Maxwell had predicted, were also involved.

It was Heaviside in 1889 who studied how the field lines around a point charge would behave when the charge moved. His mathematical analysis produced a compression in the direction of motion given precisely by the expression we have called the FL contraction, and it was an act of brilliance, by no means *ad hoc*, of FitzGerald later that year when he came to the conclusion that extended bodies *should* contract in this way, and that this would explain the null result of Michelson and Morley. FitzGerald sent a letter announcing the idea to the American journal *Science*, but apparently never bothered to find out that it had been published. He did mention his idea to Lodge, who included brief remarks about this idea in papers he published in 1892 and 1893. Lorentz came to the same conclusion as FitzGerald also in 1892, and traced FitzGerald's paper in *Science* through the mentions in Lodge's papers. He was scrupulous in giving credit to FitzGerald in his own work, and this is how the effect came indeed to be known as the Fitzgerald-Lorentz contraction.

Incidentally Hunt notes a letter of Lorentz to Einstein twenty years later suggesting that the hypothesis would have been much less easily understood as *ad hoc* if the connection to Heaviside's law of moving charges had been made clearer from the start.

From our point of view, the important aspect of the affair is that, while we are still accepting fully a unique rest frame of the Universe, we see nature seeming to deny us any possibility of finding it! This was the basis of a physics developed by Larmor and Lorentz over the next decade, but also rather separately by Poincaré, in which changes are attributed to moving systems that systematically rule out the possibility of discovering that the frame is indeed moving. Analogous to FL contraction is Larmor time dilation, discovered in 1900, in which an event that takes a time τ when the body is at rest takes a longer time $\gamma\tau$ or $\tau/(1-v^2/c^2)^{+1/2}$ in a frame moving with speed v.

More generally, a set of transformation equations was produced that did what we stressed earlier that the Galilean Transformation did not do. They are called the Lorentz Transformation (LT), although they were actually first written down by Larmor in 1897. If we start with Maxwell's Equations in the rest frame and transform them to a frame moving with constant speed using the Lorentz rather than the Galilean Transformation, then we find that the laws of electromagnetism have the same form in this moving frame, or in other words, Maxwell's Equations are true, not only in the rest frame, but also in any frame moving with constant speed. Any theory, like Maxwell's Theory of Electromagnetism, that has this property is called *Lorentz invariant.*

The form of the LT is as follows:

$$x' = \gamma(x - vt)$$
$$t' = \gamma(t - vx/c^2).$$

In these equations we note first the presence of the ubiquitous γ. We also note that, if v is small, by which we mean much less than c, then γ is very close to 1, and the LT becomes equivalent to the Galilean Transformation.

The most surprising feature of these equations is that t' is not the same as t. We took their equality for granted for the Galilean Transformation, but, of course, the Larmor time dilation formula has already alerted us to the fact that it cannot be generally true in the theory being considered. Logically, by the way, the FL contraction and the Larmor time dilation can be obtained *from* the LT, not the other way round.

Bell sums up the situation as follows:

[The theory] arose from experimental failure to detect any change in the apparent laws of physics in terrestrial laboratories, with the slowly moving orbital velocity of the earth. Of particular importance was the Michelson-Morley experiment, which attempted to find some difference in the apparent velocity of light in different directions. We have followed here very much the approach of H.A. Lorentz. Assuming physical laws in terms of certain variables (t, x, y, z), an investigation is made of how these things look to observers who, with their equipment, in terms of these variables, move. It is found that if physical laws are Lorentz invariant, such moving observers will be unable to detect their motion. As a result, it is not possible experimentally to determine which, if any, of two uniformly moving systems, is really at rest, and which moving. All this for *uniform* motion: accelerated observers are not considered.

In 1905, Einstein swept away the awkwardness of assuming that (a) there is a rest frame of the Universe but (b) there seems to be a conspiracy in nature to hide the fact. He decided that the terms 'at rest' and 'moving' were meaningless. All inertial frames were, he decided, of equal status. One might say that inertial frame A moved with respect to inertial frame B, but equally that frame B moved with respect to frame A. There was no rest frame, and since what had been thought of as the rest frame was the frame in which the ether was at rest, it seemed that the ether itself had become irrelevant and unnecessary, despite the strangeness of thinking of light as a wave without a medium to travel in. Einstein remarked on this point in his 1905 paper on special relativity.

Instead of, like Lorentz, discovering that, almost accidentally, the laws of physics happened to be the same in moving frames as in the rest frame, Einstein jumped straight in by making it a central hypothesis of his theory that the laws of physics were the same in all inertial frames—we may call this: *the principle of relativity*. This included the fact that in any inertial frame the speed of light would be c. Our previous argument that if you effectively chase after a ray of light with speed v, its speed as observed by you, or, as we should say, in the inertial frame in which you are at rest, *must* be $c-v$ is seen to be incorrect in Einstein's theory. In fact this idea was obtained essentially using the Galilean Transformation, itself supposedly *obvious*, rather than that of Lorentz. We may notice that, in particular, the simple argument does not take account of Larmor time dilation.

We may briefly survey how mechanics fares in Einstein's theory. Since, as we showed earlier in this section, Newton's Laws obey the principle of relativity using the Galilean Transformation, it seems fairly clear that they cannot obey it when, as we must, we replace this by the Lorentz Transformation. And indeed we find that we must recognize that Newton's Laws as originally given are correct only for speeds much less than that of light. (Of course, such speeds were all that were available for Newton, and, in most cases, for us, so Newton's Laws remain extremely useful and important.) So our definitions of momentum and kinetic energy must be altered for higher speeds, and in particular we meet the famous equation $E = mc^2$ for the energy equivalent of rest mass. We shall not elaborate on these important matters here.

Rather we stress that the general mathematical structure of what we may perhaps call pre-Einstein relativity goes through to Einsteinian special relativity. The Lorentz Transformation and the FitzGerald-Lorentz contraction retain their places, and they also keep their former names. Larmor time dilation is there as well, although textbook writers have almost invariably re-christened it *Einstein time dilation*. And, as Bell puts it, 'We need not accept Lorentz's philosophy to accept a

Lorentzian pedagogy.' (In other words we may use the idea of a unique rest frame if it is convenient for study of any physical problem, without necessarily believing it actually exists.)

Einstein's theory is invariably looked on as a massive step forward. Bell writes that it 'permits a very concise and elegant formulation of the theory, as often happens when one big assumption can be made to cover several less big ones'. Bell himself in no way dissents from this view. He says that: 'There is no intention here to make any reservation whatever about the power and precision of Einstein's approach.' He recognizes that there were cogent arguments for dropping the older approach which includes the ether. If the ether is unobservable, he says, why should we believe it exists? And he admits that Einstein's approach is the more elegant and simpler one. It comes, of course, with a few conceptual difficulties for each student to sort out, but the conceptual problems are nowhere near as profound as those of the quantum theory discussed in this book!

However, Bell also believes that the older theory should also be taught to students as a pedagogical device, as he considers that there are many problems that are more easily solved by thinking in terms of an ether. It is his impression that 'those with a more classical education, knowing something of the reasoning of Larmor, Lorentz and Poincaré, as well as that of Einstein, have stronger and sounder instincts'.

In a 1985 discussion with Paul Davies [17], Bell made clearer his deeper reason for, at the very least, wishing to draw attention to the pre-Einstein theory. He describes going back to this earlier theory as the 'cheapest resolution' of the problem of non-locality or non-separability or faster-than-light signalling, which was a result of his own work.

It is the case that the earlier theory centres around the rest frame of the ether in which we may say we see the events of physics occur with their real nature; Bell says that there is a 'real causal sequence that is defined in the ether'. In this frame there may be a suggestion that something is travelling faster than light, but the important point is that in this frame there is a true ordering of events. Cause must come before effect.

In other inertial frames, there may a reversal of this order; Bell says that: '[Things] seem to go not only faster than light but backwards in time'. But the important words here are 'seem to'. Bell describes this as an 'optical illusion'. The actual time-ordering is, as we have said, that in the frame of the ether.

It is rather as if we have our two children, say John and Peter, standing side-by-side in front of us. We note that John is a little taller than Peter and recognize that as a true statement of reality. At other times, when Peter is much closer to us than John, we 'observe' him to be the taller. Similarly if John is much closer than Peter, we may 'observe' him to be *much* the taller. But clearly we are not perturbed by these 'observations' for we know that there is a true sensible relation between the heights, and other 'observations' are 'optical illusions', if, of course, because of our considerable experience in such matters, of the most benign type.

Bell describes this as the cheapest resolution of the problem of non-locality. As he says, one may consider it as a revolutionary approach, or, alternatively as a reactionary approach, since it is actually putting the clock back a century or so.

Although it goes beyond what we really need, it is interesting to note that Bell actually goes rather further than we have said. It would be conventional, of course, to give Einstein's work the title of 'special relativity', but to deny this title to the work of Lorentz and the others, principally because their theory does not actually obey the principle of relativity as we defined it earlier.

In contrast Bell regards both sets of ideas as being representative of 'the special theory of relativity'. Indeed he regards them as differing in style—Einstein's rejection of the idea of absolute rather than relative motion, and philosophy—Einstein's acceptance of the principle of relativity at the outset of his work, but Bell characterizes Einstein's as a new formulation, albeit one of power and precision, of the same basic theory.

It is interesting to relate this to the views of Edmund Whittaker [23] in the second volume of his famous book *A History of the Theories of Aether* [an older rendering of 'ether'] *and Electricity*, which was published in 1953. Almost defiantly, one feels, he titles the second chapter of this book 'The Relativity Theory of Poincaré and Lorentz'. In this chapter he practically dismisses Einstein's work by saying that 'Einstein published a paper which set forth the relativity theory of Poincaré and Lorentz with some amplifications, and which attracted much attention.' (It is fair to say that he is more positive in his account of Einstein's work on quantum theory and general relativity.)

At the time of the publication of this book, Whittaker and Max Born were both professors in the University of Edinburgh, and Whittaker had asked for Born's help on some matters while writing the book, but he completely ignored Born's advice concerning Einstein's contributions to special relativity. Born did write to Einstein, expressing his hope that neither Einstein himself not anybody else would think that, because of the well-known disagreement between Born and Einstein over the interpretation of quantum theory, Born might actually have encouraged Whittaker to take this approach. Einstein, of course, dismissed this, saying: 'I myself have certainly found satisfaction in my efforts, but I would not consider it sensible to defend the results of my work as being my own "property", as some old miser might defend the few coppers he had laboriously scraped together.... After all I do not need to read the thing.'

However, in light of Bell's opinions discussed in this section, it is interesting to wonder whether Bell might have had any sympathy at all with Whittaker's views. It is, at the very least, worth noting that he wrote an article, which is included in the collections that have been mentioned, full of praise for his fellow-Irishman, FitzGerald.

We conclude this section with two more general remarks. First, in his paper Bell says that the only modern book he had been able to find that followed his approach to teaching relativity was written by the Hungarian Lajos Janossy [25] in 1971. At an important summer school in Erice in 1989, however, Bell met David Mermin, an outstanding solid state theoretician who also, as we shall see later in the book, made very important contributions to the foundations of quantum theory and to quantum information theory. Mermin reports Bell's skepticism when Mermin told him that he himself had also written a book [26] along these lines, but Bell later wrote to Mermin to say that he had looked up the book and agreed that it was exactly of the type that Bell had recommended.

The other remark is more technical. We mentioned that the one-way speed of light has never been measured. As well as the purely technical question, there is a more theoretical problem: preliminary to any such measurement is the problem of clock synchronization. We need to be able to say that two clocks at different points in space yield the same value of time for a particular event. In Einstein's special relativity, such problems are removed by his initial postulate that the speed of light is the same in all inertial frames. This makes any discussion of the one-way speed pointless, but it does provide a standard way of synchronizing clocks.

Franco Selleri (Figure 8.16) [27] suggests that having no knowledge of or interest in the one-way speed is 'not a pleasant situation, because one-way [speeds] seem to be rather natural properties... given that light and objects go from one point to another in well-defined ways in nature'. What, though, he asks, if, encouraged by Bell's argument discussed in this section, we reject this postulate of Einstein? Note that, in doing this, he actually goes beyond what Bell suggests, by rejecting Einstein's formalism as well as his philosophy. Without this postulate, Selleri deduces a generalization for the Lorentz Transformation in which the relationship between x and x' is unchanged, but the relationship between t and t' is generalized to

$$t' = (t/\gamma) + e_1(x - vt),$$

where e_1 is a quantity called the 'clock synchronization factor' that we may vary as we wish to give us different results.

If we choose $e_1 = -v\gamma/c^2$, a little calculation shows that we recover Einstein's special relativity, but other choices give a range of other theories. And while with that particular choice of e_1, the speed of light is, of course, c in all directions, for other choices it varies with the direction the light is travelling in.

Fig. 8.16 Franco Selleri [© Zentralbibliothek für Physik, Renate Bertlmann]

As an example, if e_1 is put equal to 0, the speed of light travelling in the same direction as the dashed frame is equal to $c^2/(c + v)$; in the opposite direction it is $c^2/(c - v)$. An elementary calculation tells us that if light travels a distance s in one direction and then the same distance in the opposite direction to return to its starting-point, the time taken is $(s/c^2)\{(c + v) + (c - v)\}$ or just $2s/c$, so its two-way speed is just c, which is really, of course, where we started.

We might wonder whether any experiment may be used to decide between different choices for e_1, but Selleri shows that none of the following types of experiment give us any information: Michelson-Morley type experiments; occultation of the satellites of Jupiter, where the light coming from the satellite is interrupted when it goes behind the planet; stellar aberration, or the apparent angle of the light reaching earth from a star; or radar ranging of a planet, where we reflect a radar pulse from a planet.

Selleri shows, though, that there are many advantages in choosing $e_1 = 0$. In particular, we achieve *absolute simultaneity*. Since we have $t' = t = \gamma$, two events that take place at different points in the undashed frame are also simultaneous in the dashed frame. This does not mean, of course, that time is absolute. Rather it means that a clock at rest in the undashed frame is seen by an observer in the dashed frame to run slow, but a clock in the dashed frame is seen by an observer in the undashed frame to run *fast*. In other words both observers agree that motion relative to the undashed frame slows clocks. Similarly all observers agree that motion relative to the undashed frame gives an absolute length contraction.

Selleri suggests that the choice $e_1 = 0$ is very natural if we wish to take as postulates Lorentz contraction and time dilation, but to avoid Einstein's own postulate that the speed of light is the same in any inertial frame.

Those who are used to special relativity will probably feel that this theory lacks the beautiful but subtle conceptual symmetry between the undashed and dashed frames. It must be admitted, though, that it may actually be physically more straightforward. Certainly it is interesting to see how study of the foundations of one of our great twentieth-century theories of physics, quantum theory, throws up problems of interest and subtlety in the other, relativity.

References

1. A. Whitaker, 'John Bell and the Most Profound Discovery of Science', *Physics World* **11** (12) (1998), 29–34.
2. A. Whitaker, 'John Bell in Belfast: Early Years and Education', in: R. Bertlmann and A. Zeilinger (eds), *Quantum [Un]speakables: From Bell to Quantum Information* (Berlin: Springer-Verlag, 2002), pp. 7–20.
3. R. Jackiw and A. Shimony, 'The Depth and Breadth of John Bell's Physics', *Physics in Perspective* **4** (2002), 78–116.
4. R. A. Bertlmann, *Anomalies in Quantum Field Theory* (Oxford: Clarendon, 1996).
5. M. Born, *Natural Philosophy of Cause and Chance* (Oxford: Clarendon, 1949).
6. Bell's papers on quantum theory have been collected in J. S. Bell, *Speakable and Unspeakable in Quantum Mechanics* (Cambridge: Cambridge University Press). The first edition was published in 1996, and papers written between 1996 and his death were added in the second edition of 2004. A selection of his papers in all areas of physics was published as M. Bell, K. Gottfried and M. Veltmann (eds), *Quantum Mechanics, High Energy Physics and Accelerators* (Singapore: World Scientific, 1995); the papers on quantum theory from this collection were included in *John S. Bell on the Foundations of Quantum Mechanics* published in 2001 with the same editors and publisher. In the present book, all quotes from Bell's papers will be found in these books.
7. B. J. Hiley and F. D. Peat (eds), *Quantum Implications: Essays in Honour of David Bohm* (London: Routledge and Kegan Paul, 1987).
8. J. T. Cushing, *Quantum Mechanics: Historical Contingency and the Copenhagen Hegemony* (Chicago: University of Chicago Press, 1994).
9. D. Bohm and B. J. Hiley, *The Undivided Universe: An Ontological Interpretation of Quantum Theory* (London: Routledge and Kegan Paul, 1993).
10. P. Holland, *The Quantum Theory of Motion* (Cambridge: Cambridge University Press, 1993).

11. J. M. Jauch, *Are Quanta Real: A Galilean Dialogue* (Bloomington: Indiana University Press, 1973).
12. L. Gilder, *The Age of Entanglement* (New York: Knopf, 2008).
13. B. d'Espagnat, 'The Quantum Theory and Reality', *Scientific American* **241**(1979), 158–81.
14. B. d'Espagnat, *Conceptual Foundations of Quantum Mechanics* (Menlo Park, CA: Benjamin, 1971).
15. B. d'Espagnat, *Veiled Reality: An Analysis of Present-Day Quantum Mechanical Concepts* (Reading, MA: Adison-Wesley, 1995).
16. N. Herbert, *Quantum Reality: Beyond the New Physics* (New York: Doubleday, 1985).
17. P. C. W. Davies and J. R. Brown, *The Ghost in the Atom* (Cambridge: Cambridge University Press, 1986).
18. J. Polkinghorne, *Rochester Roundabout: The Story of High Energy Physics* (Harlow, UK: Longman, 1989).
19. D. Wick, *The Infamous Boundary: Seven Decades of Controversy in Quantum Physics* (Boston: Birkhäuser, 1995).
20. J. Bernstein, *Quantum Profiles* (Princeton: Princeton University Press, 1991).
21. B. d'Espagnat, 'My Interaction with John Bell', in: R. Bertlmann and A. Zeilinger (eds), *Quantum [Un]speakables: From Bell to Quantum Information* (Berlin: Springer-Verlag, 2002), pp. 21–8.
22. B. J. Hunt, *The Maxwellians* (Ithaca, NY: Cornell University Press, 1991).
23. E. Whittaker, *A History of the Theories of Aether and Electricity: The Modern Theories 1900–1926* (Edinburgh: Thomas Nelson, 1953).
24. M. Born (ed.), *The Born–Einstein Letters: Friendship, Politics and Physics in Uncertain Times* (London: Macmillan, 2005), pp. 192–9.
25. L. Janossy, *Theory of Relativity Based on Physical Reality* (Budapest: Académiaia Kiado, 1971).
26. N. D. Mermin, *Space and Time in Special Relativity* (New York: McGraw-Hill, 1968).
27. F. Selleri, 'Bell's Spaceships and Special Relativity', in: R. Bertlmann and A. Zeilinger (eds), *Quantum [Un]speakables: From Bell to Quantum Information* (Berlin: Springer-Verlag, 2002), pp. 413–28.

9
Experimental philosophy: the first decade

Clauser and Shimony

We left John Bell with his two important papers produced in 1964, and published (in reverse order) in 1966 and 1964, respectively. They had been written in a clear but non-provocative way, and they had been published in accessible journals, but perhaps not in journals where they would naturally demand most attention. And it seemed unlikely that Bell would have pushed his ideas further if they had been ignored. He did not publish in the area at all for a further five years, until 1970 when Bernard d'Espagnat organized the Varenna summer school on the foundations of quantum mechanics mentioned in the previous chapter. Bell's paper in the proceedings for this conference differed from his previous work only slightly by including the effects of additional hidden variables in each apparatus.

Bell was to publish around twenty-five papers on the foundations of quantum theory, but the great majority were invited papers presented at various conferences, which is not to say, of course, that they did not contain important new ideas.

After his first papers in this area, virtually the only time he actually submitted papers to a journal was to *Epistemological Letters*, which has been described by John Clauser [1] as an ' "underground" newspaper', published by l'Institut de la Méthode of the Association Ferdinand Gonseth, the circulation of which was limited to a 'quantum subculture'.

This journal was perhaps unique in the 1970s in that it stated openly that it would accept papers on hidden variables and similar topics. Thirty-seven issues of this periodical appeared between 1974 and 1984, with length varying between 4 and 89 pages. The intention of the editor was stated as follows: '"Epistemological Letters" are not a scientific journal in the ordinary sense. They want to create a basis for an open and informal journal allowing confrontation and ripening of ideas before publishing in some adequate journal.' Here Bell was to swap ideas with some of the others mentioned in this chapter, confident that the more 'respectable' or less flexible members of the scientific community would presumably not be reading his work. Incidentally the unofficial editor of the papers on hidden variables was Abner Shimony, whom we shall meet shortly.

In the years after 1966, the question must have been: Bell had avoided making the kind of splash that would have antagonized the powers-that-be, but in doing so had he missed the opportunity of interesting or converting or subverting those who might be prepared to respond to his work, and even to contemplate carrying out the relevant experiments? In this chapter we shall meet several important figures as we study the development of what can be called experimental philosophy between about 1968 and 1979, but there are two of supreme significance in that, without them, Bell's work would probably never have received any attention at all!

The first is Abner Shimony [2]. Shimony (Figure 9.1) had what might be described as an ideal background for investigating the foundations of quantum theory—enthusiasm for both philosophy and physics, and indeed doctorates in both.

Shimony was born in 1928, and had early intellectual interests; at high school he was already a convinced evolutionist, and since this was Memphis Tennessee, that belief caused certain problems—he had to change geometry class as a result of evolutionary arguments he put forward in the lessons! He entered Yale University in 1944 as a prospective physics major, but in fact graduated in 1948 in philosophy and mathematics. His reason for the change was that he was interested in everything, and he felt that philosophy gave a view of the whole world, though he remained drawn to the rigour of mathematics and the enormous

Fig. 9.1 Abner Shimony [© Zentralbibliothek für Physik, Renate Bertlmann]

achievement of physics, and indeed he took courses in classical mechanics, quantum electrodynamics and special relativity. Only later did he feel that the view of the whole world provided by philosophy could end up being a little superficial.

His PhD, which was awarded in 1953, also from Yale, was in philosophy, in fact in inductive logic. In this work he was guided by Rudolf Carnap, one of the very most influential philosophers of the twentieth century. Carnap was a central figure of the famous Vienna Circle, the aim of which was to endorse logical positivism. The avowed purpose of logical positivism itself was to provide a language that could be used rigorously in the natural sciences, and Carnap himself, incidentally, had a degree in physics. Logical positivism advocated obtaining knowledge by logical analysis from experience, and eliminating metaphysics, ideas that cannot be reduced to experience and should therefore be regarded as meaningless. Carnap was unsurprisingly amazed that Shimony nevertheless was to declare himself a metaphysician!

During these years, Shimony was obliged to study probability defined in terms of reasonable degrees of belief, but he also read widely on probability in physics. Like Bell, he read Born's *Natural Philosophy of Cause and Chance*, and he also revived an interest in classical statistical mechanics and quantum mechanics. (Statistical mechanics can be defined as the area of physics in which properties of gases, liquids or solids are described in terms of the statistical behaviour of their constituent atoms of molecules.) Influenced by the famous philosopher Charles Sanders Peirce, he became convinced that probability must be meaningful for individual cases, not just for ensembles. He also thought that an opening in physics should be kept for indeterminism, either through uncertainty in the initial conditions of a system, or through genuine lack of determinism (or *stochasticity*) in the equations of physics. He had always retained an interest in the philosophy of physics, and now, becoming convinced that physics was the basic science, he decided that he would like to do a second doctorate in this subject.

After two years in the army, Shimony was able to do this, starting a PhD in physics at Boston University in 1955. His thesis work with Wigner (Figure 9.2) was in the field of statistical mechanics, but during this period he also read Wigner's writings on the foundations of quantum mechanics, and so he moved on to work in the same area, work that included study of von Neumann's book. He did not finish the PhD until 1962, but even before then, in 1959, he had been appointed to the philosophy faculty at MIT. Here he finished his second PhD, of course, but also lectured on the foundations of quantum mechanics, and during this period he also attended his first conference on the topic.

However, once he had obtained his PhD in physics, he moved to Boston University because here he had what was for him the ideal position—a joint appointment in physics and philosophy, and here he has stayed for almost half a century. He immediately built up his interest in

Fig. 9.2 Eugene Wigner pictured with Leo Szilard [courtesy of George Marx]

the foundations of quantum theory by attending a major conference in the area in 1963, which was attended by many of the giants of the subject including Podolsky, Wigner, Dirac and Bohm, and in the same year he published his first paper on quantum mechanics.

This paper was very much inspired by Wigner. At this period, several physicists had claimed to be able to show that the measurement problem, as we described it in Chapter 4, was a result of an idealization of the concept of measurement, and that if one put in a more complete account of the measurement process the problems would evaporate. They started with an ensemble of systems, each with a superposition of eigenfunctions, by which is meant a linear combination such as $a_1\phi_1 + a_2\phi_2$. They then claimed to show, by simply following the Schrödinger equation, that the wave-function of the ensemble of systems would develop into what is technically called a mixture, where each system would have a wave-function *either* ϕ_1 *or* ϕ_2. Obtaining a mixture implies that a definite measurement result has been obtained for each system; we can say this is E_1 or E_2 if the wave-function ends up as ϕ_1 *or* ϕ_2, respectively. (Thus these physicists are claiming to obtain the results usually obtained by collapse, which is totally distinct from normal Schrödinger evolution, but doing it following the normal quantum formalism.)

In early 1963, Wigner published a paper [3] making it clear that the measurement problem was a completely general phenomenon, and it appeared totally independently of the detailed model of measurement. He showed clearly and convincingly that a superposition can *never* be transformed into a mixture by the use of the Schrödinger equation, because of the mathematical linearity of the equation. As he admitted, this was not a new argument even at the time, but it seems that it constantly needs restating, more recently by Bell as we shall see in Chapter 14.

Shimony's paper [4], published later in the year, and in the same journal as that of Wigner, *American Journal of Physics*, which is a journal more concerned with physics scholarship and teaching than with research, was essentially a development of Wigner's own paper. Shimony considered the two main arms of the orthodox approach to quantum measurement. First he considered von Neumann's version, which he took to imply that collapse of wave-function was caused by consciousness. Shimony commented that this was not supported by any psychological evidence, and that it was difficult to reconcile with intersubjective agreement between different observers. (Is it possible that *any* observer can collapse a wave-function, and the result will be appropriate for all the other observers?)

Then he considered the Bohr approach, where the experimental arrangement is all-important in deciding the course of a measurement. Shimony agreed that this approach did provide useful general ideas of how measurement might fruitfully be discussed, but felt that its weakness was that it gave up any attempt at reaching an *ontological framework*. (The word 'ontology' relates to how things actually are, as distinct from 'epistemology', which is concerned with our knowledge of things and how we acquire it. Thus logical positivism and complementarity, different as they, of course, are in general, may both be more concerned with epistemology than ontology. It is a mark of Shimony's refusal to expel metaphysics from his approach to science that he required an ontology.) Shimony concluded that, successful as quantum theory has obviously been, there may be a requirement for a new formulation of the theory.

This was a substantial paper, and clearly showed that Shimony was very much prepared to consider new ideas in the foundation of quantum theory, even those that might challenge the status quo in a dramatic way. Incidentally, in this paper Shimony acknowledged Wigner for 'encouraging [his] philosophical investigations of problems in quantum theory', and also for his making several valuable specific suggestions.

The other person of the greatest significance for the work in this period is John Clauser [5]. Clauser (Figure 9.3) was much younger than Shimony, being born in California in 1942. (This meant that, while Shimony had a permanent position in Boston before starting the Bell work, Clauser did not, and that fact was to have major repercussions on Clauser's career.) Clauser's father and uncle, as well as many other family members, all obtained degrees from Caltech. His father and uncle graduated in physics, but they were then attracted into the relatively new and exciting field of aircraft design. They went to work for Theodore von Kármán, a very important pioneer in the fields of aerodynamics, aeronautics and astronautics, and founder of the Jet Propulsion Laboratory at Caltech. Von Kármán himself was a protégé of Ludwig Prandtl, famous for his work on fluid dynamics and its application to aerodynamics.

Fig. 9.3 John Clauser [© Zentralbibliothek für Physik, Renate Bertlmann]

At the end of the war, the Clauser family moved to Baltimore, where John's father was appointed chairman of the Aeronautics Department at Johns Hopkins University, where he had to create the department from scratch. His greatest achievement was working out how to put earth-orbiting satellites into space, calculating how high they would have to fly, how many stages would be required and so on. John spent much of his time in the laboratory, marvelling at the wonderful equipment, and learning from his father. He became an expert at electronics—designing some of the first video games, taking part in science projects and winning prizes at the Natural Science Fair, so it might have seemed natural for him to have studied engineering or computing at university.

His father had achieved much in engineering; he was to become Vice Chancellor of the University of California at Santa Cruz, and later to return to an endowed chair in engineering at Caltech itself. Yet he always tried to understand the physics of fluid flow behind the engineering phenomena. He also saw the importance of obtaining a broad basic education, and felt that if you understand the basic principles, you can do anything! Later John was to feel that he was probably right. His father also perhaps felt that there was an academic snobbery in universities, engineers being further down the pecking order than pure scientists, and the upshot of all these arguments was that John Clauser entered Caltech to study physics, graduating with a BA in 1964. His background and talents meant that he was clearly destined to be a superb experimental physicist.

Clauser moved to Columbia University in New York for his MA and PhD, which he obtained in 1966 and 1969, respectively. His PhD project was in astrophysics and his thesis adviser was Pat Thaddeus. Originally the idea was to put a radio telescope in a U2, the reconnaissance aircraft, or so-called spy-plane. However, for his PhD Clauser eventually carried out one of the first (in fact the third) measurement of the cosmic microwave background, which had been discovered by Arno Penzias and Robert Wilson in 1965, and which is now considered to be the best evidence for the Big Bang. (For this achievement, Penzias and Wilson were awarded the Nobel Prize for Physics in 1978.) Clauser and Thaddeus based their work on observation of the CN radical in interstellar space, and they published papers giving their results in 1966 and 1969. Their result for the temperature of the background radiation is close to the accepted value today.

However, during this period Clauser also built up a great interest in, but also a healthy scepticism towards quantum theory. In fact this interest and also the scepticism went right back to his father who, interested in physics as we have said, had studied quantum theory but found it profoundly unsatisfactory. Clauser's own position was not greatly different.

It may be said that Clauser [1] has described himself as 'a simple experimental physicist (who spends much of his time mucking about in cutting oil leaks and chasing down vacuum-system leaks and electronics and software bugs)'! Actually this is enormously far from the truth. He would certainly admit that he is not a good abstract mathematician: he is not skilled at algebraic manipulations with no clear physical meaning or implication, and in a sense neither would he probably want to be. Other theoretical physicists would delight in such tasks, of course. And in no sense would he take pleasure in seeing, for example, Maxwell's Equations reduced to a compact form that might be mathematically simple and beautiful. Rather he would like each equation to be expressed in physical terms, and to provide a direct model of one aspect of the phenomena.

But on the other hand, given a model or something to visualize, Clauser is superb at doing calculations, solving problems and coming up with new ideas. And as his work has shown, he has a remarkable ability to analyse ideas and arguments, to criticize fruitfully the erroneous conceptions of others and to present coherent schemes of his own.

Turning to quantum theory, he found the conventional accounts difficult, partly perhaps because of the abstractness of the usual formalism, but more because he found practically no logical connection between the broadly wave-based picture given in the textbooks, and the general scheme of localized measurement results. It would be easy to sum up this difficulty as a quite justifiable unwillingness to accept the picture of measurement presented in the usual accounts. Perhaps, though, it went further than that—to an unhappiness with the dichotomy between physics as it is experienced in real space and the theoretical models in an entirely different space, the abstract space or *configuration space* of the Schrödinger equation. Just to give an example of the lack of transparency of the theory, the configuration space for a system of two particles in quantum theory has six dimensions not three!

Generally unhappy about quantum theory, during the mid-1960s Clauser looked into the works of those who had created the standard interpretation—Bohr and von Neumann, and those critical of this interpretation—EPR and Bohm. Clearly he was ripe to receive any arguments, ideas or suggestions that might help him to undermine the theory.

From these brief biographies of these two up to the fateful year of 1964, it is clear what enormous resources together they brought to the crucial task of responding to Bell and testing quantum theory. Shimony brought in a wide knowledge of mathematics and physics, but also a mature and critical approach to philosophy. His worldview had sufficient metaphysics for him to be comfortable with the idea of hidden variables, he had given much thought to the foundations of quantum theory and he had already questioned whether quantum theory needed at the very least a new formalism.

Clauser brought in excellent experimental ability, and in particular the confidence to accept totally new practical challenges. He was a very strong conceptual thinker, and he already had a strong antipathy towards quantum theory.

Without either of them, Bell's arguments would probably have withered on the vine. Either one of them on his own might have achieved a certain amount. Together they were to provide a substantial, though certainly not complete or final, theoretical and experimental response to Bell, and to those whose work had led up to that of Bell, while, looking forward, they were to stimulate all those who would walk further in their own footsteps.

Preliminaries and planning

Shimony received a copy of Bell's local realism paper almost immediately on its publication. Bell produced preprints of this paper to distribute at his talk at Brandeis, and somebody in the audience, knowing of Shimony's interest in the topic, sent him a copy. Shimony's instant impression was not good; the typing was bad, and in these days before regular use of photocopiers, a stencil had been used to produce multiple copies, and the purple ink on Shimony's copy smeared over his hands. This led him to suspect that the paper was 'kooky'—all physicists frequently receive typescripts from unknowns claiming to do wonderful things, usually to disprove or revolutionize our present version of relativity or quantum theory or particle physics or, often, all three!

Every so often something worthwhile may emerge. Many people will know the story of G. H. (Godfrey) Hardy, the Cambridge mathematician receiving in 1912 or 1913 a long seemingly kooky list of mathematical results from an unknown Indian clerk, Srinivasa Ramanujan [6]. Some of the results were known, some were actually not correct, others were new and, Hardy judged, must be exciting and brilliant. It seems that several other British mathematicians had also received copies of Ramanujan's work and had put them straight in the bin, but Hardy thought twice and then invited him to Cambridge. There Ramanujan quickly became recognized as an excellent and important mathematician, and he was elected a Fellow of the Royal Society (FRS) in 1918. Sadly, though, in wartime Britain he was not able to obtain the foodstuffs his religion required him to eat, and he was to die, perhaps from malnutrition, in 1920 aged only 32. But the story does show that positives do occur from the seemingly kooky—though not often!

However, Shimony soon realized that Bell's work was, in fact, both extremely important and exceptionally interesting to Shimony himself. Immediately he took up Bell's challenge—to study any experimental evidence already available to determine whether, in fact, it was capable of deciding for or against local realism, and, if this was not the case, to design and perform (or actually to persuade others to perform) an experiment that could do so. It should be pointed out that there was at that time no question of considering experiments where the detectors in the two wings of the experiment are adjusted during the flight of the particles being analysed. Thus technically we are talking about separable realism rather than local realism. However, following conventional usage, we shall usually use the latter rather than the former term.

Were there already any experimental results that could decide the matter outright? The most obvious possibility lay in a reanalysis carried out by Bohm and Yakir Aharonov [7] in 1957 of some experimental results obtained by Chien-Shiung Wu and her graduate student, Irving Shaknov [8], in 1950. (Wu (Figure 9.4) is always known as 'Madame Wu' and she was to become more famous in 1956 when she led the group that confirmed experimentally the theoretical suggestion of Chen Ning Yang and Tsung Dao Lee that parity was not conserved in the weak nuclear interaction. Sadly Shaknov was killed in the early 1950s in the Korean War.)

Wu and Shaknov's experiment studied the decay of *positronium*, a composite of an electron and a positron. The positron had been predicted by Dirac, emerging as an unexpected part of his extremely important relativistic theory of the hydrogen atom, which he published in 1928. (It will be recalled that the earlier Schrödinger theory was non-relativistic.) The positron was found experimentally by Carl Anderson in 1932, and Anderson was confident enough in his own experimental work to announce his discovery, despite the fact that he had never heard of Dirac's prediction!

All of the positron's properties, such as its mass, were exactly the same as those of the electron, except that its charge was positive rather than negative, so the electron is written as e^- while the positron is written as e^+. The positron is said to be the *anti-particle* of the electron, and

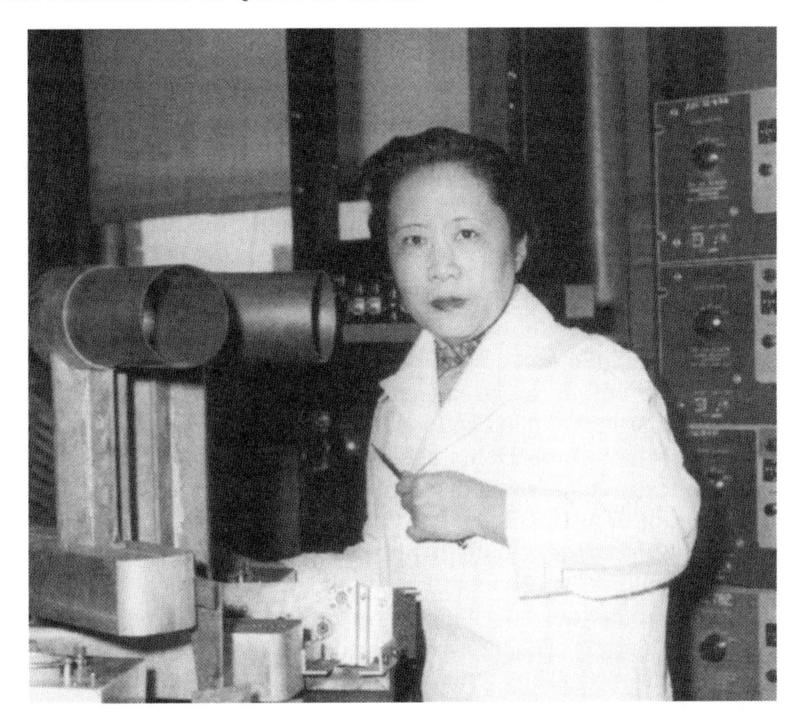

Fig. 9.4 Chien-Shiung Wu [courtesy of Professor Noemie Benczer-Koller (Rutgers University)]

this implies that when an electron and a positron meet, they annihilate each other, their energy being used to create two photons. (There must be two photons rather than just one because, as well as energy, momentum must be conserved. This can be achieved with two photons but not with one.)

However, annihilation is not instantaneous; instead the electron and positron can be thought of as performing a type of 'dance of death'. More prosaically we say that they form a compound called *positronium* or e^+e^-, which has a lifetime of around 10^{-7} s (for the case where spins of electron and positron are aligned). Interestingly this is long enough for the electromagnetic spectrum of positronium to be measured. Theoretically positronium is very much analogous to the hydrogen atom, so it is not surprising that the spectrum of positronium is analogous to that of the hydrogen atom as obtained by Bohr or from the Schrödinger equation, the only difference resulting from the fact that the mass of the proton in the hydrogen atom is, of course, much greater than that of the positron.

Wu and Shaknov examined these photons which, because of the quite high masses of the electron and the positron, were in the gamma ray region. Their object was to test various predictions of quantum electrodynamics (QED, the theory that studied electromagnetism from a quantum rather than a classical point of view). Most importantly QED predicted that if one of the photons was polarized vertically, the other would be polarized horizontally.

In their experiments a source of positrons is covered with a foil thick enough to guarantee that all the positrons are annihilated as they move away from the source, and this source is placed at the centre of a lead sphere. A narrow hole is drilled along a diameter of the sphere, and coincidences between photons detected at either end with specific polarizations are recorded. Anthracene crystals are used to analyse the polarizations of the photons. With this method, Wu and Shaknov were able to confirm the particular prediction of QED mentioned in the previous paragraph. They did, though, only consider, and indeed for their purposes only needed to consider, two orientations of the polarization of their photons—either along or perpendicular to their direction of travel.

Such was the interesting and important work of Wu and Shaknov. It was completely unnecessary for them even to recognize that, according to the laws of quantum theory, the two photons were, in fact, in an entangled state, and it was Bohm and Aharanov seven years later who pointed out that this was the case—in fact Wu and Shaknov had unknowingly produced the first source of entangled photons. Rather than saying that the polarization of photon 1 is horizontal and that of photon 2 vertical, so the state of the system can be represented as $|h\rangle_1|v\rangle_2$, or alternatively that the individual polarizations are interchanged so that the state of the system is $|v\rangle_1|h\rangle_2$, we should use the entangled state $\left(1/\sqrt{2}\right)\left(|h\rangle_1|v\rangle_2+|v\rangle_1|h\rangle_2\right)$.

While, as I have said, Bohr and Aharanov recognized that this was the quantum prediction, they also realized that it would be extremely important to check whether Wu and Shaknov's results actually confirmed it. There was another suggestion on the table, often called the Schrödinger–Furry hypothesis. Schrödinger, in his response to the EPR paper of 1935, and Wendell Furry, in two papers published in the following year, raised the possibility that, once the entangled particles were sufficiently far apart, the wave-function of the particles might not remain in the entangled form, but might move to one or other of the direct product states we looked at in the previous paragraph. The hypothesis could be regarded as an example of a hidden variable theory, as it did provide extra information over what was entailed by the entangled wave-function.

Furry pointed out that his motivation for this suggestion was the absolute opposite of that of Schrödinger. Schrödinger was broadly a supporter of EPR, and he approved of the suggestion as it removed what he considered a blemish on the theory. Furry announced himself a supporter of Bohr, and while he recognized that the change from the entangled state to the unentangled was a perfectly safe working hypothesis in most physical situations, in EPR-type situations it produced a different answer to that obtained from quantum theory and therefore, he assumed, was unacceptable.

It is interesting to realize that, if such a hypothesis were to be entertained today, and mainly under the influence of Bell's approach to these matters, immediate further questions would be asked:—how far apart must the individual particles be for this change to occur? And more basically, just how would it happen? A rough answer to the first question might be 'when the wave-functions of the particles no longer overlap', but it would seem just as difficult to answer the second question as to discover why wave-functions should collapse on measurement.

However, the fact was that, by re-examining Wu and Shaknov's data in 1957, Bohm and Aharanov were able to deduce that the Schrödinger–Furry hypothesis could not be maintained. It was certain that the wave-function kept its entangled form no matter how far the individual particles were separated. Of course this was a victory for quantum theory against what might seem more like common sense! Shimony, though, a decade or so later, had to conjecture whether the results ruled out not just the Schrödinger–Furry hypothesis but *any* hidden variable theory, and the situation was made more complicated by Aharanov appearing at MIT, where Shimony was still based, and assuring him with total confidence that indeed his and Bohm's work did rule out all types of hidden variable theory.

Shimony gradually realized, though, that this could not be the case. As we have said, Wu and Shaknov had carried out measurements only at two angles—the polarizations in the two wings of the apparatus either parallel or perpendicular. Yet these were the cases where Bell had shown that the quantum mechanical results could easily be explained by hidden variable models. Thus there was no way that experiments using only these angles could comment on the absence or presence of hidden variables.

In fact there was more to it than that, and the person who convinced Shimony of this most clearly was Michael Horne (Figure 9.5), who is another especially important character in our story. By this stage, Shimony had just taken up his new position at Boston University. Horne himself had been born in 1943 in the state of Mississippi, and he graduated in physics from the University of Mississippi, but he very much wished to carry out his PhD work in a major university, and to work in a fundamental area of physics. He came to Boston in 1965 and worked for some time in the foundations of statistical mechanics, and it was to ask for a thesis problem in this area that he appeared at Shimony's door in 1968.

Shimony realized that Horne was a student with the desire and ability to study the fundamental questions of physics and he showed him the two important papers of John Bell. Horne himself was exceptionally interested in these papers, and at once commenced working with Shimony to plan a genuine test of Bell's work.

Fig. 9.5 Michael Horne [© Zentralbibliothek für Physik, Renate Bertlmann]

Horne was able to show—it was included, with much other material on this topic, in his PhD thesis, which he was to write in 1970—that the Wu–Shaknov scheme would not work even if more angles were included. The problem was that, since the experiment used gamma particles of high energy—about 0.5 MeV each, the polarizers we may be familiar with from working with optical photons, Polaroid or calcite prisms, are no use; the photons just go straight through. Because of this, Wu and Shaknov had used so-called Compton polarimeters to study the polarization of the photons.

The so-called Compton effect had been discovered by the American physicist Arthur Compton (Figure 9.6) in 1923, and it was this experiment that was finally responsible for convincing the great majority of physicists that electromagnetic radiation of any wavelength, whether it be, for example, light, X-ray or radio wave, had a particle-like as well as a wavelike nature, the former being displayed when it interacted, the latter when it propagated.

We have stated that it was Einstein who first spoke of this *wave–particle duality* in 1905, when he discussed the photoelectric effect, and he used the idea of what he called the *light quantum*, which much later was called the *photon*. However, although Einstein received the Nobel Prize for physics in 1922 for his photoelectric equation, rather strangely even at that time extremely few physicists actually believed in the idea of the photon, even though it was the basis for the equation which they accepted!

Compton's work studied the interaction between an X-ray and an electron, and he showed that the details of the results could only be

Fig. 9.6 Arthur Compton (right) with Vannevar Bush in 1940. Both Compton and Bush were highly involved in the American military effort during the Second World War. [Courtesy of Lawrence Berkeley National Laboratory]

explained by treating the X-ray as a particle, and analysing the interaction with the electron as a straightforward Newtonian collision in which kinetic energy and momentum both had to be conserved. With the Compton effect understood in this way, it was difficult to deny the existence of wave–particle duality, and the effect played quite an important role in the emergence of Heisenberg–Schrödinger quantum theory in the next few years.

Wu and Shaknov were able to study the polarization of their photons by causing them to interact in a Compton-like way with electrons, and studying the resultant behaviour of the electrons. However, Horne was able to show that, while entanglement means that the wave-functions of the photons are completely correlated, the transfer of correlation between the photons and the electrons in this way is so weak that the resultant correlation between the electrons was too small to give any hope of discriminating between quantum theory and hidden variables. In fact Horne performed what was to become the classic way of showing that particular quantum predictions or experimental results could not rule out hidden variables; he was able to construct a hidden variable model that duplicated the results!

It was clear that what was required was an experiment involving entangled low-energy light photons, so that more conventional methods could be used for studying the polarization—Polaroid or calcite prisms or, as Clauser was to use, a pile of plates. In the last few months of 1968 and the first few of 1969, Shimony and Horne travelled to many universities searching for ideas for obtaining entangled beams of low-energy photons, as well as picking up hints of methods of measuring the polarization.

The great breakthrough was when they heard of the experiment of Carl Kocher and Gene Commins [9] at the University of California at Berkeley. Kocher and Commins had studied the polarization of a pair of low-energy photons produced in a calcium cascade. In a cascade, an atom in a stable level receives energy via a photon—a photon in the ultraviolet range in Kocher and Commins' case. It absorbs the energy and moves up to an excited state, but this excited state is unstable, and the atom returns to its ground state in two stages, each time emitting a photon. The two photons are entangled; in this case they will *either* both be polarized horizontally *or* both be polarized vertically, so the state can be represented as $\left(1/\sqrt{2}\right)\left(|h\rangle_1|h\rangle_2+|v\rangle_1|v\rangle_2\right)$.

Clearly this seemed an ideal setup for testing Bell's ideas. Kocher and Commins did indeed cite the EPR paper and Bohr's response to it in their paper, and this paper was indeed intended to check the theory in a general way. However, they were presumably unaware of Bell's work as they did not give a reference to it, and, just like Wu and Shaknov, they took readings only at 0° and 90°, so their experiment was totally unable to decide on the possible existence of local hidden variables. Actually the experiment was initially planned as a lecture demonstration on a trolley, and it was only when it turned out to be more complicated than had been expected that it became Kocher's thesis topic. It certainly was not intended to solve major problems!

So Shimony and Horne, and actually at more or less the same time but unbeknown to them, Clauser, reached the understanding that no experiments had yet been done to check whether Bell's inequalities were obeyed, even for the case where settings in the two wings of the experiment were *not* changed during the transit of the entangled particles. There was certainly no need to proceed immediately to the case where the settings would be altered, as Bell had perhaps suggested. Clauser regarded this as 'calling Bell's bluff'.

However, Shimony and Horne now had an excellent idea of what direction to move in; they should repeat the Kocher–Commins experiment for a wide range of angles, and even better was to follow. They were told that Frank Pipkin, a highly experienced physics professor at Harvard, had exactly the type of apparatus that they were interested in. His apparatus in fact used a cascade in mercury rather than calcium, and another difference from the Kocher–Commins case was that both photons always had opposite rather than the same polarization.

The down-to-earth purpose of Pipkin's apparatus was to measure the lifetime of the intermediate state, the state produced after the emission of the first photon of the cascade, which decays back to the ground state by the emission of the second one. In any experiment of this type, a coincidence check is needed to identify photons emitted from the same atom, and in this case the time delay between the arrival of the correlated photons tells the experimenter how long the intermediate state has lasted.

Shimony and Horne now paid a visit to Pipkin and his new PhD student, Richard Holt. Their aim was to try to persuade Pipkin and Holt to use the apparatus to undertake an experiment that some might think conceptually far more interesting than measurement of an atomic lifetime, but which others might think a complete waste of time! Fortunately, Holt was interested enough to take on the task, which he optimistically thought of as a fairly straightforward experiment that could be finished reasonably quickly so that he could get back to his 'real' project.

Everything now seemed set fair for the four would-be collaborators. Shimony had intended to present a paper at the spring 1970 meeting of the American Physical Society in Washington DC, but the deadline was missed. However, Shimony remarked that it was hardly likely that anybody else would be working on such an obscure problem, and he intended to move directly to writing a full paper for a journal. That was until he saw the *Bulletin* of the Society that gave the programme and abstracts for the meeting, and came across an abstract from Clauser...

Clauser himself was not sent a copy of Bell's paper, but some time in 1967 he came across a copy in the library of the Goddard Institute for Space Studies. As a quantum agnostic, he was excited and stimulated by Bell's ideas, but as a sceptic he first checked the work by trying to construct counter examples—hidden variable theories that *were* local. He could not find any and thus became convinced that Bell was right.

From now on, it is fair to say that, for Clauser, astrophysics took a back seat. He was to become obsessed with experiments that could decide between local realism and quantum theory—first ascertaining whether such experiments had already been carried out, and if not, hoping to design and perform the first to find the truth of the matter. And while, for example, Bell and Shimony would have been delighted if quantum theory had proved to be wrong for these particular experiments, but in their heart of hearts thought it unlikely, Clauser was convinced that the experiments would uphold local realism and find that quantum theory was wrong. He was to make comments such as: 'If we get the quantum mechanical result, I'll quit physics', and he took out a bet with Aharanov with odds of two-to-one against quantum theory.

Holt incidentally took a different view. He was to say, jocularly, that he had to get the quantum mechanical result, because if he didn't, he might get a Nobel Prize, but stuffy Harvard would not give him a PhD! The other main participator, Horne, was probably even less optimistic than Shimony or Bell of showing that quantum theory was wrong for these experiments; he felt that the theory had been so supremely successful so far that it was practically certain that it would not fall at this hurdle.

Clauser followed much the same path as Shimony and Horne. He too realized that the Wu–Shakov results needed to be at a greater range of angles, and he visited Wu at Columbia to see whether she had obtained further data but it had remained unpublished. This was not the case, but it does seem that Clauser raised Wu's interest, and she and her two graduate students were actually to publish a test of Bell's ideas in 1975, though, as we have already suggested, the results could not be conclusive; we shall return to her work briefly later in this chapter.

It must be remembered that Clauser was in a completely different position from Shimony. Shimony had an established position, and, though he may obviously have had limited time to work on the project himself, the arrival of Horne and the agreement with Holt and Pipkin must have improved things enormously. Clauser, on the other hand, as a graduate student, was naturally expected to spend all his time on astrophysics. Instead he was putting a lot of effort into thinking about quantum theory and its perceived limitations, and planning the crucial experiment that he hoped would bring it down.

He travelled quite widely, looking for information, but also giving talks that were actually mostly job interviews. He himself would admit that he was naïve, failing to realize that most physicists were likely to regard the work as a waste of time; they believed that quantum theory was as sound as a rock and that Bohr had seen off EPR decades earlier.

Clauser regarded Pat Thaddeus, his thesis advisor, along with his father, as one of his main influences, but Thaddeus must have become highly irritated and disappointed with his admittedly brilliant graduate student. He certainly thought Clauser was totally wasting his time on quantum theory, and wrote job 'recommendation' letters for him along the lines—don't appoint him because he may do quantum mechanics experiments, and it is all junk science!

Like Shimony and Horne, Clauser became aware of the work of Kocher and Commins. In his case this came about when, hearing of experiments being performed at MIT on crossed-beam scattering experiments with alkali metal atoms, a technique that he thought could be useful for his own project, he visited MIT to give a seminar on Bell's work. By coincidence, Kocher had just taken up a postdoctoral appointment at MIT, and when Clauser's host asked him whether his experiment had performed the test that Clauser was interested in, Kocher replied that it certainly would—that was precisely why they had done it! However, when Clauser checked up their paper, he found that this was very definitely *not* the case. In the first place, as we have already said, data had only been taken at 0° and 90°, and Clauser also found that the efficiencies of the polarizers were far too low to provide a conclusive test.

Becoming, like Shimony and Horne, convinced that no experiments had in fact been performed that actually tested Bell's work, Clauser decide to take the bull by the horns, and he wrote to Bell, Bohm and de Broglie asking them if (a) they knew of any experimental tests; (b) they thought a repeat of Kocher–Commins at a greater range of angles and with improved polarizers would be useful; and (c) how important such tests would be. All said 'no' to (a) and 'yes' to (b), while Bell was very enthusiastic in response to (c); he made the comment reported before that there was a 'slim chance of an unexpected result, which would shake the world'. Wick [10] has reported that Clauser's letter was the first response to his 1964 paper that he had received.

It was thus that Clauser submitted the paper for the Washington meeting.

Shimony was obviously nonplussed to find that another physicist had trodden more or less the same route that he had imagined was too arcane for anybody but himself to have been interested in. To Horne, he confessed: 'We've been scooped', and they were forced to work out what course to take now—ethically, tactically. Some of their colleagues suggested they should pretend that they had not seen Clauser's abstract, and go ahead themselves with publication in a mainstream journal. They could, of course, have published a more substantial paper than that of Clauser with a reference to Clauser's own work, but with a note that they had performed the same work independently. (When this is done, referees and readers of the paper may choose to believe or disbelieve the claim as they wish!)

In the end, it was Wigner who encouraged Shimony to phone Clauser to tell him that they had come to exactly the same conclusions as he had, and to see if he might be willing to join forces. This is what Shimony did, but at first Clauser was not keen. Apart from the fact that he had moved a step ahead of Shimony and Horne with his publication in the *Bulletin*, he would fairly have thought that he had made great strides already on his own, and could see no advantage in sharing the glory he genuinely thought might emerge from the project with two other people he knew very little about.

However, he had to rethink his position when Shimony told him of Pipkin and Holt ready and waiting to carry out the experiment, and he had to consider the possibility of missing out altogether. He was very keen to carry out the first experiment to test local realism against quantum theory, and it now seemed to him that the best way for him to achieve this was for him to carry out the experiment with Holt.

Clauser, Shimony, Horne and Holt all met up at the Washington meeting, and obviously got on well because they made fairly detailed plans to push the project forward together. This was to be the start of an immensely fulfilling relationship for Shimony, Horne and Clauser, if, unfortunately, not quite so much for Holt. The first immediate step was to write a joint paper for *Physical Review Letters* (*PRL*), which would present suitable generalizations of Bell's work, and would describe a suitable experiment to test these ideas.

It must be remembered that Clauser was still a graduate student, and in fact was at the stage of finishing off his PhD, so one can imagine how busy he was! At last, though, he completed his thesis and had two weeks free before he had to undergo its defense. He was able to utilize this time by visiting Holt's laboratory, and also staying with Shimony in Boston, where Clauser, Shimony and Horne made a good start on the paper for PRL. At first it was thought that it would be just the three of them who would be authors of this paper, but Holt was able to make a particularly important contribution, to be explained in the next section, and so he was included and the paper became Clauser, Horne, Shimony and Holt; this was to be a very important paper, always known as CHSH.

(Yet for the rest of the decade, it was Shimony, Clauser and Horne who worked together on theory and analysis, as distinct from the actual experiments. There are several examples of this. An important paper of 1974 made significant additions to Bell's work. It was authored by Clauser and Horne, but Shimony contributed a lot to the study, declining to be a joint author only because he did not think his contribution was significant enough. And in 1978, Clauser and Shimony published a review of progress made up to that point; Horne had been invited to be a co-author but declined because of lack of time. In neither case, though, was Holt part of the team.)

In fact there was to be one further throw of the dice. Clauser was offered a position at the University of California at Berkeley to work as a postdoc under Charles Townes (Figure 9.7) studying the early universe by radio astronomy. Berkeley, of course, was where the Kocher–Commins experiment had been performed, where Commins was still in a senior position, and most importantly where the experimental equipment still lay. It was natural for Clauser to ask if he might be allowed to resurrect the equipment and use it for a conclusive test of Bell's ideas.

Commins was totally negative. As has been said, the Kocher–Commins experiment had not been carried out in a spirit of open-minded inquiry; rather it had originally been intended just as a student demonstration. Commins himself made clear his opinion that such experiments were

Fig. 9.7 Charles Townes [from Denis Brian, *Genius Talk* (New York: Plenum, 1995)]

entirely pointless, that entanglement was a basic and non-controversial part of quantum theory, and that Bohr had defeated EPR decades earlier. Clauser has commented that most faculty at Berkeley felt much the same. Even if Commins had been less hostile to the proposed experiment, though, it is still unlikely that he would have jumped at Clauser's suggestion. After all, when you hire a postdoc to work on radio astronomy, you obviously expect him to work on radio astronomy, not on an entirely different topic of his choice!

At this point we meet yet another hero of our story, for Townes was sufficiently interested in Clauser's suggestion to take his side. Townes was an extremely influential figure in American physics. Born in 1915, his work in the development of the maser and the laser had been responsible for him being awarded the Nobel Prize for Physics in 1964. (Just as 'laser' stands for 'light amplification by stimulated emission of radiation', in 'maser', 'light' is replaced by 'microwave'.) Most of the people who worked in lasers and astrophysics in the country were his protégés, including Thaddeus.

So when Townes suggested to Commins that Clauser's proposed experiment looked interesting, Commins had little option but to agree. Notionally Clauser was to be allowed to spend half of his time on this project, the other half being spent on radio astronomy. In addition, Commins allowed a new graduate student, Stuart Freedman, to work with Clauser. Freedman knew nothing about the subject up to this point, but he thought it sounded interesting, and he joined in whole-heartedly. Incidentally, like Holt, it is possible that all of them—Freedman, Commins, Townes and even Clauser—may have drastically underestimated the time that the experiment would take. After all, it might seem that all that was required was a little upgrading of the Kocher–Commins equipment and taking some measurements. This was to turn out to be very much not the case!

It seems dubious that Clauser spent much time, if indeed any, on radio astronomy. Every so often Townes would give him a ring commenting that not much of the work he was really being paid for was being done, but he seems to have been willing to turn a blind eye, And fortunately Howard Shugart, the head of the atomic beams group, presumably with the encouragement of Townes, was prepared to keep him on the payroll as a nominal member of that group. Even with the best will in the world, such flexibility could never be allowed today; money would be obtained for a particular project, and would have to be spent on that project. Even in the 1970s, it was only possible because Clauser was technically a joint employee of the University of California at Berkeley (UC), and Livermore Berkeley National Laboratory (LBL). Financially the relations between UC and LBL were complicated; the atomic beams group was funded through LBL by a blanket funding, so that Shugart had an enormous amount of discretion over how he spent the money.

One result of Townes' generosity or foresightedness was, of course, that Clauser and Holt had become, not possible collaborators on the same apparatus, but competitors—friendly competitors perhaps, but there was no way that Clauser wanted to come second!

Clauser was an extremely keen sailor, and in fact he lived on his racing yacht. Thus he decided that the best way of organizing his journey back to California was to sail to Galveston in Texas, truck the boat to the West Coast, and then sail up along the Californian coast. Though the latter stages of this journey were disrupted by a hurricane, communication with Shimony was maintained as he sailed down the East Coast. Shimony would arrange for the latest material for the CHSH paper to be sent to the marina where Clauser would be stopping next, and Clauser would comment and argue by phone. By the time he reached Berkeley, the paper was finished. It was submitted to the journal on 4 August 1969 and published on 13 October.

It is interesting that, after he had arrived in California [1], Clauser took the opportunity to visit Caltech to discuss his ideas with Richard Feynman. Feynman was entirely negative towards Clauser, effectively throwing him out of his office. For him, it was inconceivable that what Clauser was suggesting could have even the smallest amount of merit. All it could do would be to demonstrate Clauser's lack of competence and professionalism. We shall see much more of Feynman in Chapter 16.

CHSH—Clauser, Horne, Shimony and Holt

The 'four-man paper' by Clauser, Horne, Shimony and Holt [11] played an extremely important role in the understanding and testing of Bell's seminal ideas. It was in a sense the bridge between theory and experiment. In his paper on local realism, Bell comments that: 'The example considered has the advantage that it requires little imagination to envisage the measurements involved actually being made.' This is quite true, and of course the supreme advantage of Bell's work over that of, for example EPR, Bohr in his response to EPR, and Bohm, is that, rather than floundering in thought-experiments, easy to discuss but impossible to carry out, armchair philosophy or vague suggestions of where problems with quantum theory might lie, Bell's work did lead to experiments that could actually be performed.

For all that, there was a vast amount to be done before one could even set about constructing an experiment. Clauser, the supreme experimentalist (though, as we insisted earlier, that by no means exhausted his talents) was even quite critical of some of Bell's assumptions, at least insofar as they could be followed in practice. Bell, he says, being a theorist, assumed an ideal apparatus, an idealized configuration and an ideal preparation of an entangled state. His result, therefore, Clauser says, must apply only to ideal systems, not to real ones! ('Damn theorists!', he remarks [1].)

The problem he saw in Bell's analysis, at least from the point of view of generalizing to real experiments, is Bell's preliminary argument in which, following EPR, he argues *from* locality *to* deterministic hidden variables. From Clauser's point of view, the difficulty is that this argument does assume that the two photons are indeed travelling in precisely opposite directions and so their polarizations are precisely correlated. As Clauser says, even an infinitesimal departure from this ideal situation would mean that Bell's assumption is no longer true, and the whole proof collapses. Fortunately Clauser was able to show that this first stage of Bell's argument was by no means essential.

In CHSH, a small amount of departure from perfect correlation is allowed. This is called δ; the authors explicitly state that they 'avoid Bell's experimentally unrealistic restriction that for some pairs of parameters there is perfect correlation (i.e. $\delta = 0$)', but they are able to show that the main result is independent of the value of δ. As well as being a useful generalization of Bell's proof, this argument is also important as a contribution to understanding what is essential and what is inessential for Bell's result to go through. We shall take this argument further later in this chapter.

CHSH was also important for introducing the famous inequality to determine whether experimental data agree with local realism. Here we write it in somewhat different notation. We write two settings of the polarizer in the first wing of the experiment as $\hat{a}$ and $\hat{a}'$, and analogous settings in the second wing as $\hat{b}$ and $\hat{b}'$. A and B are the results in each wing of the apparatus, and the units are arranged so that each of them takes the values ± 1. Then $P(\hat{a},\hat{b})$ is the average value of AB for settings $\hat{a}$ and $\hat{b}$, the average being taken over all values of λ, which is the hidden variable. (We are writing in terms of a single hidden variable, but can generalize to many such variables very easily.)

Then the inequality is $-2 \le S \le 2$, where

$$S = P(\hat{a}, \hat{b}) + P(\hat{a}, \hat{b}') + P(\hat{a}', \hat{b}) - P(\hat{a}', \hat{b}')$$

Local hidden variable theories *must* obey the inequality. A wide variety of values can be put in for $\hat{a},\hat{b},\hat{a}'$ and $\hat{b}'$, but an interesting and useful choice is to have all the vectors in the same plane, and to choose $\hat{a} = 0$, $\hat{b} = \theta$, $\hat{a}' = 2\theta$, $\hat{b}' = 3\theta$, with θ varying.

The quantum mechanical expression for $P(\alpha,\beta)$ is just $\cos\{2(\alpha-\beta)\}$, for any α and β, so it is easy to show that S becomes $3\cos(2\theta)-\cos(6\theta)$ for the quantum case. Then S is shown as a function of θ in Figure 9.8. We note that S disobeys the local hidden variable inequality for wide ranges of θ. It is just on the boundary of the permitted region for $\theta = 0°$ and 90°, and there is a wide band around 45° where it obeys the inequality easily, but for values of θ in ranges above 0° and below 90°, it violates it by a considerable amount. This is another way of seeing what Bell had already observed—that quantum theory does not conform to local realism.

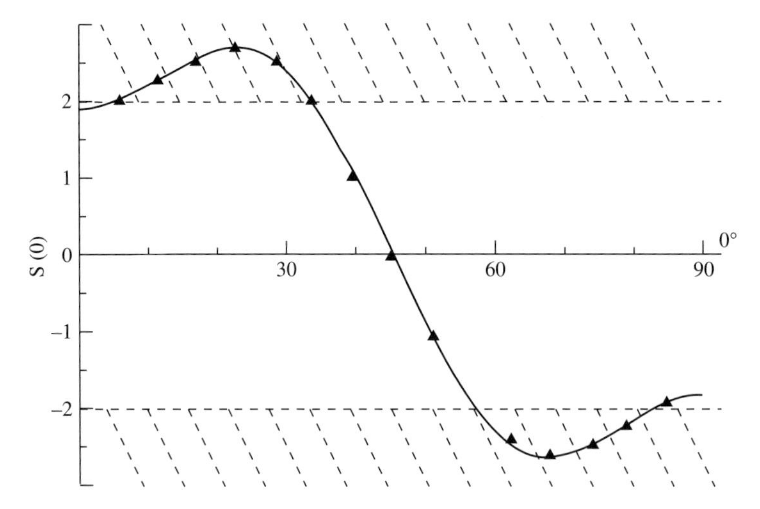

Fig. 9.8 $S(\theta)$ as described in the text. According to the CHSH inequality, S should not lie in the hatched region. The solid line is the prediction of quantum theory which is clearly in the hatched region for large values of θ. The triangles are the experimental values of Aspect and collaborators as discussed in Chapter 10. [A. Aspect and P. Grangier, in P. Lahti and P. Mittelstaedt (eds), *Symposium on the Foundations of Modern Physics: 50 Years of the Einstein-Podolsky-Rosen Gedanken-experiment* (Singapore: World Scientific, 1985)]

It can be noted that S violates the inequality most strongly if $\theta =$ 22½° or 67½°; in these cases it is equal to $2\sqrt{2}$ and $-2\sqrt{2}$, respectively. With the choice of $\hat{a}, \hat{b}, \hat{a}'$ and $\hat{b}'$ just given and $\theta = 22½°$, in all cases the angle between the polarizers, between *either* $\hat{a}$ or $\hat{a}'$ and *either* $\hat{b}$ or $\hat{b}'$ is either 22½° or 67½°. The merits of this case were discovered by Freedman, and these angles are often called Freedman angles.

The general inequality should probably most fairly be called the CHSH inequality. Often in fact it is called the CHSH–Bell inequality, and even more often just the Bell inequality, which is a little misleading. By using Bell's perfect correlation equation, it is possible to recover an equation from Bell's earlier paper, which can be called a genuine Bell inequality, but this equation is less symmetric than the CHSH inequality, and in any case it does not seem that Bell intended it to be used as a test for local realism. I am not sure that Bell went much further than recognizing that if experiments on entangled systems supported quantum theory, then that implied that they violated local realism.

With the CHSH inequality, one can examine experimental results in two ways, one after the other. First, one determines whether the inequality is obeyed. Clearly that provides direct evidence on whether the experiment supports local realism. It is important to recognize that this procedure is entirely independent of any quantum theoretical considerations. Neither the way in which the inequality is produced nor the examination of the experimental results using the inequality involves any use of the quantum formalism. If at this point it is found that local realism is violated, then it is an entirely separate matter to see whether the results agree with quantum theory, within, of course, the possible errors of the experiment.

Another area of crucial importance that CHSH opened up was the actual analysis of the experimental results. Bell blithely assumed for his theoretical purposes that each member of each entangled pair could be investigated. He was thinking of experiments involving electron spin, and it was natural to think of electrons being detected whether they had 'spin up' or 'spin down'.

Real life, as studied by CHSH, was cruelly different. They were, of course, discussing measurement of the direction of polarization of photons rather than the direction of spin of electrons. In this case, and for normal polarizing devices, photons with polarization parallel to a given axis are passed and continue on to the detector, but those with polarization orthogonal to this axis are absorbed. We may call this a *one-channel analyser*, since only one beam of photons emerges and may be detected.

However, this is a minor problem concerned with a fact that caused immense difficulties for CHSH, and, under the name *detector loophole*, continues to do so today. This is that photon detectors are exceptionally inefficient. In the kind of experiments being suggested, in only one in about a million cases are both members of an entangled pair detected. In most cases even where both photons have parallel polarization, still neither will be detected. When a photon is detected in one wing of the apparatus, we don't know whether its partner was polarized parallel to the axis and just failed to be detected, or orthogonal to the axis and was absorbed. And of course if a pair of photons are both orthogonal, there is no possibility of any detection. And the geometry of the experiment means that in most cases neither photon reaches a polarizer in any case! But the only useful data are for cases where a photon is detected simultaneously in each wing of the apparatus.

To solve this problem, CHSH proposed to carry our the measurement in each setting four times: once with polarizers in place in each wing of the experiment as we have taken for granted up till now, once with each polarizer in turn removed, and once with both polarizers removed. With this procedure the required analysis is straightforward, but it is essential to note that it uses an assumption—we shall always call such assumptions *auxiliary assumptions*, and we shall need to discuss their significance later. The assumption in this case is that the probability of both of a pair of entangled photons being detected is entirely independent of whether polarizers are present in either wing, or, in cases where they are, the direction of either axis of polarization. In other words, the probability of a photon being detected can be treated as a constant, and since, in fact, we will be dealing with ratios of numbers of events in different cases, this constant can be ignored.

We can then think of the number of simultaneous detections with both polarizers present giving us the number of pairs of photons both polarized parallel to the appropriate axis. When we remove the polarizer in the first wing, we are seeing those pairs of photons for which the polarization is parallel to the axis in the second wing, but it may be parallel or orthogonal to the axis in the second wing, and the situation is analogous when we remove the polarizer in the second wing. Finally when we remove both polarizers, we may see all pairs of photons whatever their directions of polarization. Of course, as we have said, when we say we 'see' them, in fact we detect only a very small fraction of them, but, because of our auxiliary assumption, it is the same fraction every time. So it is very easy to write down expressions such as $P(\hat{a}, \hat{b})$ in terms of ratios of our coincidence count rates for the four different experimental cases, and then we can also write down the CHSH inequality in terms of these ratios.

That is all straightforward, but now we must come back to our auxiliary assumption. Is it actually the case that the possibility of detection of an individual photon does not depend 'where it has been' or in other words whether it has passed through a polarizer and, if so, the direction of the axis of the polarizer? An immediate response would probably be: why should it depend on these things? Indeed there may seem no reason, and the assumption is of a kind that would be made automatically in any normal experimental context. Indeed one might say that far less likely assumptions are regularly made in such contexts.

However, we must face up to the fact that we are discussing here a very special experiment designed to answer a very deep question. It is also, as we shall see, a question on which different people have very strong feelings—some very much for local realism, some very much for quantum theory. Such people will certainly pick up on any doubt in the experimental setup.

A more strictly scientific comment might go as follows; we saw in, for example, Bell's analysis of von Neumann's 'proof' and the other 'no-go theorems' that assumptions that seem totally natural when made in a fairly abstract way may turn out to be not just wrong, but even to seem rather silly, when examined in more specific circumstances.

It can also be said that the whole point of hidden variable theories is that different members of an ensemble, which quantum theory would characterize as identical, may behave very differently in any given context because they have different values of a hidden variable. When one puts it this way, it is at least more plausible that photons with different histories might behave differently in a particle detector. We shall return to the vexed question of auxiliary assumptions later.

Here we take up the last essential point in CHSH that translated theoretical vision to experimental practicality. In Bell's theoretical analysis, the entangled particles came out in exactly opposite directions, one might say, for example, along the positive and negative z-axes. Of course if the experiment was restricted to this case only an infinitesimally small number of pairs would be included. Obviously we need to include photons emerging at considerable angles to any central axis and then to use wide lenses to focus the emergent photons onto the polarizer and detector. Thus entangled pairs that end up being detected are not necessarily travelling in precisely opposite directions, and this does complicate the analysis.

Actually it is easier to understand what happens for the case Bell considered theoretically, which, of course, is not the one which CHSH were planning to use in their experiments. CHSH were going to use photon polarization, while Bell had spin angular momentum, and the advantage of discussing the latter case is that it can all be explained in terms of angular momentum. For the cascade in calcium that Kocher and Commins had used, and that Freedman and Clauser also intended to use, the initial state has $J = 0$, the intermediate state has $J = 1$ and the final state has $J = 0$, so we can call it a 0-1-0 cascade. Here J is the total angular momentum of the various atomic states, in units of $\hbar$ of course. For this cascade the overall change in total angular momentum is 0, so the total angular momentum of the two photons emitted in the cascade must also be 0.

The awkward point is that this angular momentum is the sum of the spin angular momentum of the two electrons, which is what we are interested in, and their orbital angular momentum, which is essentially the angular momentum related to the actual paths of the electrons, and which we are not interested in. Fortunately when the two photons emerge in precisely opposite directions, the orbital angular momentum is 0, so we can say directly that the total spin angular momentum of the two electrons is 0, or that the spins of the two electrons are in opposite directions. We can say that the correlation is 100%. This is what we have taken for granted up to now.

We can mention that two other types of cascades can be used in this type of experiments; we can write them as 0-1-1 and 1-1-0. In these cases, the total angular momentum of the atomic system changes through the process by $\hbar$, so, in the case where the electrons emerge in opposite directions and the electrons have no orbital angular momentum, the two electrons have a total spin angular momentum of $\hbar$, so both spins must point in the same rather than opposite directions. Again we can talk of 100% correlation.

However, when we allow the electrons to emerge in beams of considerable angular width, perhaps 30° half-width, the orbital angular momentum will not be 0, and quite a complicated calculation is required to find the effective correlation.

That, of course, as we have stressed is for study of electron spin not photon polarization, which is the case for which we actually require information. However, for this case, although the picture is not as obvious as the one we have been looking at, the actual calculations are broadly analogous. It was Holt who was able to show the other authors a good approximate method of carrying out the procedure, they followed this prescription and the reward for Holt was to have his name added to the list of authors of CHSH. It turned out the correlation for beams of this width was only slightly less than 1, about 0.995. As we have said, Holt's method was an approximation, but when Shimony had time to carry out the calculations exactly, he obtained exactly the same answer, and he included this work in his Varenna paper.

We have assembled the various points required for the bridge between theory and experiment, and we now conclude this section by making two general points. First there is a footnote to the paper saying that the paper was an expansion of Clauser's paper presented at the Washington meeting, but adding that Shimony and Horne had reached the same conclusions independently. This was a way of putting things that seems to have been fair to all.

Secondly, and this is an aspect of a much more general point, it is interesting that the paper was published in *PRL*. It should be explained that, while the American journal *The Physical Review* (*PR*) had been publishing papers in physics since 1893, and since the Second World War it had certainly been the leading physics journal in the world, in 1958 the American Physical Society, which publishes the journal, launched a new journal, *PRL*, which was intended to provide much quicker publication for short but very important pieces of work. The period of time between reception of the paper and its publication might be two or three months (two for CHSH) as compared to perhaps eight or nine months for an paper in *PR*. Many other journals in different areas of science took similar steps at roughly this time.

Almost immediately *PRL* became what it still remains—by far the most prestigious journal publishing only papers on physics. (The American journal *Science* and the British journal *Nature* are at least as prestigious as *PRL*, but they publish material from all areas of science.) Speed of publication may be almost a red herring. Authors will happily wrangle with the Editor of *PRL* for many months if not years to obtain publication in that journal, rather than cut their losses and settle for publication in a lesser regarded journal, although they might achieve the latter much faster.

So having said all that, it is interesting to note that CHSH was indeed published in *PRL*. It is true that, from the nature of some suggestions made by one of the referees, the authors were fairly sure that that referee must have been Bell, who obviously would have supported publication, but the paper must also have received support from at least one other referee, probably more. In any case most of the papers presenting the result of measurements on Bell's work were also published in *PRL*. We shall return to this point later in the chapter.

The experiments of the first decade [12]

All the main issues had been sorted out in CHSH, and Holt in Harvard, and Clauser and Freedman in California could move onto designing and constructing their experimental systems. This was far from straightforward, of course.

Clauser's original idea had been essentially to redo the experiment of Kocher and Commins, using basically the original equipment but upgrading substantially the polarizers. Of course the experiment would use a much greater range of angles.

In practice, though, only very minor components of the Kocher and Commins apparatus remained in the setup eventually used by Clauser and Freedman. As they studied each part of the original equipment they came to the conclusion that they could do better. They were able to purchase new photomultipliers, but mostly Clauser did what he excelled at, looking through the 'junk' left as previous experiments had been dismantled, and picking out what would best suit his purpose. Every piece of equipment—lamps, lenses, filters, polarizers, photomultipliers—had to be chosen with great care so that performance would be maximized at the wavelengths of the photons involved in the experiment.

Clauser was actually one of the few postdocs at Berkeley essentially working on his own programme. Most people in his position were working fairly much under the direction of a particular professor, and Townes and Shugart were unusual in allowing a postdoc to follow independent ideas. But he still had to generate a whole range of new skills for the experiment. His PhD work had involved him in such areas as vacuum systems, microwaves, plumbing, electronics and computing, but while the last two in particular are important for any experimental work in physics, he and Freedman had to develop expertise in other areas such as optics, atomic physics, photomultipliers and lamps. In total the experiment took them about two years to build and test, and another two months to run.

In the experiment, calcium atoms in a beam emerging from an oven are exposed in a vacuum chamber to radiation from a deuterium (D_2) arc. This radiation has been passed through a filter allowing only a narrow range of wavelengths that would excite the calcium atoms from the lowest energy level in Figure 9.9, which is the ground state and has $J = 0$, to the highest energy level in Figure 9.9. The wavelength required is 2.275×10^{-7} m, and lies in the near UV. This energy level has $J = 0$, and is at the top of the cascade. The cascade is then back to the ground state in two stages. The intermediate level has $J = 1$, so we see that the cascade is indeed of type 0-1-0. The two photons produced in the cascade are entangled and have wavelengths 5.513×10^{-7} m, which is in the green, and 4.227×10^{-7} m, which is in the purple.

The photons then pass through two symmetric optical systems, consisting of a primary lens, of half-angle about 30°, a wavelength filter, to select the photons of the desired wavelength in each wing of the apparatus, a rotatable and removable polarizer, a further lens and a single photon detector or photomultiplier.

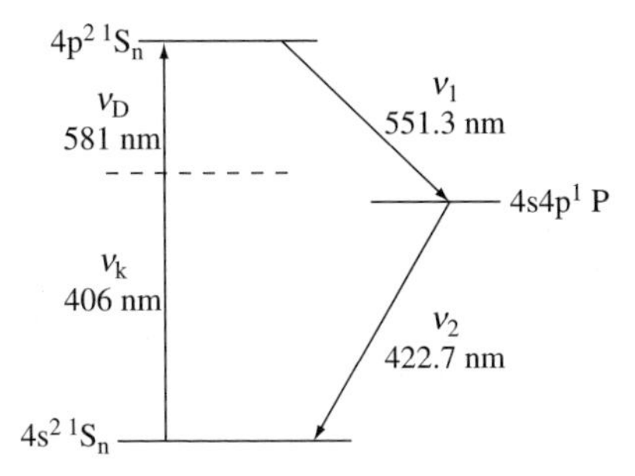

Fig. 9.9 The atomic cascade used by Freedman and Clauser, and also by Aspect and Grangier. Atoms are pumped to the upper level by the absorption of photons of frequency ν_K and ν_D, and return to the ground state emitting photons of frequency ν_1 and ν_2, which have correlated polarizations. [A. Aspect and P. Grangier, in P. Lahti and P. Mittelstaedt (eds), *Symposium on the Foundations of Modern Physics: 50 Years of the Einstein-Podolsky-Rosen Gedanken-experiment* (Singapore: World Scientific, 1985)]

Because of the very low proportion of initial excitations to the upper energy level that would lead to a photon being detected in each photomultiplier, Clauser was determined to construct a system that had a very large number of such excitations, in other words a very large system with a very powerful oven, a strong deuterium arc and, of course, very large polarizers, but also polarizers that were extremely efficient. It may be remembered that Clauser had identified inefficient polarizers as one of the weakest points of the Kocher–Commins experiment, and he recognized that Polaroid sheets could not be efficient enough.

Instead Clauser and Freedman used so-called *pile-of-plates* polarizers, which are indeed, as the name suggests, just series of glass plates. The idea used is that when a beam of light reaches a plate of glass making an angle nearly equal to what is called the Brewster angle with the beam, the only light reflected is polarized parallel to the surface. (The Brewster angle refers to Sir David Brewster, a nineteenth-century Scottish physicist, and for a typical glass it is about 56°.) This means that the transmitted light consists of both perpendicular and parallel polarizations, but the former is dominant. If then the light meets a long series of such plates, at each plate light with parallel polarization is lost, and what emerges is, to as high an approximation as one wishes, a beam of perpendicularly polarized light. The pile of plates is, of course, acting exactly as a polarizer, the only other point needing to be mentioned is that, compared to our previous explanation of a polarizer, the words 'parallel' and 'perpendicular' must be interchanged. In the apparatus of Clauser and Freedman, the thickness of each plate was 0.33 mm, and each pile consisted of ten plates.

Clauser and Freedman needed their equipment to perform the same basic task, generating photons and recording their polarization, for a very large number of cases—different angles of the polarizer in each wing of the apparatus, and also the case where one or both of the polarizers was absent. To achieve this practically automatically, they concocted a highly ingenious system using a so-called Geneva mechanism that rotates one or other of the polarizers by 22½°, then waits for 100 s for a run of the experiment, before performing another rotation. Throughout the process the angle between beam and plate is maintained very close to the Brewster angle.

The Geneva gear system has been used for centuries to transform regular circular motion into a sequence of discrete rotations. It is used, for example, in film projectors. In the mechanism, a slotted wheel rotates continuously, and as it turns it engages and disengages a peg on a smaller wheel.

Each polarizer could also be folded completely out of the optical path and the whole sequencing was achieved by an old telephone relay that Clauser had saved from 'junk', and that ordered one or other of the polarizers to move through 22½°, or to be removed from the beam. With 100 s for each experimental run, the total time for all the runs was 280 hours.

The photons in each wing of the experiment then passed into the photomultipliers. The art of counting single photons was still in its infancy, and Freedman in particular spent a lot of time with the manufacturers, RCA, helping to improve their product and maximize its effectiveness in this particular experiment. The phototubes used the photoelectric effect, utilizing the incoming photons to obtain an output of electrons, which was then amplified and sent onto the coincidence counter. The phototubes were particularly sensitive to the occurrence of many other types of process also giving off electrons. This is the so-called *dark current*, and to reduce this to tolerable levels it was necessary to cool the phototubes using a mixture of dry ice and alcohol.

The coincidence counter had to take account of the fact that, because of the lifetime of the intermediate state, Level 1, which is about 5×10^{-9} s, the second photon will be detected after the first one by this amount. The length of the coincidence window was actually larger than this, 8.1×10^{-9} s. To check for accidental coincidences, an arbitrary delay of 5×10^{-8} s was inserted in one line, and the number of spurious coincidences was determined; actually the number was so small as to be irrelevant. Information from these counters was sent to the computers, which, in 1970s fashion, produced reams of paper tape from which the results of the experiment could be determined.

We briefly note the different efficiencies of different parts of the apparatus. The efficiencies of the polarizers are high, 96–97%, but those of the phototubes are as low as 0.13 and 0.28%. We must also remember that the probabilities of the directions of the photons being within the angles of the initial lenses, and so even reaching the polarizers is comparatively small. Overall the probability of an individual photon that is produced as part of the cascade being finally detected at the photomultiplier is about one in a thousand, so, as we said before, the probability of both members of an entangled pair being detected is about one in a million.

We now turn to the results, which could be analysed in two different ways. First, with R being the coincidence rate with both polarizers present, R_1 and R_2 the rates with polarizers 2 and 1 respectively removed, and R_0 the rate with both polarizers removed, we can focus our attention on $R(\phi)$, the dependence of the coincidence rate on ϕ, the angle between the polarizers. In fact $R(\phi)/R_0$ gives us the central result for the experiment, which can be compared with quantum theory or any other theory.

However, as we have already said, before that we should investigate what the experiment tells us about the CHSH–Bell inequality. A general analysis of this point requires values of R, R_1, R_2 and R_0, but a particularly neat way of studying the result is to remember that the maximum difference between quantum theory and the requirements of local realism occurs when $\phi = 22\frac{1}{2}°$ or $67\frac{1}{2}°$. Using these angles, the CHSH–Bell inequality can conveniently be expressed in terms of a parameter δ, which is give by

$$\delta = \left| R(22\tfrac{1}{2}^0)/R_0 - R(67\tfrac{1}{2}^0)/R_0 \right| - \tfrac{1}{4}.$$

Then the inequality is $\delta \leq 0$.

Let us now see what results Clauser and Freedman obtained. The first point is that δ, as given just above, has the value 0.050 ± 0.008. Clearly the inequality is strongly violated, which is evidence against local realism. Indeed we can say that it is violated by about six standard deviations ($0.050/0.008 > 6$). Moreover this value of δ is very close to the quantum mechanical prediction of 0.051 ± 0.007. The more general forms of the inequality are also violated. When we turn to $R(\phi)/R_0$, we find that there is absolutely no evidence of any departure from the predictions of quantum theory.

Stuart Freedman was awarded his PhD in May 1972 for this work, and it was also published, again in *PRL*, under the names of both Freedman and Clauser. This paper was submitted in February and published in April 1972. The conclusion stated by the authors are as follows: 'We consider these results to be strong evidence against local hidden-variable theories.' Clauser in particular must have been desperately disappointed to come to such a conclusion. It must be remembered, of course, that the authors made the CHSH assumption that the probability of detecting a particular photon is independent of whether it has passed through a polarizer, and if so the orientation of the polarizer.

Meanwhile in Harvard, Holt had constructed his own apparatus, along the same lines as that of Clauser and Freedman, of course, but differing sharply in some ways. For a start, instead of calcium, Holt was using mercury. Mercury has a number of different isotopes (versions of the same atomic species with different numbers of neutrons), and Holt was using the comparatively rare mercury-198, which meant that he had to separate out this isotope and then distill it into his main glass tube.

His equipment was on a much smaller scale than that of Clauser and Freedman. It was very much a table-top experiment rather than one taking up a whole laboratory. Once his mercury vapour was in his glass tube, an electron gun was used to excite the atoms to the top of the cascade. From there the two entangled photons were emitted, and, to polarize them, rather than the pile-of-plates that was required in California because of the massive size of the equipment there, Holt used calcite. Detection and analysis were much the same as used by Clauser and Freedman.

What was startling about Holt's results, which he had obtained before those of Clauser and Freedman, was that they strongly disagreed with quantum theory and were comfortably in the region agreeing with local realism. His value of δ, to be compared with that of Clauser and Freedman, was -0.034 ± 0.013, well within the range permitted by local realism. Because of the rather different values of the experimental parameters, the quantum prediction in his case was not actually very far from the local realism range, being equal to 0.016, but it was still a very long way from the value he actually obtained.

Naturally Holt and Pipkin were extremely concerned about these results. It is one thing to be open-minded in abstract terms about the possibility of finding violations of quantum theory. However, to announce such violations in the open literature only to be found later to have been wrong would certainly be to finish one's career on a note of farce. On the other hand, of course, to announce it and be proved right would be to have an exceptionally good chance of making a trip to Stockholm to receive a Nobel Prize!

Probably in their heart of hearts they did not really believe that quantum theory had been proved wrong in such a comparatively simple way. (It would be interesting to know what Clauser would have done if he had been in their position—almost certainly published, one would think!) Eventually they decided—almost perhaps by default—to search high and low for a possible source of error in their experiment, and in the absence of finding such an error, to hang on and wait from the results from California.

When the results of Clauser and Freeman became known, Pipkin and Holt must have realized that the game was up. Holt wrote up the work and was awarded his PhD in 1972, and in 1973 the pair distributed a preprint, without any intention of submitting it for publication. The preprint would at least tell others what had been done, and leave them to ponder on where there might have been a systematic error. One suggestion is that the tube containing the mercury vapour was curved, and thus it may have been optically active and rotated the polarization.

Obviously this would have had a catastrophic effect on the results of the experiment.

Clauser felt that the overall state of affairs—one result for local realism, one against—was patently unsatisfactory, and decided to repeat Holt's experiment. He was able to borrow suitable equipment, though he ended up using the isotope mercury-202 rather than the mercury-198 that Holt had used. It turned out that the particular atomic transitions used in this experiment made the work far more laborious than that of either himself and Freedman or Holt, involving around 400 hours of taking readings. In the end, though, the results, which were published in 1976, were much the same as those for the experiment on calcium—disagreement with local realism, and agreement with quantum theory.

Even before this experiment, Clauser [1] had tested a very important assumption of CHSH, and of course of the experiments based on it, and in doing so, he had also made a contribution to the fundamental understanding of quantum theory and quantum electrodynamics, which can be defined as the theory of photons. It will be realized that throughout our discussion, we have been glibly speaking of photons travelling around, entering polarizers, being detected or failing to be detected, in other words behaving exactly as particles. And the coincidence structure of the detection in either wing of the apparatus gives general credence to this idea. But of course at the same time at the very heart of quantum theory is the fact that light also has a wavelike nature. Could it be that this wavelike nature has some bearing on the results of the experiment?

The best piece of apparatus to check on this is the half-silvered mirror. When a reasonably bright light (many photons) is incident on such a mirror, a fraction of the light is transmitted and a half reflected. But when we reduce the light to a single photon, is it the case that the photon is *either* transmitted *or* reflected? Clauser's experiment used *tagged photons*; in other words he detected a photon in one wing of the apparatus, and studied phenomena in the other wing only during the appropriate coincidence window. He used, in fact, two silvered mirrors and four photomultipliers, and his results showed clearly that light did indeed behave as photons in this type of experiment.

There was one more physicist heavily involved in this first round of experiments, and he had everything to gain from the divergence of results between Freedman and Clauser on the one hand and Holt on the other. He was Ed Fry (Figure 9.10), who, as a recently appointed untenured assistant professor at Texas A&M (Agricultural and Mechanical) University, heard a talk on Bell's Theorem and CHSH in late 1969. Immediately, and unaware of what Clauser would call the stigma associated with such experiments, he determined to become involved. In fact, it was possible for Fry to design an experiment that was much more elegant than the earlier ones. Narrow-band tunable dye lasers had just become commercially available, and one of these could be tuned to a wavelength of 5.461×10^{-7} m, which was an emerald green colour and

Fig. 9.10 Ed Fry [© Zentralbibliothek für Physik, Renate Bertlmann]

just the right wavelength to trigger a cascade in mercury-200. Use of lasers should make the experiment much easier to perform.

Sadly Fry's grant applications were soundly rebuffed. It was generally thought ridiculous even to consider testing quantum theory; it was bad enough that money was being wasted in Berkeley and Harvard without further profligacy in Texas!

Holt's results in particular drastically changed the situation, and Fry received enough money to go ahead with the experiment, together with Randall Thompson, his research student. Because the laser provided light at a precise wavelength, it stimulated mercury atoms to the top of the appropriate cascade, not to any other excited state. This greatly increased the production of the required entangled pairs and enabled much higher rates of data accumulation. The whole experiment was carried out in around 80 minutes, rather than the hundreds of hours of Holt, and of Clauser and Freedman.

In fact, although Fry and Thompson used the atoms of mercury-200 in their experiment, they did not actually separate the isotope, but just used naturally occurring mercury. It was the precise wavelength of the laser that singled out atoms of the desired isotope, and in the two wings of the experiment, optical filters were used to eliminate all but the wavelengths desired.

The short duration of the experiment minimized errors due to variation of the properties of the various components of the apparatus during longer runs, and also gave every opportunity for Fry and Thompson to search for and remove any systematic errors. Their results were published in 1976, and they found δ, as defined above, to be 0.046 ± 0.014, in clear violation of the requirement for local realism, but very close to the value given by quantum theory, 0.044 ± 0.007.

During this decade, experiments examining Bell's inequality were carried out in other areas of physics. The first area was positronium annihilation, the kind of experiment that both Clauser, and also Shimony and Horne considered very seriously, but then ruled out for providing a definitive test. Nevertheless Wu decided to proceed ahead with such an experiment, carried out with her research student Leonard Kasday and also John Ullman.

As we said before, the reason for the rejection of this method by Shimony and Horne, and by Clauser was that, because there were no good polarizers for the high energy photons produced in this experiment, the interaction of the photons with electrons, or Compton effect, had to be used, and the resultant correlation between the behaviour of photons and electrons was too weak for useful results to be obtained. Both Horne and Bell had been able to invent hidden variable models that reproduced the quantum mechanical results predicted for the experiment.

Kasday, though, was able to analyse his results by making two additional assumptions. First, he assumed that it was possible in principle to construct an ideal polarization analyser. Second, he assumed that the results in an experiment using ideal analysers and the results obtained in a Compton scattering experiment are correctly related by quantum theory. This was despite the fact that quantum mechanics cannot actually handle the kinds of polarization that actually occur in these experiments.

With these assumptions made, the experiments of Kasday, Ullman and Wu agreed with quantum theory and were in violation of local realism. Kasday gave an account of their results at the Varenna meeting that we mentioned previously. However, it was generally agreed that the assumptions of these authors were far more serious, and much less easy to justify than those of the cascade experiments, and little attention has been paid to these results. Three other experiments along the same lines were performed in this decade, and rather similarly to the cascade experiments, two agreed with quantum theory and disagreed with local realism, while the other one did exactly the opposite.

The other experiment, carried out by Mohammad Lamehi-Rachti and Walther Mittig and published in 1976, was actually closest to Bell's original conception since it used spin rather than photon polarization as the basis of its measurements. The authors studied spin correlation in proton–proton scattering; in their experiment a proton from an accelerator hit a target containing hydrogen, liberating another proton, the two protons are entangled and their spins are investigated. The results of this experiment again agreed with quantum theory and violated local realism, but again the authors needed to use a considerable number of auxiliary assumptions in their analysis, broadly analogous to those used in the positronium experiments, so again it is difficult to give a great deal of weight to these experiments. General analysis at the end of the 1970s would be based almost entirely on the cascade experiments.

Putting the theory on firmer foundations

During the 1970s, both Bell and, working closely together, Clauser, Shimony and Horne attempted to make the theory behind Bell's theorem more rigorous, to find out, for example, what assumptions are essential to the theorem, and, as Bell put it, 'to be rather more explicit and general about the notion of locality, along lines only hinted at in previous publications'.

In 1974 Clauser and Horne [13] published an important paper, always known as CH. They had actually been thinking out the issues involved over the previous couple of years, and, as we said before, Shimony had given much advice that was acknowledged in the paper, though he thought that this was not enough to justify including his name among the authors.

The first point that they stressed was the place or lack of it of determinism. From the case of perfect correlation Bell had argued from locality to determinism. Because CHSH realized that this perfect correlation was a theoretician's artefact that could not be used in any experiment, they did not take over this argument, but they nevertheless still assumed that determinism held in any case.

However, it was soon noticed by Bell and others that the theorem was true even if determinism was not assumed, and even in cases where it definitely did not hold. A model including hidden variables cannot reproduce the quantum theory results even if the evolution of the hidden variables is probabilistic. (A more technical word for 'probabilistic' is 'stochastic'.) Then CH asked themselves—what *is* the characteristic feature of systems for which Bell's theorem holds? They rechristened the term 'local hidden variables theory' as 'objective local theory', or OLT, partly perhaps for precision, mostly probably because the words 'hidden variables' still very much had Clauser's 'stigma' attached to them. But that was just a change of words; what did the words actually mean?

CH defined 'an object', that is to say something that may be a part of an objective theory, as something that may be put 'in a box', or more pompously, enclosed within a Gaussian surface in space-time. The idea of locality is covered by ruling out action at a distance, and stressing in addition that a measurement reveals local properties of this object.

They stress that the use in CHSH of such terms as 'the photon being incident on the polarizer' or 'passage through the polarizer' is actually illegitimate, as such processes are unobservable, and indeed do not actually appear in all theories. Theoretical analysis should rather stress the event of detection, since this occurs independently of any particular theoretical framework.

CH found it necessary to say more about the measurement process. They assumed that the experimenter had free will, and so could choose arbitrarily what properties of the object were to be measured. So if the correlated properties of two objects in different boxes are measured at the same time (or more generally with a spacelike separation between the two measurement events) the absence of action at a distance implies that the experimenter's choice of property measured in one box (or indeed the result obtained in this frame) cannot affect the results obtained in a measurement in the second box.

They found that the simple sets of precepts listed above leads to a general formulation of Bell's theorem. The new inequality they found is known as the CH inequality; OLT theories must obey the inequality but they were able to show that it is disobeyed by quantum theory. The CH inequality is totally robust; it retains its discriminatory power completely irrespective of the fact that only a very low fraction of the photons produced enter the polarizers, and of these only a very low fraction are detected, and experiments testing the CH inequality would not be vulnerable to the detector loophole. However, the very strength of the CH result also implies that it is exceptionally difficult, indeed certainly impossible up to the present time, to perform an experiment testing it.

For the cascade experiments already carried out at this time, and indeed for all experiments until comparatively recently, CH showed that, to establish a clash between OLTs and quantum theory, an auxiliary assumption is yet again required. To put it another way, without such an assumption it is possible to construct an OLT model that agrees with the results predicted by quantum theory. However, CH suggested an auxiliary assumption that is weaker than that of CHSH.

The CHSH assumption was that the probability of detecting a photon was independent of the presence or orientation of any polarizer. The CH auxiliary assumption is that the probability of detecting a photon that has passed through a polarizer is not greater than the probability would have been if the polarizer had not been there. Thus the presence of a polarizer cannot increase the probability of detection, and this is often called an assumption of *no enhancement*. Such an assumption seems eminently reasonable and it is certainly true for many important theories, and although the assumption is weaker than that of CHSH, it is still strong enough to ensure that the results of Freedman and Clauser violate the appropriate inequality and thus refute local realism.

Bell too attempted to tighten discussion of the assumptions and meaning of the central theorem. Typically his paper, entitled 'The Theory of Local Beables', was presented at a conference in Jaca in June 1975, but after that published only in *Epistemological Letters* in March 1976. In this paper he said that 'I am particularly conscious of having profited from the paper of CHSH and from that of CH.' However, Clauser, Horne and Shimony disagreed with some aspects of Bell's ideas and published a criticism in the same journal later in the year. Bell replied in 1977 and Shimony published a further comment in 1978, in which he remarked that 'The distance between the positions of Bell and CHS seems to have converged to zero', but still added that 'The latter can stake claim to have articulated the common position with greater clarity'. The debate was very instructive, and in 1985 it was republished in the mainline philosophical journal *Dialectica* with a little editorial comment by Shimony.

The purpose of Bell's paper was to present a theory 'which hardly exists otherwise, but which ought to exist'. Conventional quantum theory talks of *observables*, which may be observed. The term *beable* denotes something that exists independent of any observation, and Bell is particularly concerned with *local beables*, where his definition is rather similar to the comparable one of CH. Incidentally a minor part of Shimony's objection is just to the word *beable*. He says that this should mean 'able to be', which is confusing, and would prefer the work *existent*.

Bell provides a careful account of his theorem, restricting himself to using beables—hidden variables, apparatus settings and so on—but it is one point in particular that attracted the attention and criticism of CHS. It is usual to think of apparatus settings as choices of free will of the experimenter. Thus the decision to impose a polarization setting along, say, the x-axis is to be regarded as external to the experimental situation, not determined by anything in the past—technically we can say by anything in the backward light cone of the experimenter.

However, Bell at least considers the possibility of the choices of the observers being the result of the past experiences of the observers, or alternatively perhaps the choices being made by random number generators in either wing of the apparatus. It might be said that this has the conceptual advantage that one is not singling out experimenters as special features of the setup, totally different from the rest of the experimental situation, though of course others might put great store on human beings having free will.

Even if one takes this as an advantage, there is an accompanying problem, for if the decision of the experimenter in each wing of the apparatus is based on his or her past experience, it is easy to see that locality would be compromised where their past experiences overlap—or we might say in the overlap of their backward light cones. The choices of measurement by both observers might be determined by some event in their common past. Bell suggested that the settings might have been 'at least not determined in the overlap of the backward light cones', hoping in a sense to ground the settings in the initial conditions of the experiment, as would be done in the analysis of many physics experiments. However, under the pressure of the argument of CHS, he was forced to admit that, for these particular experiments, the only relevant initial conditions were those at the very creation of the world, and he had to admit that to bring in that idea was inappropriate!

Even from this brief summary of the ideas and discussion of Bell and of CHS, we see that the original work of Bell was of astonishing depth. It might seem at first sight just to give an important insight into quantum theory, which would be significant enough, but in fact it goes much further, telling us very fundamental and actually quite startling things about the structure and behaviour of the Universe. Not for nothing did Henry Stapp [14] call Bell's theorem 'the most important scientific discovery ever'.

Conclusions towards the end of the decade

In 1978, Clauser and Shimony [12] published a comprehensive and extremely clearly written review of the background to the topic and the progress made so far—the review is always called CS. They were able to say, 'One can assert with reasonable confidence that the experimental evidence to date is contrary to the family of local realistic theories', and even more strongly that 'Experimental results evidently refute the theorem's predictions for these [local realistic] theories and favour those of quantum mechanics.'

To reduce the scope for those who might be inclined to retort that it was obvious that such results would be obtained, with the implication that Bohr had been right all along and the experiments might better have been left unperformed, CS then stressed that 'It can now be asserted with reasonable confidence that either the thesis of realism or that of locality must be abandoned', and that 'The conclusions are philosophically startling: either one must totally abandon the realistic philosophy of most working scientists or dramatically revise our concept of space-time.'

Despite this 'reasonable confidence', CS did discuss the detection loophole in detail, largely along the lines of CH. They said comparatively little about the locality loophole—for what they did say, see later. They did mention also, though, another difficulty with the cascade experiments, which had been mentioned by CH.

Bell's original conception had been of a two-body decay. The initial spin-zero particle decayed into the two spin-½ particles; by conservation of momentum, these two particles had to move off in opposite directions and this was the basis of the resulting analysis. For the cascade, in the decay two photons were created, but the initial atom did not disappear; it remained, though in a different atomic state. Thus there were three particles at the conclusion of the process, the two photons and the atom; it can be called a three-body decay. The total final momentum, like the total initial momentum, must be 0, but this final momentum of the system must be the sum of those of each of the three particles, so the photons will not necessarily move off in opposite directions. Thus one photon may enter the first polarizer but the other may miss the second polarizer, complicating the analysis enormously.

CS put forward their conclusion with reasonable firmness but still with moderation. Though they made their philosophical statements, it was still in essence the result of a set of experiments that they were stating. As such, and as compared to the situation later in the 1980s, the work received comparatively little attention, positive or negative; in particular it was not attacked strongly by those supporting local realism.

Prominent among the latter was Franco Selleri. He has told of visiting Bell at CERN and starting to tell him of the problems with the cascade experiments, only for Bell to pull out a great sheaf of notes on exactly this matter. Bell, who would, of course, been delighted if local realism had been upheld, was well aware of the experiments being inconclusive. However, he was also well aware that he had encouraged Clauser in particular to risk his career perhaps by carrying out the experiments, and could not help feeling that it would be a slap in the teeth to say at the end that they were not good enough.

In fact there were fairly good reasons to suspect that imperfections or errors in the experiments would lead to Bell's inequality for local realism being obeyed in the experiment when it should have been disobeyed, rather than the other way round. This is because quantum theory predicted higher correlations between the results in the two wings of the experiment than allowed by local realism, and it would be expected that errors in the experiment would lower rather than increase correlations. It was for this reason, rather than a rather vague reference to the general success of quantum theory, that CS were inclined to disregard the two experiments, including that of Holt and Pipkin, that agreed with local realism.

In 1981, after writing that 'those of us who are inspired by Einstein' would like it best if quantum mechanics were wrong in sufficiently critical situations, Bell [15] was forced to add that:

> The experimental situation is not very encouraging. It is true that practical experiments fall far short of the ideal, because of counter inefficiencies, or analyser inefficiencies, or geometric imperfections, and so on. It is only with additional assumptions, or conventional allowance for inefficiencies and extrapolation from the real to the ideal, that one can say the equality is violated. Although there is and escape route there, it is hard for me to believe that quantum mecanics works so nicely for inefficient practical set-ups and yet is going to fail badly when sufficient refinements are made.

Bell was much more concerned with what we have called the locality loophole; indeed, as we said, he mentioned this in his earliest paper in this area. He continues:

> Of more importance, in my opinion, is the complete absence of the vital *time* factor in existing experiments. The analysers are not rotated during the flight of

the particles. Even if one is obliged to admit some long range influence, it need not travel faster than light—and so would be much less indigestible. For me, then, it is of capital importance that Aspect is engaged in an experiment in which the time factor is introduced.

Indeed, as early as 1976, Alain Aspect had published a detailed proposal of experiments in which the orientations of the polarizers would be changed during the flight of the photons. These experiments would be the most important in the 1980s, and they are the subject of our next chapter.

CS also mention Aspect's suggestion towards the very end of their paper. Of the authors, though, it would seem that Clauser at least is not very perturbed by the locality loophole. Much later he was to write [1] that theories that take advantage of this loophole are

rather improbable, indeed almost 'pathological'. To accept the locality loophole as a reasonable possibility, it is first necessary to believe that a pair of detectors and analysers that are several meters apart are somehow conspiring with each other, so as to defeat the experimenter... To be sure, it may seem that a belief in such logically possible conspiracies requires a certain degree of paranoia. A logical consequence of admitting to this paranoia is then an associated recognition that every piece of equipment that has been in your laboratory for more than a few microseconds might possibly participate in such a grand conspiracy against you, whereupon no effort on your part can rule out this logical possibility! However, if all experimentalists were similarly paranoid, then experimental physics, in general, would seem to be a pointless endeavor.

We now discuss a different side of research on quantum foundations in the 1970s, the extent to which the 'stigma', as Clauser [1] calls it, that was a result even of contemplating the fundamental nature of quantum theory, let alone working on it, remained. He reports that many of his famous colleagues at Columbia, and also many of Holt's advisers at Harvard, warned them not to get into the subject.

It was certainly true that the stigma was there right up to the time of Bell's papers, but it is usually assumed that the fact that Bell's work led to experimental tests surely if slowly changed the situation for the better. It is certainly the case that CHSH and the accounts of the main experiments were published in *PRL*, the most prestigious physics journal, while CH was published in the leading journal for regular papers, and CS in a leading review journal. So it does not seem that achieving publication of such views was any longer insuperable.

Building a career, though, was rather different. Bell himself, whatever his courage in publishing the first ideas, was desperate to avoid being cast as a crank. He wished to preserve his own reputation, but he also hated the thought that other scientists in CERN might regard his work as somehow undermining the common purpose. The latter idea was preposterous, of course, as a different view of the most fundamental aspects of quantum theory could scarcely affect its application to particle physics, but Bell was right to recognize the extreme sensitivity of most scientists to such topics.

We have already seen Bell's reluctance in these years to publish anything except the bare minimum in regular journals, and Clauser gives another good example of a sensitivity on Bell's own part. In 1982 the two of them were joint winners of 'The Reality Foundation Prize', but, before accepting it, Bell wrote to Clauser checking whether the award was a 'quack' one and would harm his reputation as a serious scientist. Only when Clauser assured him that this was not the case did he agree to accept the award.

What about any effect on the careers of those leading the events of this chapter? For a start, Shimony, of course, was in a stable position at the beginning of this work. He had actually given up tenure when he moved from MIT to Boston University but soon regained it, and he continued in his excellent career straddling philosophy and physics. On his retirement in 1997, colleagues in both these disciplines combined to organize a conference on *Foundations of Quantum Mechanics*, and to publish the papers from this conference in two volumes of a *Festschrift* in his honour [16]. This book contained contributions from many of those discussed in this book, and a substantial number of other physicists and philosophers.

Horne became a professor at Stonehill College, a small mainly teaching establishment just outside Boston, where he still works as Chair of Physics. For a number of years he had one day a week free of teaching and, of course, weekends and college vacations, and he used this time highly profitably working in the Physics Department of MIT. We shall meet him again later in the book.

Fry was also lucky in obtaining tenure at Texas A&M, in fact obtaining tenure during the course of the work described earlier in this chapter. His record on paper was not outstanding and it seemed certain that he was about to be denied tenure by the appropriate committee. However, one of his colleagues called in Frank Pipkin, Holt's thesis adviser. Pipkin spent a day in Fry's laboratory and sent a glowing report to the tenure committee, which responded positively. Forty years later it is clear that this response was the correct one as Fry is now Chair of the Physics Department with an excellent research record. Again we shall meet him later in the book.

Stuart Freedman and Dick Holt also made careers in academia. Freedman had spells at Stanford, Princeton and Chicago before returning in 1991 to Berkeley where he now has the Luiz W. Alvarez Chair in Experimental Physics. In his work, he exploits precision techniques from atomic and nuclear physics in the study of fundamental questions in particle physics, and his work has implications for astrophysics and cosmology. He claims to have carried out an enormous number of null experiments—experiments ruling out exotic effects that had been suggested, of which those described in this chapter were merely the first.

Holt felt that he suffered from the stigma that Clauser claimed came from work on quantum fundamentals. However, he has ended up as professor in the Physics and Astronomy Department at the University of Western Ontario in London, Canada. He returned to the cascade type of experiment, using it for its original purpose of finding atomic lifetimes; the data are for use on astrophysics, to the extent that the title of the group is Laboratory Astrophysics. He also measures atomic energy levels and obtains the wavelengths of lasers with extreme accuracy.

Clauser is the one person discussed in this chapter whose academic ambitions seem to have been ruined by the stigma against those working on Bell-type problems. In the mid-1970s, he applied for numerous academic positions, but met total rejection. Clearly while there was a certain tolerance in the journals for discussion and accounts of experiments on Bell's theorem, there were still many who felt the whole topic was unnecessary, and probably even more who would prefer to employ someone in a more conventional area of physics. This was clearly unfair, as, entirely independently of the actual topic being investigated, Clauser had shown himself to be a remarkable experimentalist.

He was to have an excellent career carrying out much significant work and making many important inventions. From 1975 to 1986 he was a research physicist at the Lawrence Livermore National Laboratory performing experiments in plasma physics, the long-term goal of which being sustainable nuclear fusion. However, in 1986 he felt that Livermore was stagnating—the government had essentially closed down the project Clauser was involved with without telling anybody, and he left the laboratory.

Much of the rest of his career has been spent working on projects of his own choice, in particular the atomic interferometer; he was a research physicist at Berkeley from 1990 to 1997, while at other times he worked for his own company, but he was always funded from outside. Clauser patented a novel form of atomic interferometer, intending to use the technique as a gravity sensor, capable of finding deposits of oil, gas or water, and in the *Festschrift* for Shimony mentioned above, he published an important paper titled 'De Broglie Interference of Small Rocks and Large Viruses'. Up till then it had been taken for granted that objects of this size would not exhibit the interference effects demonstrated by light or less massive particles, but, more recently, as we shall see in a later chapter, Anton Zeilinger has taken substantial steps in this direction.

The disappointing aspect of Clauser's career is not that he has been unable to carry out cutting-edge research—he has consistently done so. It is that so dynamic a physicist, with the ability and confidence to make his own judgements, even against a stubborn and unyielding orthodoxy, the determination to carry his projects through and the ability to explain and justify his views clearly to colleagues or students, was prevented from entering a career in university teaching where he would undoubtedly have been a shining light!

The first generation of physicists and philosophers to work on Bell-type problems achieved an enormous amount, and it is important that this is not forgotten. It is my fervent hope that over the next few years, several of the people mentioned in this book will obtain Nobel Prizes for physics. It is, of course, unfortunate that Bell did not live long enough to obtain the prize. At the time of his death, even though he had vastly clarified an area of physics that had divided the two giants of the century, Einstein and Bohr, it was probably the case that his work seemed too esoteric for him to be awarded the prize. Of course if the results of the experiments on Bell's inequalities had gone the other way and quantum theory had been dethroned, things would have been very different, but non-locality on its own was not enough! As it was, even Bell himself modestly felt it would be difficult to maintain that his work had 'benefited mankind' as the Nobel statutes require. He was nominated for the prize shortly before his death, but the nomination was either unsuccessful or came too late.

Twenty years later things seem very different. The fuller appreciation of the fundamental aspects of quantum theory, and probably more importantly, the rise of quantum information theory, which has become extremely important and is discussed in Part III of the book, means that, were he still alive, Bell would undoubtedly be awarded the prize.

It should happen, and I think it will happen, that others, again in both quantum foundations and quantum information theory, will also be successful. The point I would make here is that it should be unthinkable that those discussed in this chapter who took the baton directly from Bell should not be included in this number.

References

1. J. F. Clauser, 'Early History of Bell's Theorem', in: R. Bertlmann and A. Zeilinger (eds), *Quantum [Un]speakables; From Bell to Quantum Information* (Berlin: Springer-Verlag, 2002), pp. 61–98.
2. Interview of Abner Shimony by Joan Lisa Bromberg on September 9, 2002, Niels Bohr Library and Archives, American Institute of Physics, College Park, MD USA, http://www.aip.org/history/ohilist/25643.html
3. E. P. Wigner, 'The Problem of Measurement', *American Journal of Physics* **31** (1963), 6–15.
4. A. Shimony, 'Role of the Observer in Quantum Theory', *American Journal of Physics* **31** (1963), 755–73.
5. As for Ref. [2], Interview of John F. Clauser by Bromberg on May 20, 2002, http://www.aip.org/history/ohilist/25096.html

6. R. Kanigel, *The Man Who Knew Infinity: Life of the Genius Ramanujan* (London: Abacus, 1992).
7. D. Bohm and Y. Aharanov, 'Discussion of Experimental Proof for the Paradox of Einstein, Rosen and Podolsky', *Physical Review* **108** (1957), 1070–6.
8. C. S. Wu and I. Shaknov, 'The Angular Correlation of Scattered Annihilation Radiation', *Physical Review* **77** (1950), 136.
9. C. A. Kocher and E. D. Commins, 'Polarisation Correlation of Phonons Emitted in an Atomic Cascade', *Physical Review Letters* **18** (1967), 575–7.
10. D. Wick, *The Infamous Boundary: Seven Decades of Heresy in Quantum Physics* (New York: Springer, 1995).
11. J. F. Clauser, M. A. Horne, A. Shimony and R. A. Holt, 'Experimental Test of Local Hidden-Variable Theories', *Physical Review Letters* **23** (1969), 880–4.
12. J. F. Clauser and A. Shimony, 'Bell's Theorem: Experimental Tests and Implications', *Reports on Progress in Physics* **41** (1978), 1881–1927.
13. J. F. Clauser and M. A. Horne, 'Experimental Consequences of Objective Local Theories', *Physical Review D* **10** (1974), 526–35.
14. H. P. Stapp, 'Are Superluminal Connections Necessary?', *Il Nuovo Cimento* **40B** (1977), 191–205.
15. J. S. Bell, *Speakable and Unspeakable in Quantum Mechanics* (Cambridge: Cambridge University Press, 1996, 2004); M. Bell, K. Gottfried and M. Veltmann (eds), *Quantum Mechanics, High Energy Physics and Accelerators* (Singapore: World Scientific, 1995).
16. R. S. Cohen, M. Horne and J. J. Stachel (eds), *Quantum Mechanical Studies for Abner Shimony*, Vol. 1: *Experimental Metaphysics*; Vol. 2: *Potentiality, Entanglement and Passion-at-a-Distance* (Dordrecht: Kluwer, 1997).

10
Alain Aspect: ruling out signalling

Aspect and Bell

The next hero of our story is a Frenchman, Alain Aspect (Figure 10.1), who was born in 1947 in the southwest of France. His early studies were at the École Normale Supérieure at Cachan outside Paris. This is one of the Grandes Écoles, the most prestigious institutions of higher education, which lie outside the main framework of the university system, and for which entrance is by highly competitive examination.

After completing his first doctorate, Aspect undertook his national service, volunteering to teach in the heat of Cameroon from 1971 to 1974. During this period he spent most of his spare time thinking about quantum theory. In his earlier courses he had, of course, studied the subject, but his lecturers had concentrated on the mathematics; he learned much about differential equations, but comparatively little about the physical applications, and virtually nothing about the conceptual foundations of the theory.

In Cameroon he studied quantum theory using the book *Quantum Mechanics* [1] by Claude Cohen-Tannoudji, Bernard Diu and Frank Laloë. Cohen-Tannoudji is one of the most famous of French physicists, and we shall see later he was to win the Nobel Prize for physics in 1997 for the use of lasers in trapping single atoms and ions. The book that he wrote with his two collaborators is acknowledged to be one of the most comprehensive and deep accounts of quantum theory. Aspect particularly appreciated it because it was definitely a book on physics rather than on mathematics, and also because it refrained from the usual vague remarks that Bohr had solved all the conceptual difficulties. It left decisions like that up to the reader.

In particular Aspect read the EPR paper, and he became completely convinced that it was not only correct but expressed an extremely important truth of nature. Then after his return from Africa at the end of 1974 he read the Bell paper on local realism, and as he put it, it was 'love at first sight'. He found the Bell argument not only intellectually brilliant but emotionally fascinating, and he longed to make a substantial contribution towards its experimental investigation. He found out, of course,

Fig. 10.1 Alain Aspect [© Zentralbibliothek für Physik, Renate Bertlmann]

that the first generation of experiments had been performed and had given a fairly clear message that, within what would come to be known as the loopholes, local realism had been violated.

However, that left the loopholes. Aspect paid only a little attention to the detector loophole, but he tackled the so-called locality loophole head-on. It will be remembered that in the first-generation experiments carried out in the 1970s, the polarizer settings were static, so that there was a possibility that the polarizers and detectors in the two wings of the experiment could share information by exchange of signals at speeds less than or equal to that of light. So Bell had always considered that experiments in which the settings of the polarizers were changed during the flight of the photons would be crucial. (It might be said, going back to the precise terminology we introduced earlier, that, other loopholes being disregarded for the moment, the experiments of Clauser and his friends ruled out Einstein separability in which there can be no communication between the two wings of the experiment, but not Bell locality in which there can be communication at speeds up to c.) Aspect determined to perform such an experiment. In fact he planned to do it as a so-called *these d'état*, sometimes known as a 'big doctorate' rather than the earlier 'small doctorate'. At this time this could be a really long work, and Aspect was able to persuade Christian Imbert, a young professor at the Institute of Optics at the University of Paris-South at Orsay, to act as his thesis adviser.

Imbert advised him to discuss the possible experiment with Bell, so Aspect obtained an appointment, travelled to CERN in Geneva in early 1975, and rather nervously sketched out his ideas to see how Bell would react. He was rather taken aback by Bell's first question, which was 'Have you a permanent position?' Aspect was able to answer that, because the French system was very different to that of most other countries, although a graduate student, technically he was a permanent employee.

However, Bell's warning was well meant. It was alerting Aspect to the fact, not only that the experiment would be extremely long and difficult, but that it might actually hinder him in any attempt to obtain an academic position. Indeed Aspect, like Clauser, Fry and the others before him, was to meet many who would find it ridiculous to believe that a theory of such great and wide-ranging success as quantum theory could conceivably be undone by a comparatively simple experiment. Incidentally Aspect actually did *not* expect any such thing. Although he was entranced by Bell's work, he believed that quantum theory would triumph, and, in this belief, he was, of course, in complete contrast to Clauser, who had expected local realism to triumph, but also somewhat opposed to those such as Bell and Shimony, who felt there was a chance, but probably only an outside one, that there might be an upset.

Bell, of course, was delighted that Aspect did intend to close the locality loophole and encouraged him greatly. Aspect published a short and rather general note in 1975 and then what amounted to rather a full experimental proposal in 1976, but as Bell had implied, it was a long project. It took four years obtaining funding and constructing the apparatus, it was a further two years before the full results of the experiment were published and Aspect did not get his thesis written until 1983.

The Aspect experiments

As well as hoping to rule out signalling between the two wings of the apparatus, Aspect [2] intended to make two further improvements to the first generation of experiments. He and his collaborators, Philippe Grangier, Gérard Roger and Jean Dalibard, reported that 'technological progress (mostly in lasers) was sufficient to allow a new generation of experiments'. Even compared to the experiment of Fry and Thompson in the mid-1970s, which had used lasers, progress in laser technology had been sufficiently fast that an improvement of about ten in excitation rate could be obtained, obviously leading to data being obtained faster, and results of greater accuracy being achieved. In these experiments, the laser excitation of the cascades was sufficient to provide a cascade rate of 4×10^7 s^{-1}.

Also, while the experiments in the first generation had used single-channel polarizers, which allow for detection of only one of the two modes of polarization, in some of the experiments of Aspect and collaborators, two-channel polarizers were used. These were polarizing cubes with dielectric layers that transmit one polarization and reflect the other one. When these polarizers are used, a fourfold counting system follows, and in a single run four quantities, $N_{++}(\hat{a},\hat{b})$, $N_{+-}(\hat{a},\hat{b})$, $N_{-+}(\hat{a},\hat{b})$ and $N_{--}(\hat{a},\hat{b})$ could be measured. It was then easy to calculate $P(\hat{a},\hat{b})$, which we defined earlier as the average value of AB, where A and B are the results in the two wings of the experiment, in units such that each is ±1. We have

$$P(\hat{a},\hat{b}) = \frac{N_{++}(\hat{a},\hat{b}) + N_{--}(\hat{a},\hat{b}) - N_{+-}(\hat{a},\hat{b}) - N_{-+}(\hat{a},\hat{b})}{N_{++}(\hat{a},\hat{b}) + N_{--}(\hat{a},\hat{b}) + N_{+-}(\hat{a},\hat{b}) + N_{-+}(\hat{a},\hat{b})}$$

and, for four choices of angles, either $\hat{a}$ or $\hat{a}'$ in the first wing, and either $\hat{b}$ or $\hat{b}'$ in the second, we may calculate S, as given in the previous chapter, for which local realism dictates that $-2 \le S \le 2$. One does not have to perform experiments with polarizers removed as in the experiments of the first generation. Clearly with two-channel polarizers we are much closer to the original conception of Bell.

Of course, as for the experiments of the first generation, there must be auxiliary assumptions. The most natural assumption here is that we may think of the same photons being detected in the experiment whatever the orientations of the polarizers. For different orientations, of course, a particular photon will appear in different channels. This is very much analogous to a fair-sampling assumption, which says that the detected pairs are a fair sample of the total ensemble of all emitted pairs, though the fair-sampling assumption is actually stronger and we want to keep whatever auxiliary assumption we use as weak as possible. Some justification for the assumption (though naturally not a proof—otherwise it would not be an assumption!) is that, as the orientations of the polarizers change, the sum of all the coincidence rates, which is the denominator in the expression just above, remains constant, although each of the four rates changes drastically.

However, Clauser in particular would feel that Aspects' assumptions were stronger than those of CHSH, so that, for the detector loophole, Aspects work represented a backward rather than a forward step from the experiments of the previous generation.

In their early experiment Aspect and his collaborators used these two-channel polarizers without the later switching between channels. They found $S = 2.697 \pm 0.015$, or in other words the Bell inequality is violated by more than 40 standard deviations ($0.697/0.015 > 40$). Their result was also in good agreement with the quantum prediction that was $S = 2.70 \pm 0.05$, and they checked this agreement for a wide range of angles between the polarizers. Clearly, even without moving into the switching, the steady improvement of available equipment, especially lasers, made this a particularly strong rejection of local realism.

In their next experiment, they attempted to answer Bell's call to change the settings of the polarizers while the photons were in flight so that information could not be exchanged below the speed of light. In the modified scheme, single-channel polarizers were used. In fact each polarizer was replaced by a switching device followed by two polarizers in different orientations, $\hat{a}$ and $\hat{a}'$ in one wing of the apparatus, and $\hat{b}$ and $\hat{b}'$ in the other. The optical switch was able to redirect the incident light from one polarizer to the other extremely quickly.

Switching occurred between the two channels about every 10 ns (1 ns =1 nanosecond = 10^{-9} s). We may compare this period, and also the lifetime of the intermediate level in the cascade, which is 5 ns, with the tine taken for information to pass between the two ends of the apparatus. Each switch is about 6 m from the source, so the two ends of the apparatus are about 12 m apart, and the time for light to pass from one end to the other is 12 m/(3×10^{10} ms^{-1}), which is 40 ns. It is indeed clear that the change of polarizer angle in one wing of the apparatus and a detection event in the other wing are separated by a spacelike interval; a signal travelling at the speed of light or a lower speed cannot travel from one event to the other.

The switching of the light is achieved by acousto-optical interaction with an ultrasonic standing wave in water. The standing wave itself results from interference between two counter-propagating acoustic waves produced by two electroacoustical transducers driven in phase at 25 MHz (or 2.5×10^7 s^{-1}). Light is incident on the water, and it is transmitted when the amplitude of the standing wave is 0, but deflected when the amplitude is a maximum. Each occurs twice in an acoustical period, so the optical switch works at twice the acoustical frequency.

The experiment was particularly difficult, and a number of problems emerged as the work was performed. In tests with a laser beam, complete switching had been obtained, but in the actual experiments, because of the divergence of the light beams, the switching was not complete. Also the large distance between the two switches meant that complicated optics were required to match beams with the switches and the polarizers.

Since the individual polarizers were single-channel, coincidence rates had to be measured (i) with all polarizers in place, (ii) with all polarizers removed and (iii) with a single polarizer removed on each side, as with the experiments of the previous decade. Because the divergence of the beams had to be reduced to obtain good switching, collection efficiencies were also reduced by a factor of about 40. Coincidence rates were around a few per second, not very much greater than the accidental rates of around one per second, so runs had to be much longer than in the earlier experiment, typically 12,000 s, with the time equally divided between (i), (ii) and (iii) above. This led to problems with drift, and to minimize these, (i)–(iii) were alternated every 400 s.

With these difficulties it is scarcely surprising that results were somewhat less accurate than those without switching. The relevant form of Bell–CHSH inequality was $-1 \leq S' \leq 0$ for local realism, where S' is a parameter analogous to S. Aspect and his team found $S' = 0.101 \pm 0.020$, so while the uncertainty was relatively high, the inequality was still violated by more than 5 standard deviations. The quantum mechanical prediction was $S' = 0.113 \pm 0.005$, so we are well within the experimental uncertainties, although the central values are a little way apart.

Thus the experiments give at least a reasonably good indication that local realism is not saved by prohibiting the two wings of the experiment from exchanging information. However, as Aspect himself admitted, the experiment was not ideal. First we can say that the switching was not complete; some photons were unaffected. However, it is difficult to believe that quantum mechanics would be violated given total switching, but is totally unaffected by a substantial amount of switching.

Much more important is that the switching is not random as would be ideal, but periodic. However, it should be stated that the switches in the two wings of the experiment were driven by different generators at different frequencies, and that the inevitable drift in both devices was independent. Overall the experiment could be said to have presented reasonably good if not perfect evidence that Bell's inequalities were violated, and quantum theory obeyed, even if the two wings of the experiment could not communicate with each other. In other words, to the same extent, we could say that the detector loophole had been closed, that the results of the experiment violated local realism not just separable realism as in the experiments of the previous generation.

The aftermath

Indeed this message came through quite strongly to concerned scientists and to those members of the public that were at all interested in these matters. Certainly for some reason Aspect's experiment attracted far more attention than those experiments of the previous generation; more and more people became convinced that the whole subject was important. Possibly this was because either Aspect or the university authorities at Orsay employed better public-relations officers than those at the American universities (although frankly that seems unlikely). Perhaps it was more that Aspect was based in Europe, and at least at this time, with Bell in particular working not far away, there was more interest in these matters in Europe than in America. Or perhaps it was just that interest in Bell's ideas, together with suspicion of Copenhagen, had slowly but steadily risen, and had now reached a level where new events, such as the Aspect results, created a self-sustaining level of discussion and comment. D'Espagnat's popular article that we mentioned in Chapter 8 had appeared a few years before, and this had also attracted much attention.

This had a number of consequences. First, it was, of course, Aspect's experiments that were discussed, and to an extremely large extent his was the only name even mentioned. At one level this was natural—

certainly his experiments were more powerful than the earlier ones. This was in the normal way of things, as experimental equipment steadily improves and new techniques become available, just as, for example Fry and Thompson were able to use lasers for their experiment while only a few years earlier this had not been possible. However, this did have unfortunate consequences—Aspect certainly had made major advances and richly deserved his prestige, but it was a pity that, through nobody's fault, this rather came at the expense of the previous generation of experimenters.

There was another consequence of the attention drawn to the Aspect experiments, and in particular to the fact that the limitations of the conclusions drawn were nearly always dropped or just forgotten, not by Aspect himself, but as very often reported. The result was frequently stated bluntly that local realism had been disproved.

This other consequence was that the principled opposition to this idea, which had been muted at the results of the previous generation, came to the fore. The opposition came from scientists whom I shall call, perhaps a little loosely, the realists, who, for deep-seated philosophical reasons, believed extremely strongly in the traditional virtues of local realism. They were convinced that, if all loopholes were plugged, experiments would undoubtedly show that Bell's inequalities are obeyed and quantum theory disproved. This loose grouping included Franco Selleri, Emilio Santos, Tom Marshall and more recently Caroline Thompson.

In 1983, shortly after the Aspect results were announced amidst considerable publicity, the first three [3] produced a very strong riposte, trenchantly titled 'Local Realism Has Not Been Refuted by Atomic Cascade Experiments'. In this paper they argued that 'a great deal of wishful thinking' lay behind the assertion that the experiments reported in this chapter and the previous one had agreed with quantum theory and violated grossly the predictions of any local realist model. The authors used a model including an object with wavelike as well as particle-like properties created in an atomic emission, and they were able to produce a consistently local and realistic theory of atomic cascades that fitted the experimental data as well as the quantum results did. Their theory used a single hidden variable representing an angle; it also involved some enhancement, in the terms used by CH in the previous chapter, though the paper attempted to justify this enhancement, and in any case Thompson has argued since that a slightly different model does not require enhancement.

It was at this stage that, with the locality loophole seemingly closed, the realists laid great stress on the detector loophole. Yuval Ne'eman [4], a highly accomplished elementary particle physicist, in fact one of the pioneers of group theoretical methods in particle physics, said that:

> Here and there some of our colleagues are still fighting a rear-guard action—finding another gap in the Aspect text, placing their hope in yet another awaited

neutron interferometry experiment... These are highly intelligent scientists—they would not have detected the remaining gaps otherwise. Faithful to their own creed, they are now forced to invent complicated classical alternatives, each managing to hold on to at least one classical concept: either locality or determinism. My personal impression is that these ideas resemble the attempts to hold on after Copernicus to Ptolemaic Astronomy, by inventing more and more sophisticated epicycles within epicycles. It can be done and was done. However, placing the sun in the centre was simple... And yet there is no doubt that these workers in this conservative struggle fulfill a very important role. Every opening should indeed be explored, every gap should be closed... After all, there is nothing holy about the quantum view, it just works. Should it, however, fail somewhere—I believe it would have to be replaced by something really new. My guess is that this would imply a further revolution, not a return to the 'Ancien Regime'.

Selleri, though, not surprisingly, expressed a very different point of view in a book a few years later entitled *Quantum Paradoxes and Physical Reality* [5]. Selleri's career and beliefs are discussed in an interesting article by Olival Freire [6] on the growth of research on quantum foundations. (Also included are accounts of Hans-Dieter Zeh, Bell, Clauser, Shimony, Wigner, Rosenfeld, d'Espagnat and Bryce de Witt, all of who we have met or shall meet in this book.)

Selleri was awarded his PhD for work in theoretical high-energy physics (elementary particle physics) in 1958 from the University of Bologna, and he returned there to a permanent position ten years later after performing important work in the same area. However, somewhat disillusioned by the political and cultural climate in Bologna, he soon moved to a new university in Bari, where he has remained ever since. He also became frustrated with the abstract nature of particle physics, and pinned down the problems on the fact that quantum theory itself was abstract and poorly understood. He thus shifted his research to this area, and it was Selleri who suggested to the Italian Physical Society that the Varenna summer school that we have heard so much about should be organized.

As well as his scientific beliefs, Selleri takes seriously wider social responsibilities. He has said to Freire that:

I always thought that it is our duty to build a science that can be communicated to everybody. And at those times I was thinking in terms of the working class... If the only way to understand what I'm doing is to study differential equations or Hilbert space, there is too high a threshold.

In *Quantum Paradoxes and Physical Reality*, he explained that the discussion is from a point of view generally sympathetic with the realistic and rationalistic outlook of Einstein, de Broglie and Schrödinger, though taking care 'to reconstruct faithfully the true opinions and insights of many other physicists who, from different perspectives than ours, have made essential contributions to the shaping of the new theory'.

Selleri's position involves three claims. The first, reality, says that the basic entities of atomic physics actually exist, independently of human beings and their observations. The second, comprehensibility, says that it is possible to comprehend the structure and evolutions of atomic objects and processes in terms of mental images formed in correspondence with their reality. The third, causality, says that physical laws should be formulated so that at least one cause can be given for any observed effect. It is clear that at least the first two take as given ideas that one would usually say would be decided by the experiments on Bell's theorem. The third claim is not as firm as determinism, which would say that there is definite connection between present and future. Rather the connection is objective but still probabilistic.

Locality is also of great importance for Selleri, though he uses what he calls probabilistic Einstein locality, based on his definition of causality rather than strict determinism. Because of this importance, Selleri must be ambivalent towards the work of Bohm. On the one hand it did show that the Bohr–von Neumann position was not unassailable, but on the other its non-locality spoke against it. Selleri remarked that it took the efforts of physicists such as Einstein, de Broglie and Bell, working in a hostile cultural environment, to make plausible a causal completion of quantum mechanics, but says that it is the dualistic approach of Einstein and de Broglie that is the real talisman for realists—dualist in the sense of encompassing real waves and real particles for photons and electrons alike.

What about the Bell experiments, which are usually thought to be the death knell of local realistic theories? Not surprisingly Selleri makes use of the auxiliary hypotheses in his defence of these theories. He comments that: 'A logical refutation of these hypotheses is all that is required to restore full agreement between Einstein locality and quantum predictions... The situation is thus fully open to debate, and opposite claims reflect more than anything else old-fashioned views biased by ideological choices which obfuscate logical thinking.'

Indeed for Selleri, internal analysis of the working of quantum theory can be only part of the discussion. He says that quantum mechanics is a philosophically committed theory, and its inner mathematical structure leads to empirical predictions that contradict simple consequences of very general physical ideas, ideas, in fact, so general that they are actually philosophical ideas. He says that one must consider the role of society, culture and prejudice in the birth of quantum theory.

As examples he gives Born, who he says was attracted to acausality as early as 1920, well before his interest in quantum theory; the younger generation of quantum physicists, such as Heisenberg and Jordan, who had grown up in the years of the First World War, the Russian Revolution and class struggles in Germany, and thus held strong political convictions, which were likely to have philosophical implications in physics; and Bohr, whom he claimed was strongly influenced by such prominent Danish philosophers as Søren Kierkegaard and Harald Høffding, and lived in an environment saturated with the ideas of Danish existentialism, which was engaged in polemics against rationalism and the philosophical vigour of nearby Germany.

Selleri argues that while every physical theory contains some objective content that is unlikely to be fundamentally altered, it also contains some content that is logically arbitrary, that may be historically determined by religious prejudice, cultural tradition or power structure. This content, which he believes includes the non-realism and acausality of quantum theory, may well be modified or jettisoned completely in future theories.

Selleri applauds attempts to reconsider critically the foundations of modern physics from a vision of the world devoid of religious considerations, and anthropomorphic and anthrocentric (human-centred) illusions. He believes that an accurate historical analysis would show that profound distortions had been introduced, and a fundamental role had been played by the ideological and philosophical ideas of the Copenhagen and Göttingen schools, and he suggests that the often presumed falsification of local realism 'depends on the historically determined, logically arbitrary, ideological contents of the quantum paradigm and not on unchallengeable empirical evidence'.

Selleri is prepared to argue from the success of sciences such as astrophysics, geophysics, molecular biology and psychology, which strongly oppose, he says, Bohr's views on reality to the conclusion that: 'In a theoretical and conceptual sense, one can thus already speak of an existing falsification of the Copenhagen-Göttingen paradigm.' Rejection of the proposition 'objective reality exists', he believes, and this is at the heart of his argument, entails a highly damaging exclusion from the cultural heritage of everyone but a cultural elite of experts. 'Today', he wrote (in 1990), 'we may confidently declare that the roads to a causal completion of quantum mechanics lies open before us.'

Since 1990, as we shall see in the following chapter, there has been considerable progress in closing the loopholes. Yet such is the passion of Selleri and the other realists that one feels that it will not be unless and until every possible hope for local realism is totally extinguished that they might be prepared to renounce their deeply held beliefs. (And of course Selleri's approach to special relativity discussed in the last section of Chapter 8 is an attempt to cope with the situation if the worst does actually come to the worst and quantum theory triumphs.)

It is a remarkable fact that quantum theory, basically a physical theory designed to explain experimental facts, reaches such philosophical depths, and a tribute to Bell and his successors that their analysis unlocks these fundamental debates about the Universe.

As a footnote to Aspect's work, though, we must mention an argument produced in 1986 by Anton Zeilinger [7], whom we shall meet a great deal through the rest of the book. Zeilinger pointed out an awkward coincidence in the parameters used in Aspect's experiment. In either wing of the experiment, the polarizer in use changes every 10 ns, and the polarizers are a distance 6 m from the source. Suppose that a signal passes from the polarizers in wing 1 of the apparatus to those in wing 2, and let us also suppose that it travels at speed c, which is reasonable enough. Then the time taken to travel from one end of the apparatus to the other will be just 40 ns, as we calculated above. In that time, the polarizer being used in wing 1 will have changed from say polarizer A to B, back to A, B again and finally A again. In other words when the signal reaches the polarizers in wing 2, the message that the signal provides is exactly right. It may be said that the switching of the polarizers is effectively irrelevant!

One can choose whether to take this argument very seriously, but overall it helped to make the point even clearer that, for any unambiguous closure of the locality loophole, for all the great advance of Aspect's experiment, the switching between the polarizers in each wing of the experiment should be totally random rather than periodic. Such an experiment would not be carried out for 16 years after Aspect's work, and when it was performed, it was under the guidance of Zeilinger.

In the years since the experiments described in this chapter, Aspect has continued as one of the foremost workers in the world in the area of quantum optics. In 1994 he returned to the Institute of Optics at Orsay, and he is also Research Director of the French National Centre for Scientific Research, the largest fundamental science agency in Europe.

References

1. C. Cohen-Tannoudji, B. Diu and F. Laloe, *Quantum Mechanics*, 2 vols (New York: Wiley, 2006).
2. A. Aspect, 'Bell's Theorem: The Naïve View of an Experimentalist', in: R. Bertlmann and A. Zeilinger (eds), *Quantum [Un]speakables; From Bell to Quantum Information* (Berlin: Springer-Verlag, 2002), pp. 119–53.
3. T. W. Marshall, E. Santos and F. Selleri, 'Local Realism Has Not Been Refuted by Atomic-Cascade Experiments', *Physics Letters A* **98** (1983), 5–9.
4. Y. Ne'eman, 'EPR Non-separability and Global Aspects of Quantum Mechanics', in *Symposium on the Foundations of Modern Physics: 50 Years of the Einstein-Podolsky-Rosen Gedankenexperiment* (Singapore: World Scientific, 1985), pp. 481–95.
5. F. Selleri, *Quantum Paradoxes and Physical Reality* (Kluwer, Dordrecht, 1990).

6. O. Freire, 'Quantum Dissidents: Research on the Foundations of Quantum Theory', *Studies in the History and Philosophy of Modern Physics* **40** (2009), 280–9.
7. A. Zeilinger, 'Testing Bell's Inequalities with Periodic Switching', *Physics Letters A* **118** (1986), 1–2.

11
Recent developments on Bell's inequalities

Zeilinger, Greenberger, and Gisin

In this chapter we meet three more heroes of our story, Anton Zeilinger, Daniel Greenberger and Nicolas Gisin. In some ways perhaps they did not have to be quite as heroic as the previous generations of experimenters on Bell's inequalities, since we are talking about the late 1980s and the 1990s in this chapter, when, following the work of Aspect, thinking about and performing experiments on the foundational aspects of quantum theory was broadly accepted and even applauded among the community of physicists. Another difference from previous decades was that work on Bell's inequalities had spawned the new subject of quantum information theory, which was rapidly becoming a hot topic, and this was another reason for the change in attitude to foundational work itself. Indeed experimentally, and perhaps particularly in the work of Zeilinger and Gisin, there was an almost seamless move from working on quantum foundations to working on quantum information theory, to a considerable extent the same equipment and techniques being used in both. In this chapter we shall concentrate on the foundational work, leaving the study of quantum information to Part III of the book.

To start with Anton Zeilinger (Figure 11.1), he was born in Ried im Innkreis in Austria in 1945. His father was a chemistry professor, but, though as a child Zeilinger enjoyed visiting his father's laboratory, his own interests were to be in what he considered to be the most fundamental areas of science. He studied physics and mathematics at the University of Vienna from 1963 to 1971, working for his PhD under Helmut Rauch (Figure 11.2), who was well known for his research on the physics of neutrons and their interactions with solids. From 1972 to 1981, Zeilinger was a research assistant, still under Rauch, at the Atomic Research Institute in Vienna.

He was lucky that during this period Helmut Rauch [1] was one of two scientists who independently invented the neutron interferometer, the other being Sam Werner at the University of Missouri. Interference of light had been known for a hundred and fifty years, the two names

Fig. 11.1 Anton Zeilinger [© Zentralbibliothek für Physik, Renate Bertlmann]

Fig. 11.2 Helmut Rauch [© Zentralbibliothek für Physik, Renate Bertlmann]

most associated with the initial demonstrations being Thomas Young and Augustin Fresnel, and at this time interference was looked on quite naturally as proof that light was a wave. A hundred years later the waters were muddied by Einstein's argument that light had a wavelike as well as a particle-like nature, and by the photoelectric and Compton effects that demonstrated this particle-like nature experimentally. They were further muddied by de Broglie's suggestion of 1923 that entities such as electrons, until then thought to be indisputably particles, also had a wavelike nature, this suggestion being experimentally confirmed for electrons by the demonstration of interference as well as diffraction.

It was obvious that, in principle, other particles such as protons and neutrons should also demonstrate interference, but, of course, to say that is very different from actually constructing a working device for neutrons. To do so was the achievement of Rauch and Werner. Rauch's neutrons were produced in a reactor at the Laue-Langevin Institute at Grenoble, where the actual experiments were carried out. The neutron interferometer has many interesting features from the point of view of this book, and for convenience we will discuss the instrument, its construction and some of its applications in the following section.

Here we just mention that during these years Zeilinger played an important part in these experiments, particularly those we shall meet that involve the rotation of the neutron spin. Yet it may be only a little over-dramatic to suggest that there seem to have been forces pushing or pulling him in the direction of the work that would make him famous, work on Bell's inequalities and quantum information theory. In 1976, Rauch received an invitation to attend a conference on the foundations of quantum theory at Erice in Sicily. At this time the first generation of experiments on Bell's inequalities had been completed, and clearly the main thrust of the meeting would be on this area of work, so Rauch suggested that Zeilinger attend in his place, and that he should learn all he could about entanglement and Bell's work, in the hope that such work might in the future be carried out in Vienna (as it certainly has been!).

The meeting was a small one, invitations only having been sent to around twenty physicists centrally involved in the work, and among those attending were Bell, Shimony, Horne and Pipkin. Zeilinger's own presentation would have been extremely interesting for the delegates, but it would have been very much outside the main thrust of the meeting; he was talking of a single particle effect (as will be clarified in the next section of this chapter) while all the other papers presented concerned two-particle entangled states. Horne, for example, spoke of the CH paper that had recently been published.

Following Rauch's instruction, Zeilinger made every effort to find out about EPR, Bell's inequalities, entanglement and so on, and his particular informant was Horne. However, there was much more to this than merely passage of information; the two men took to each other immediately, and became firm friends, forming a close relationship that would blossom over the years, and eventually become central in the creation of science of enormous importance.

Even at the scientific level, there was an added ingredient to the relationship. As we said earlier, Horne had taken up a position at Stonehill College outside Boston. This was a highly respected college for teaching, but with no emphasis on research. Horne definitely hoped to continue in research himself, but, after the time writing CH with Clauser, with all the complication of the various loopholes, he felt that it was difficult to take the Bell work much further, and that it was time to develop new interests. As we have said, Clauser and Shimony asked him to take part in the writing of what was to become CS, but he declined because of lack of time.

In fact Horne had already moved into a new field, that of working with neutrons. Both he and Shimony were fascinated by the neutron interferometer, and in fact they fairly quickly prepared a paper on the rotation of neutron spins. Unfortunately, they then discovered that, not only had a theoretical paper been published by Herb Bernstein in Amherst, but the experiment had actually been performed by Rauch's group, which included, as we have said, Zeilinger himself. However, Horne was not put off by this setback, and he was still keen to work in such a fundamental area of physics. He knew that Cliff Shull, the pioneer of neutron diffraction, was based at MIT and used the MIT reactor for his work.

Shull himself had been born in Pittsburgh in 1915, and he gained his bachelor's degree at Carnegie Tech, which is now Carnegie Mellon University, before obtaining a PhD from New York University in 1941. His thesis work used a Van de Graaf electron accelerator to perform the so-called electron-double-scattering (EDS) experiment giving a direct demonstration that the electron had a spin. This, of course, had been 'known' for around fifteen years but by less direct means. Shull spent the war years working for the Texas Company on problems associated with the production and use of petroleum fuels, but at the end of the war he moved to the Clinton Laboratory (now Oak Ridge National Laboratory) to work with Ernest Wollan on the pioneering studies on nuclear diffraction. He was to move to MIT in 1955, and he stayed there for the remainder of his career.

Much later in 1994 he was to receive the Nobel Prize for his work, sharing the prize with Bertram Brockhouse. Both parts of the prize related to work on neutrons; it may be said that Shull showed where the neutrons were, and Brockhouse showed what they were doing. Both men had had to wait a record period of 48 years between doing the work and getting the award. Shull was 79 at the time of the award, and it was particularly unfortunate that Wollan, who Shull considered should have shared his part of the prize, had died in 1984. Shull had retired in 1986 but fortunately survived until 2001.

Back in 1975 Horne introduced himself to Shull, told him of his work on entanglement, and asked him if he could take part in the neutron diffraction work in Shull's group. Shull readily agreed, and so, for the next ten years, every Tuesday, the day that Horne had no classes at Stonehill, and every vacation, Horne spent his time at MIT taking part in all aspects of the work, and being one of the authors on a very large number of publications.

So when he and Zeilinger met in Amherst in 1976, as well as Zeilinger learning from Horne about Bell's inequalities, and since Zeilinger was then vastly more experienced that Horne in work with neutrons, he was able to impart a great amount of knowledge about the neutron interferometer. Horne, in turn, of course, told Zeilinger of the work in Shull's laboratory, and in the following year Zeilinger applied to spend the year of 1977–8 at MIT as a postdoctoral fellow supported by a Fulbright scholarship. This was to be only the first of many visits over the next decade or so; as Zeilinger rose through the academic ranks in Vienna, he made many very long visits to MIT, working with Shull and the rest of his group, but especially with Horne.

In fact from 1980, we may say that there was a third member of the informal group, Danny Greenberger [2]. Greenberger (Figure 11.3) was somewhat older than the other two, having being born in 1933 in the Bronx. In 1964 he was to return to New York to the City College, which had just opened a graduate school with a programme in physics, and he has remained there since then, but before that he had travelled fairly freely. He graduated from MIT in 1954, then went to the University of Illinois where he was awarded a Masters degree in 1956, and he then began doctoral work under Francis Low, a very well-known high-energy physicist. However, Low then moved to MIT, and Greenberger moved with him, so his PhD thesis was written at MIT, though the actual degree came from Illinois in 1958. In this work, Greenberger used the group

Fig. 11.3 Danny Greenberger in centre; also pictured Mary Bell and Jack Steinberger [© Zentralbibliothek für Physik, Renate Bertlmann]

theoretic methods of analysis for elementary particles that had recently been invented.

Two years in the Army followed, during which Greenberger spent most of his time working in a physics research laboratory. He was trained as a cryptanalyst, which naturally he never imagined would be useful in his later work. Surprisingly much later he became involved in quantum cryptography, an important topic in Part III of this book. After the Army, he spent a year at Ohio State University, and then won an NSF Fellowship to work at Berkeley from 1961 to 1963 as a postdoctoral fellow with Jeffrey Chew, another extremely influential particle physicist, before his return to New York.

Greenberger thus had a very strong background in high-energy particle physics, but he also had an intense interest in quantum theory and its fundamental nature, and in particular in the relationship between quantum theory and general relativity. This was by no means an easy thing to do experiments on! An initial idea was to analyse the hyperfine structure of atoms—the extremely small energy differences between states with different orientations of the nuclear spin—and thus to discover the gravitational effect on a nucleon. Greenberger and Ken Rubin obtained a small grant to study this, but it turned out to be much *too* small!

Greenberger's thoughts then turned to free neutrons. It seemed eminently sensible to concentrate on neutrons; they were definitely quantum objects with a quantum mechanical spin, but they were also affected by gravity—the ones studied in Grenoble and MIT were thermal neutrons, travelling slowly enough that the distance they fell under gravity was detectable. Greenberger visited Shull at MIT in the early 1970s to suggest the experiment. The neutron interferometer had not yet been invented, but Shull did think that the experiment might be possible on his own equipment; unfortunately, though, he told Greenberger that the reactor was down for a couple of years. Greenberger was not unduly worried, as it seemed unlikely that anybody else would have such an outlandish idea. (This is rather reminiscent of Shimony and Horne being confident that nobody else would contemplate using cascades to investigate Bell's inequalities!)

This was just the time that the neutron interferometer was invented, and Greenberger was rather shocked to read next year a proposal by Roberta Colella, Al Overhauser and Sam Werner to perform the experiment (known as COW after their initials), and even more shocked next year to read that they had actually performed it! All was not lost, however, because Greenberger spotted a problem with their work, and when he wrote to Overhauser to point this out, Greenberger was asked to visit Purdue to give a talk. This was to lead to a very useful collaboration and friendship. In 1979, Overhauser was asked to talk about the COW experiment at a neutron conference in Grenoble, but he asked Greenberger to go instead. Here he met nearly all the people doing coherent neutron optics, mostly with the neutron interferometer, for the first time, but particularly Horne and Zeilinger, with whom he got on particularly well.

The three of them—Greenberger, Horne and Zeilinger—were to form a very close grouping. Greenberger spent a sabbatical leave in MIT in 1980, and for the next seven years or so made frequent visits there. Zeilinger too, as we have seen, made many visits, some of them quite long, and the three of them, with Shull and occasional contributions from Shimony and Herb Bernstein, performed a great deal of important work. During work time, their talk was solely of single particle physics with neutrons, but away from the laboratory they spent a lot of time discussing entangled pairs and Bell's theorem.

Actually they did make an occasional contribution to the Bell type of work; for example, Zeilinger, as mentioned in the previous chapter, pointed out the problem in Aspect's experiment in 1986, and Horne and Zeilinger wrote an interesting paper for a 1985 conference in Joensuu in Finland. In the next chapter, though, we shall see how GHZ, for Greenberger, Horne and Zeilinger, was to become as famous a set of initials as CHSH.

Shull, however, was due to retire in 1986. He had hoped that Zeilinger could be his successor, but MIT preferred to discontinue the type of work. There was no immediate break—the three of them, together with Bernstein, were awarded an NSF grant, so some collaboration could continue, and in Vienna Zeilinger obtained a Fulbright Fellowship for a long visit from Greenberger. However, it was still a sad ending to the work on neutrons, but more positively it was an opportunity for the three comrades to move more completely into the study of quantum theory and quantum information theory, an opportunity they were to accept with open arms.

Let us first, though, say a little about Nicolas Gisin (Figure 11.4). He was born in Geneva, Switzerland in 1952, so he is somewhat younger than most of the people we have met in the book so far. After obtaining degrees from the University of Geneva in both physics and mathematics, he was awarded his PhD in 1981 for a thesis that covered aspects of quantum theory and statistical physics. In fact it was very much a thesis from the New Quantum Age, since it studied an unorthodox interpretation of quantum theory that did not include the collapse of the wave-function. The Louis de Broglie Foundation, suitably commemorating the famous unorthodox French physicist by rewarding the questioning of quantum dogma, awarded the equally unorthodox newcomer a prize. It is striking that much earlier than this he had met John Bell at CERN and had been greatly impressed, and it is interesting to surmise that this meeting may have affected the whole of the rest of Gisin's professional life!

Gisin then spent a period as a postdoctoral fellow in the University of Rochester, New York, where he worked with some of the greatest

Fig. 11.4 Nicolas Gisin [© Zentralbibliothek für Physik, Renate Bertlmann]

names in quantum optics—Leonard Mandel, whom we shall meet later in this chapter, and Emil Wolf, who, with Max Born, wrote perhaps the most celebrated book on optics in the history of physics.

At this point one might have expected Gisin to take up a position in a university. Instead, though, he made a career move that was to turn out to be one of genius. Shimony had apparently advised him that a spell in an industry based on optics would stand him in extremely good stead, enabling him to apply the most advanced technological techniques to the study of the most fundamental theoretical and conceptual questions, and this is exactly how things were to work out.

Gisin took up a position in Alphatronix, a start-up company dedicated to providing fibre optic instrumentation to the telecommunications industry. At first he was head of software, but he soon became responsible for the interface between software and hardware. Then after four years in that company he joined a Swiss software company that was developing an image processing package. During this period, he had the energy and enthusiasm to continue, in his own time, his theoretical research into unorthodox approaches to quantum theory. Both Zeilinger and Gisin combine enormous experimental flair with the capability for deep conceptual thought and mathematical fluency. (In addition they each have a major hobby. Zeilinger's is music—he plays the bass—while Gisin's main hobby has been hockey, which he played at top national level.)

In 1988, Gisin was asked by the University of Geneva to become the new head of the optics section of its Group of Applied Physics. At this time the section was virtually wholly dedicated to support of the Swiss telephone company, now Swisscom. With his experience in telecommunication and fibre optics, and perhaps his plans for future experiments putting this technology to the service of the foundations of quantum theory and quantum information theory, this relationship must have been exactly to Gisin's taste.

However, to build up the section, he added two new research fields, optical sensors and quantum optics. Again Gisin was able to straddle comprehensively just about any divide. The telecommunication and sensing work has been extremely successful, many patents having been produced, together with a great deal of technological transfer to Swiss and international industry. The quantum optics division has been able to do fundamental work in the study of quantum theory, as we shall see later in this chapter.

The neutron interferometer

As has been mentioned, the neutron interferometer was important in the early scientific experiences of G(reenberger), H(orne) and Z(eilinger), and it was responsible for their coming together. However, even apart from that historical significance, it deserves attention in this book because, more than any other device, it has succeeded over the past 35 years or so in turning thought-experiments, which had often been at the heart of the conceptual development of quantum theory, even though it was assumed that they could never be performed in practice, into real experiments actually carried out in the laboratory.

Such an achievement may conceivably be something of a double-edged sword for supporters of orthodox approaches to quantum theory. On the one hand it may serve to bring to experimental reality the seemingly abstract arguments of the pioneers of quantum theory. On the other hand, though, experiments may be performed that can lead some of the unorthodox brigade to claim demonstration of failures in, or even contradictions to, the traditional approaches. We shall see examples of both responses to experiments using the neutron interferometer in this section.

We shall first say a little about the device itself and its mode of construction. It can be mentioned that the neutron interferometer was developed from an earlier device, the X-ray interferometer, which had been constructed by Ulrich Bonse and Michael Hart in 1965. X-rays are a component of the electromagnetic spectrum, and so it is natural to assume that they may be made to interfere, though it was still a major feat to demonstrate that they actually did so. The wavelength of an X-ray is around 10^{-10} m, which is also roughly the distance between two atoms in a typical solid. Thus a solid clearly provides a possible system for demonstrating X-ray interference.

From de Broglie's work, it is clear that neutrons also have a wave-like nature. In fact, as we have said, neutrons emerging from a nuclear reactor are so-called thermal neutrons, neutrons at room temperature; from the most basic laws of statistical mechanics, it is easy to calculate the expected energy of these neutrons, and, from the de Broglie relationship between energy and wavelength, it is thus also easy to obtain a rough value for the wavelength of these thermal neutrons. In fact it is again about 10^{-10} m, a fact that suggests the possibility of constructing a neutron interferometer in an analogous way to the X-ray interferometer.

However, Greenberger [3] points out that the general consensus was that the only reason that the X-ray interferometer worked was because part of the X-ray beam was absorbed in the crystal. In the analogous neutron case, there would be no such absorption, so no one tried to make a neutron interferometer until Rauch, together with Bonse, with what Greenberger calls 'an act of courage', invested the required time, money and effort to do so. Happily the device they constructed did indeed work.

The neutron interferometer and the X-ray interferometer have the same basic structure shown in Figure 11.5. In each there are three active regions. The top view is shown in Figure 11.5a. The incoming beam of X-rays or neutrons is split into two diverging beams in the first region at A. In the second region, the beams are reflected at B and C, respectively, so that they subsequently converge and interfere in the third active region at D. X-rays or neutrons may be detected at E and F.

In principle three separate regions could be used for the three active regions, but in practice it is essential that a single crystal of perfect periodicity is used for all three regions, and there must be perfect lining-up of the atoms in all three regions. The solution thought up by Bonse and Hart for the X-ray interferometer, and taken over by Rauch and Bonse for the neutron interferometer, was to make all three slabs from a large single crystal of silicon, cut as shown in Fig. 11.5b to give

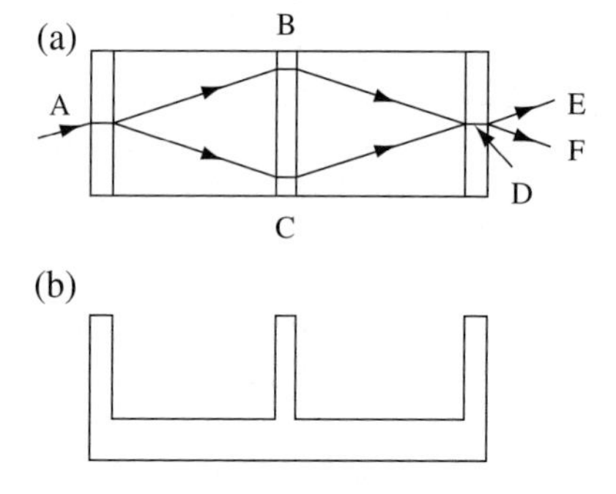

Fig. 11.5 The neutron interferometer

three slabs or 'ears' with enough of the crystal left to retain structural support.

The length of single crystal must be 10 cm or more, and in 1965 production of a perfect single crystal of this size, with perfect periodicity and no faults at all, was only possible because of tremendous advances in the technology of crystal growth demanded for the solid state semiconductor industry in the twenty years or so since the end of the Second World War. Thus fundamental science reaped a great advantage from technology, but we must remember that semiconductor technology would not have been possible without the development of quantum theory and its application to the theory of solids from the 1930s to the 1950s, so this is an excellent example of the mutual cooperation between pure science and highly lucrative technology.

Construction of the neutron interferometer was a major achievement. The distances ABD and ACD had to be the same within a minute amount of possible error. Thus the thickness of the ear and the distances between them had to be correct to about 10^{-4} cm or about 10,000 atomic layers. The crystal had to be supported with great care to avoid distortion under its own weight, and it also had to be shielded from temperature gradients and vibration.

In fact the problem of vibration is much more difficult for the neutron interferometer than for the X-ray case. While X-rays move at the speed of light, at least in between the ears, and thus pass through the interferometer in about 10^{-9} s, thermal neutrons travel much more slowly and take about 10^{-4} s. Thus for the neutron case, the slightest vibration on this timescale will completely destroy the interference pattern.

It was Greenberger [3] in 1983 who produced the most clear and relatively elementary account of the neutron interferometer. He explained that we may think of each neutron as being represented by a wave-pulse of about 10^{-3} cm, and the cross-section of the beam as being about 1 cm^2, so the wave-pulse is very much analogous in size to a small postage stamp! The two beams passing through B and C may be a few centimetres apart, while a typical reactor produced roughly one neutron per second, so neutrons usually pass through the system individually, rarely interacting with each other. Thus we may think of the neutrons very much as single particles confined within the interferometer as they pass through it. Conceptually the picture is very clear.

In contrast Greenberger points out that it would be immensely difficult to achieve for electrons what is done for neutrons in the neutron interferometer, essentially to confine them within the device itself. For an electron interferometer, the two beams could be separated by at most a fraction of a millimetre, not centimetres as in the neutron case.

From any classical point of view, the neutrons would thus appear to have been produced very much as particles. Yet at D in Figure 11.5a, interference *is* seen. Thus it is clear that the neutron has sampled or travelled along both of two well-separated paths, which is, of course, the essence of wavelike behaviour. This is the type of effect that was discussed from the early days of quantum theory, but usually it was deduced from more complicated experiments, rather than being directly demonstrated. Thus a great advantage in clarity is achieved by working with neutrons.

As usual with interference experiments, a wedge of suitable material may be inserted into one of the beams, and the interference pattern shifts in phase. As the width of the wedge is increased, the pattern changes in a way that may be calculated exactly.

Another nice feature of the neutron interferometer is that, as Greenberger showed in detail, we can find out exactly what happens if we attempt to determine which of the beams the neutron went along. The debate between Bohr and Einstein confirmed that if you do detect a neutron (or photon or electron) in one particular beam, the interference pattern is destroyed. In the 1920s, the argument was satisfactory in general terms, but abstract rather than specific. For the neutron interferometer, in contrast, Greenberger was able to produce an exact theoretical model, in which the detailed process by which the destruction of the interference pattern takes place gives a very nice insight into how the laws of quantum theory manifest themselves in actual physical situations.

Another interesting experiment that can be performed with the neutron interferometer is to pass one of the beams through a partial absorber of neutrons. Classically one would expect that some neutrons will indeed be absorbed, but that the properties of those not absorbed, and so are eventually detected, will be unchanged. Quantum mechanically, though, we must describe the situation in terms of the resultant wave. The amplitude of the wave emerging from the absorber will be reduced, and so the interference pattern will be very much changed by the absorber, although we would be inclined to say that the neutrons that *do* emerge should have been unaffected by the absorber!

We now come to the experiments that were mentioned in the previous section. The first, performed by the team including Zeilinger, covers the rotation of the electron spin. Quantum mechanical spin is a totally non-classical object, and one of the strange predictions resulting from this fact is that if a neutron is rotated through a full 360°, its spin-state changes sign. It takes a double rotation of 720° to return it to its original state. The reason for this is that the spin of the neutron, like that of the electron and the proton, has a quantum number of ½; particles whose spins have integral quantum numbers do return to their original state under a rotation of 360°.

The neutron spin is involved in another experiment that confirms predictions that had been suggested by a thought-experiment twenty years earlier. This experiment uses a polarized incident beam in which the spins of all the neutrons are in the same direction. A beam-flipper is then included in one of the beams; this rotates the directions of all the spins in that beam. The beams are then allowed to interfere at D, and the results agree exactly with those obtained from theory, provided the theory used makes a certain assumption. This is that *no* collapse of state wave-function takes place when the beams are separated at A.

This is actually a slightly tricky point, as it is very natural to describe the splitting resulting in sending neutrons with different directions of spin in different directions as 'measuring' the z-component of the spin. This would imply a collapse of the state function to *either* spin-up in the beam travelling from A to B *or* spin-down in the beam travelling from A to C, rather than the typical quantum mechanical linear combination of both states. Wigner had pointed out in 1963 that there was actually no measurement and therefore no collapse, and so in turn the beams may, at least in principle, interfere when they reach D. Though this was generally accepted, Wigner, like everybody else, had assumed that it would be impossible to keep the path lengths along the two beams exactly the same, so he thought the idea must remain as a thought-experiment, and it was very pleasant to see the thought-experiment become an actual experiment using the neutron interferometer.

The experiment that Greenberger had been particularly interested in was one in which the beams are oriented with one vertically above the other. Thus there is a very small difference in gravitational potential between the two beams, and this in turn leads to a small but measurable difference in the interference pattern. Experiment and theory agreed perfectly, and this was the first direct experimental confirmation that quantum theory could be applied in discussions of gravity.

So far this has all been good from the point of view of those holding orthodox views on the interpretation of quantum theory: some of the more sensitive points in the argument have been confirmed and demonstrated explicitly. However, as we have said, there is also ample opportunity for opponents of the orthodox position to claim support for their own position from experiments on the neutron interferometer. One such was Jean-Pierre Vigier, who had worked with Bohm in the earliest days of the Bohm model, and who remained a scourge of all those happy with Bohr's ideas.

The general idea was to confound the orthodox view that the presence of interference shows the neutron must have sampled both paths through the interferometer. It claims to show that in fact it had gone down a particular path. There are actually two levels of such a claim. The less revolutionary approach, the *Einweg* or *one-way* position, which was actually held by Einstein, says that the neutron went along one beam or the other, although it is not possible to say which. The more revolutionary, the *Welcherweg* or *which-way* position, goes further by saying that one can actually say which beam the neutron passed along.

Rauch and Vigier together wrote a paper along these lines; it may be taken for granted that Rauch did the experiment while Vigier did most of the analysis. In the experiment, neutrons were sent through the interferometer singly, and there is a spin-flipper in each beam, operating at frequencies f_1 and f_2, respectively. When a neutron flips its spin, it exchanges one photon with one of the flippers, the energy of the photon being either hf_1 or hf_2. Experiments show that if f_1 and f_2 are equal, the interference pattern remains unchanged, but if they are different, then the pattern oscillates at a frequency given by the difference of f_1 and f_2.

What does this experiment actually tell us? Rauch and Vigier argued that the photon must have been exchanged *either* in the first beam *or* in the second beam, since there can be no half photons. This, they said, shows that each photon must have travelled along a particular one of the beams, although they do not claim to know which—the *Einweg* position. They specifically commented that this supports the position of Einstein, and is in direct opposition to that of Bohr. Indeed on the basis of further experiments, Vigier was prepared to move from the *Einweg* position to the *Welcherweg* position.

Much discussion ensued, but in the end the situation was clarified. It would now be fairly generally accepted that the Copenhagen view was an over-simplification. Actually there is a duality relation between predictability or the *Welcherweg* type of information, and the visibility or intensity of the interference pattern. The higher the predictability, the lower the visibility and vice versa. So while Vigier may be able to obtain the information he claims, it must always be at the expense of this interference pattern.

The last experiment [4, 5] we consider was based on an earlier theoretical argument, and it is a violation of a Bell's inequality, but not in this case between properties such as the polarization of two different particles, but between the spin and spatial parts of the wave-function of a particular neutron. The detection of neutrons is very much easier than that of photons, so at first sight it might be suspected that the experiment might be able to close the detector loophole. Unfortunately, though, only half of the neutrons produced in the experiment travel along the path that allows them to be used in the Bell type of analysis, so obviously we cannot assume fair sampling, since the neutrons that were not sampled could possibly have had completely different properties from those used in the experiment.

Overall we see the tremendous usefulness of the neutron interferometer. Not only could it give direct evidence of some of those central parts of the accepted quantum mechanical story that hitherto had been assumed on the basis of more complicated phenomena, but it could also clarify some of the controversial points of the argument.

Parametric down-conversion

In the first section of this chapter, we showed that, by the end of the 1980s, Zeilinger had brought his work on neutron interferometry to a conclusion, and Gisin had moved into academia, albeit an aspect of academia where he would be expected and would certainly wish to maintain the closest contacts with industry. They would thus be free to tackle the experimental demands of quantum foundations theory and quantum information theory. Greenberger and Horne were ready to engage the same area from a more theoretical point of view.

In fact the first major advance came initially from others. This was the use of parametric down-conversion to produce pairs of entangled photons far more copiously than the method of cascades used in almost every experiment up to this date. Parametric down-conversion had been discovered by David Burnham and Donald Weinberg as early as 1970. It occurs only in very special crystals such as β-barium oxide (known as BBO), lithium iodate and barium borate. When light from a high-frequency laser is shone on the crystal, at first all that happens is that light emerges from the other side, but as the intensity of the light is increased, a halo appears surrounding the crystal containing all the colours of the rainbow.

The explanation is that for these crystals there is mathematically a term in the electric dipole moment that is proportional to the square of the applied electric field rather than the field itself. (The crystals are called *non-linear* crystals.) This term naturally becomes more important as the field increases because the square of the field increases much faster than the field. What happens is that, for a very small fraction of the incident photons, the photon breaks down into two photons, and the two emerging beams are called the signal beam and the idler beam. Naturally the energies of the two photons must add up to that of the original photon, and since $E = hf$, this tells us that thc frequencies obey the same rule. Thus the frequencies of both emerging photons are less than that of the original photon, and so it is natural to call the process down-conversion. Momentum must, of course, also be conserved in the down-conversion process.

It was not until the later 1980s, with the increased interest in matters such as entanglement following the publicity gained by the Aspect experiment, that various people suggested that these pairs of photons might be used in similar experiments. Probably the first was Leonard Mandel, who had been born in Berlin in 1927 but had emigrated with his family to England, where he gained a PhD from Imperial College London. He was a faculty member at Imperial College until 1964 when he emigrated once again to the United States, where he joined the University of Rochester in New York. His initial research interest was in cosmic rays, but in the late 1970s he took up the study of quantum optics, working with lasers. Around 1987 he realized that the down-converted photons could be used to test Bell's inequalities.

At about the same time, Yanhua Shih, who was a research student of Carroll Alley at the University of Maryland, came to the same conclusion. Shih had been born in 1949, and took his bachelor's degree from

Northwestern University in China in 1981, followed by a master's degree and a doctorate from University of Maryland in 1984 and 1987, respectively. He subsequently took up a faculty position in 1989, moving from the College Park campus to the Baltimore County campus, where he now leads a major group in quantum optics.

Alley and Shih, and independently Mandel with his own research student, Zhe-Yu Ou, carried out the following experiment. The two beams emerging from the down-converter were reflected onto a beam-splitter from opposite sides. (A beam-splitter is a slab of glass with its top surface half-silvered. If it is set up making an angle of 45° with the incident beam, half of the energy in the beam goes straight through the beam-splitter, while the other half is deflected through 90°.) The beams then travel onto detectors, and as always one looks for coincidences at the two detectors.

The increased intensity of the down-conversion source relative to the atomic cascade means that Bell's inequalities may be violated by several standard deviations (three for Alley and Shih, six for Ou and Mandel). Another extremely good point about this new type of experiment is that it is a genuine two-body decay. At the conclusion of the down-conversion there are only two bodies present, the two photons. You will remember that for the cascade case, the atom is still present at the end of the decay, in its final decayed form of course, and this complicates the analysis because at the conclusion of the process, energy and momentum must be shared between the three bodies, the two photons that are actually observed and the atom.

So there are good things about this type of experiment. Unfortunately there is also a rather bad thing. You may have noticed that I did not mention in my above account of the process that the photons produced in the down-conversion are entangled. This was not an oversight! The kind of process used in these experiments can be called a type-I scheme in which both the output photons have the same polarization. Because we therefore cannot distinguish between the two photons using polarization, it turns out that, unless we use a special trick, the state of the two photons is not entangled at all. Rather it is a simple product state in which the behaviour of each of the two photons is entirely independent of the behaviour of the other. (When we use the word 'behaviour' we mean what measurements may be performed on the photon and what results we may get.)

Mandel and the others did apply such a trick: they considered only those pairs of photons that take different exits from the beam-splitter and thus travel away from each other, ignoring those pairs that take the same exit from the beam-splitter. Provision of this extra 'label' for each of the photons—which exit they took from the beam-splitter—in fact makes the final state the pair of photons entangled, and, as we have said, the usual type of analysis can now be used.

However, the procedure is rather unsatisfactory. We are ignoring half of the photon pairs produced, and it is perfectly conceivable that those pairs we are considering may differ in fundamental ways (apart from the exit they take from the beam-splitter) from those we are ignoring, so any assumption of fair sampling is highly dubious. Ignoring half the pairs of photons produced certainly means we can be making no progress at all towards closing the detector loophole.

For this reason it was excellent news when Paul Kwiat and other members of Zeilinger's group [6] effectively solved the problems inevitable with a type-I down-conversion scheme. They were able to achieve a type-II scheme in which the polarizations of the two photons are different and so a genuinely entangled state is produced. The group was working in Innsbruck, by the way; Zeilinger had moved to a senior position here in 1990, before moving back to Vienna in 1999.

The source was around ten times brighter than those produced by type-I schemes—coincidence detection rates were as high as 1500 s^{-1}, and so experiments using the source enabled Bell's inequalities to be violated by around 100 standard deviations, the highest figure so far! The experiments were simpler than those using the type-I down-conversion, as no beam-splitter and no mirrors were required. They also led to a higher detector efficiency, because it was possible to have larger collection angles than for previous experiments, and so the experiments made a major step towards closing the detector loophole.

As we have said, by this time, the emphasis of the physicists involved was perhaps less on Bell-type work, important as this still was, and more on establishing the basic methods of quantum information theory, and the method of type-II down-conversion would be equally central in both areas.

Closing the locality loophole

Another important achievement from Zeilinger's group came in 1998, when Gregor Weihs (Figure 11.6) and co-workers [7] finally closed the locality loophole. Fifteen years after Aspect's work, there had been major technological improvements in experimental equipment, and what had been state-of-the-art for Aspect was now practically archaic!

In the new experiment, locality was achieved by ensuring that the measurement stations in each wing of the experiment were sufficiently far apart that completely independent data registration took place in each wing. Weihs and his colleagues demanded that, if a individual measurement was defined as beginning when the decision of what measurement direction would be chosen in one wing of the experiment, and ending with the actual registration of the photon, then the measurement must be carried out sufficiently quickly that, during the time of the measurement, no information can travel to the other wing

Fig. 11.6 Gregor Weihs [© Zentralbibliothek für Physik, Renate Bertlmann]

of the experiment. The two wings of the experiment were separated by 400 m, so the time taken for light to travel between wings was 400 m divided by the speed of light, which is about 1.3×10^{-6} s. The time taken by a measurement, then, must be rather less than this.

In each wing, the direction of the polarization measurement was determined by extremely fast electro-optic modulators controlled by a physical random-number generator sampled every 10^{-7} s. The random-number generator was equipped with a light-emitting diode that illuminated a beam-splitter, the outputs of which were monitored by photomultipliers. The subsequent electronic circuit set its output to 0 or 1, depending on which photomultiplier it had most recently received a pulse from. Thus the time taken for a measurement at each observer station was around 10^{-7} s, must less than the 1.3×10^{-6} s allowed.

It was also essential that the photons reached the two wings at the same time within an exceedingly small leeway, as otherwise, of course, it would again be possible for information to be passed from the wing at which the photon arrived first to the other wing. In the Weihs experiments, the photons were sent to each wing of the experiment through optical fibres that differed in length by less than 1 m, so the photons were registered simultaneously within 5×10^{-9} s.

A last requirement for locality was that measurements in the two wings should be synchronized—otherwise, it would be impossible to correlate results taken in the two wings, but apart from the synchronization all procedures should be totally independent in the two wings. To ensure these requirements, before each experimental cycle, a synchronization pulse of precision 3×10^{-9} s was sent to each wing, but subsequently all information on times and results of detection were stored in an individual computer in each wing. The files were analysed for coincidences only long after all measurements had been completed.

In these experiments not only was the locality loophole convincingly closed, but Bell's inequalities were violated by about 30 standard deviations.

Quite recently, Clauser, Aspect and Zeilinger have shared the 2010 Wolf Foundation Prize for physics for 'their fundamental conceptual and experimental contributions to the foundations of quantum physics, specifically an increasingly sophisticated series of tests of Bell's inequalities or extensions there of using entangled quantum states'. This is an extremely prestigious prize, and its award often leads on to the award of a Nobel Prize.

Gisin and the experiments at Lake Geneva

An extremely interesting set of experiments was carried out by Wolfgang Tittel and co-workers in Gisin's group [8] towards the end of the previous century. These experiments owed everything to Gisin's experience in telecommunications and quantum optics as well as in quantum theory, and to his excellent relations with the Swiss telephone company. It seems unlikely that any other physicist in the world could have organized and led them.

We saw in the previous section that in the Innsbruck experiment the photons travelled hundreds of metres. Gisin wanted them to go much further than that, more than 10 km, and in this experiment asked for and received enthusiastic help from the telephone company to achieve this aim. The source for the photons was at a telecommunications station in Geneva, and the photons were sent along installed standard telecom fibre under Lake Geneva to analysers in Bellevue and Bernex. The distances from Geneva to Bellevue and Bernex are 4.5 and 7.3 km, respectively, while Bellevue and Bernex themselves are 10.9 km apart as the crow flies; in fact, taking into account the places where the cable could not follow a straight path, the actual length of the cable between Bellevue and Bernex was around 16 km. The photons were detected by photon counters at Bellevue and Bernes, and the signals, which were now classical after detection, were transmitted back to Geneva where the coincidence electronics was located.

Compared with laboratory experiments, the absorption in the wires meant that the number of coincidences decreased by a factor of 10, and the visibility of the fringes also decreased to 86% compared with 94% for a similar experiment carried out in the laboratory. This loss of 14% may be attributed to detector noise, to the rather large bandwidth of the down-converted photons, which led to birefringence and chromatic dispersion in the interferometers, and also possibly to slightly unequal transmission amplitudes for the interfering paths.

Nevertheless these experiments were still able to demonstrate violations of Bell's inequalities by about 25 standard deviations, and they gave results agreeing well with quantum theory. This work was important both in its right and perhaps even more so because it led to similar studies in quantum cryptography, one of the main branches of quantum information theory, which requires the maintenance of quantum correlations over long distances. For example, the 86% fringe visibility would be just about high enough to make quantum cryptography possible.

Although this result was the clear prediction of quantum theory, it was far from obvious that this result would necessarily have been found, as quantum theory had never been tested at distances anything like these. There were alternative possibilities. As early as 1970, Hans-Dieter Zeh had suggested the idea of environment-induced decoherence, according to which particular photon states become correlated with particular elements of the environment, and this prevents the photon states from interfering. Wojciech Zurek has also promoted this approach, which is now widely recognized as playing an essential part in bringing about our classical world.

Another possibility is that of spontaneous collapse. The Schrödinger–Furry hypothesis discussed in Chapter 9 was an early example of such an idea, though more recently, authors such as Philip Pearle, and Gian-Carlo Ghirardi, Alberto Rimini and Tullio Weber (GRW) have suggested much more sophisticated schemes to attempt to solve the measurement problem. In the experiments of Tittel and colleagues, this idea of spontaneous collapse would imply that the linear combination of states such as $\left(1/\sqrt{2}\right)\left(|h\rangle_1|v\rangle_2+|v\rangle_2|h\rangle_2\right)$, where $|h\rangle_1$ denotes photon 1 has horizontal polarization, $|v\rangle_2$ denotes photon 2 has vertical polarization and so on, which is so essential in the idea of entanglement, decays or decoheres spontaneously, over the long distances travelled by the photons, to *either* $|h\rangle_1|v\rangle_2$ *or* $|v\rangle_1|h\rangle_2$.

The Lake Geneva experiments cannot show that these types of process do not take place. Indeed, as we have said, environment-induced decoherence certainly does take place, while the theory of GRW is also important, and both will be discussed in Chapter 13. But what the experiments *do* show is that neither of these effects or possible effects disturbs the predicted quantum correlation over the distances involved in the experiment.

Gisin described two other experiments carried out with the same setup in a beautiful article [8] written for *Quantum [Un]speakables*. In these experiments he was considering the possibility that the collapse of the wave-function should be taken as real, rather than as hypothetical and used only conceptually just to obtain general agreement between experiment and theory.

First, he considered what would happen if the measurements at the two detectors were made at exactly the same time, to within as small an error as is unavoidable. He was actually repeating the Innsbruck experiment described in the previous section, but the very large distance between the two detectors meant that he could estimate the minimum value of the speed of any signal between the two detectors even more dramatically. The demands of relativity mean, of course, that the meaning of the phrase 'at exactly the same time' must be thought out carefully. He decided to use the reference frame in which the massive parts of the apparatus were at rest, and assumed that the collapse took place at the detector in each wing of the experiment. In these experiments, the distances travelled by the two photons were accurate to about 1 in 10^7, and it was possible to establish that, if there were a signal between the two detectors, it must be travelling at a speed of at least some million times the speed of light.

Another experiment was to allow one of the detectors to be travelling at a high enough speed that each detector in its own reference frame will effectively detect its photon before the other one. (As we saw earlier this is a possible situation relativistically.) It turns out that the speed needed is comparatively low, only about 50 ms^{-1}. If one takes seriously the idea that collapse is real, it is very difficult to describe this experiment in a straightforward unique way, since it seems that each detector must cause a collapse before the other one, and it seems conceivable that the interference pattern might disappear. However, this is not the case; whatever the relative speed, the pattern is still visible.

Other experiments on Bell's theorem and the detector loophole (with a detour on Bell and Bertlmann)

In this section we first briefly mention a few other suggestions for testing Bell's theorem. First, in 1990, John Rarity and Paul Tapster demonstrated the violation of a Bell's inequality for an entangled photon pair based not on different directions of polarization, but on phase and momentum. This idea had been suggested independently by Horne, Shimony and Zeilinger. James Franson has suggested experiments using entanglement in energy and time, and such experiments along these lines have been carried out, though other have argued that there is a local hidden variable model that may predict the results obtained.

A rather unusual suggestion for a test of Bell's inequalities was made by Peter Samuelson, Eugene Sukhorukov and Markus Büttiker. This is to use mesoscopic conductors, which are intermediate in size between microscopic and macroscopic conductors. They suggest that a superconductor could generate entangled states by injecting pairs of electrons into different leads of a normal conductor.

For the rest of this section we shall be discussing attempts to close the detector loophole. The difficulty of doing so using photons was spelled out by Philip Pearle, who argued that to contradict the Bell inequality conclusively using photons, the detector efficiency would have to be at least 84%. Philippe Eberhard was actually able to show that this could be reduced to 67% by considering a range of quantum states, but even this is well beyond detectors available so far.

Very many suggestions have been made of ways of closing the loophole. A few have used photons but with special tricks to avoid the problem stated in the previous paragraph. Kwiat and colleagues who included Eberhard have suggested using non-maximally entangled photons, for which the requirements on the detectors are somewhat lower. Santos, who had drawn attention to the problem that the final state of the atomic cascade is a three-body rather than a two-body state, suggested an experiment with an extra measurement, that of a recoil atom, which may provide extra information on each pair of entangled photons so that no auxiliary assumptions are required.

However, the majority of suggestions have involved the use of more massive particles, for which detection rates can be much higher than those for photons. One interesting idea [9], which was suggested by Shimony together with T. K. Lo as early as 1981, but was described in considerable detail by Ed Fry, Thomas Walther and S. Li in 1995, was to analyse correlated spins in pairs of atoms obtained by dissociation of a beam of dimers containing two atoms of sodium or mercury. Quite a number of proposals have used pairs of atoms entangled by use of a cavity filled with black-body radiation or a classical radiation field.

Many of the proposals take advantage of the possibility of using relatively massive elementary particles, protons or neutrons or especially K-mesons (kaons). In Chapter 9 we mentioned the experiments of Lamehi–Rachti and Mittig using protons, but explained that they could not be regarded as conclusive tests of Bell's inequalities, while earlier in this chapter we looked at neutrons being used in the neutron interferometer for the same task.

However, most work has been done on the use of neutral kaons. Kaons are the elementary particles with the most bizarre properties, all relating from the fact the distinction between particle and anti-particle, between K_0 and $\tilde{K}_0$, is very unusual. For charged particles, the difference between particle and anti-particle is, of course, obvious: e^- and e^+, or p^+ and p^- . Even for neutron and anti-neutron, n and $\tilde{n}$, although both have zero charge, the two are totally distinct, because in one case spin and magnetic moment are in the same direction, while in the other they are in opposite directions.

The distinction between neutrino, ν, and anti-neutrino, $\tilde{\nu}$, is more problematic. For a considerable period, during which it was assumed that neutrino and anti-neutrino had zero mass, it was thought that the two differed by one having spin along the direction of motion, the other having the spin opposed to the direction of motion. However, now that it is know that neutrino have a small but definitely non-zero mass, that distinction cannot be maintained, as a relativistic transition may change neutrino into anti-neutrino and vice versa, and it is allowable to construct theories in which neutrino and anti-neutrino are not distinguished—the so-called Majorana neutrino.

Anyhow our interest is in the kaon. The difference between kaon and anti-kaon lies in their values of the quantum number *strangeness*. The kaon has strangeness 1; the strangeness of the anti-kaon is –1. But the catch is that strangeness itself only has any real meaning over the timescale of the strong nuclear interaction, which is about 10^{-23} s. So for longer timescales particle and anti-particle cease to be distinct. Thus there are many interesting aspects of the behaviour of this system over different timescales, and so there are many ways of achieving entanglement. Much theory, experiment and discussion has ensued.

Interestingly enough, one of the main participants in this work has been Reinhold Bertlmann (Figure 11.7) [10], who has played quite an interesting and important part in Bell's life. To explain this, we shall make a brief detour to catch up with Bell's life and work. In Chapter 8 we discussed his early life and his two great quantum papers. In Chapter 9 we saw his interaction with, in particular, Clauser, Horne and Shimony, including the discussion in *Epistemological Letters*, while in Chapter 10 we noted his pleasure at the Aspect experiments.

Fig. 11.7 Reinhold Bertlmann [© Zentralbibliothek für Physik, Renate Bertlmann]

It is interesting that through this period, though he published over twenty papers on the foundations of quantum theory, virtually all had been presented at conferences and were published in their proceedings. He took advantage of these occasions to reiterate his own views, and to comment on Bohm positively, and on many worlds broadly negatively, though with the caveat that many worlds might conceivably have something useful to say about EPR.

However, he did not, except in a couple of exceptional cases, submit papers to regular journals. It was possible that he was still concerned about the stigma even for himself; he may have felt that he had raised his head above the parapet twice and that was enough. He may perhaps have felt that people at CERN would be highly negative about his concerns with quantum theory. Gisin has suggested that it was merely that he wanted to avoid the too-often sterile discussions with referees! (Gisin made this remark as a footnote to a comment that the referee reports on one of his papers discussed earlier in this chapter varied from fascination to desperation!)

What he did, of course, as well as following the quantum literature closely, was to work conscientiously and highly successfully at what he was paid for—elementary particle theory and quantum field theory. Some of his most important achievements were mentioned in Chapter 8, but these were merely the peaks among a host of contributions. But he contributed much more than that. He was called the Oracle of CERN, sharing his wisdom with the staff members and senior visitors, but also being prepared to give his time for the younger visitors. When they arrived in a new and not necessarily over-friendly environment, he would make a point of greeting them and introducing himself, and he would very often be called on to answer technical questions, to give advice and to comment on manuscripts.

To be sure he was often a little disappointed that, when he had helped to revise a manuscript, which, to be frank, might often entail a total re-writing of both the science and the presentation, he was given little or no acknowledgment when the paper was actually submitted and published. But one might see why this could happen. Bell perhaps gave the impression that he was totally relaxed in his life and work, and that ideas that the young visitor found complex and daunting he was able to assimilate with ease. They may even have felt that he would hardly *want* to be thanked.

None of this was really true. It would have been out of character to impress on other scientists how much effort he may have taken on revising their work, but, in any case, it may be assumed that no scientific work is easy, and any scientist, however apparently laid-back, deserves and appreciates appropriate credit for work performed. Bell was actually working particularly hard against a background of some ill-health—intermittent headaches sufficiently serious to stop him working or at least to stay at home for a day or so.

He was in effect doing three jobs—first his work for CERN, second his constant study of quantum foundations, and third replying to correspondents of all types, anxious to get his views and often to put over their own. He always said that he held the right not to reply to letters to be the most fundamental of human freedoms, but in practice he felt morally obliged to reply if it was at all appropriate. It is sad to think that this triple load of work may have contributed to his untimely death.

However, let us now go to 1978, when Bertlmann, a postdoc from Vienna, arrived at CERN. He certainly found CERN extremely impressive with its state-of-the-art accelerators and the large number of distinguished physicists, but, like most arrivals, he was also probably a little ill-at-ease. Bell indeed welcomed him in the common room after a seminar of the theory division, and when they discussed Bertlmann's work, it became clear that their interests and approaches very much overlapped. During Bertlmann's stay in CERN they worked together highly successfully on a programme involving what may be called 'magic moments'.

They studied systems called *quarkonium*—combinations of quark and anti-quark. In 1974 we may say that the fourth quark, the charmed quark or c, was discovered. (The first three quarks—up, down and strange or u, d and s—were included in the original 1960s scheme of Murray Gell-Mann and George Zweig.) In fact what was discovered initially was $c\bar{c}$, a combination of the charmed quark and anti-quark. In analogy with e^+e^- or positronium, which we met in Chapter 9, $c\bar{c}$ is always called *charmonium*. Its total charm is 0 since c had charm 1 and $\bar{c}$ has charm -1, and we may say that it has hidden charm, in contrast to, for example, c which has naked charm. Later different particles with naked charm were produced.

Similarly in 1977 *bottonium* or $b\bar{b}$ was produced, where b is the fifth quark, the bottom quark. (The sixth quark, t or top quark, was not to be produced for some years afterwards because of its surprisingly high mass.)

It was particularly because at low energies the quarks are confined and their states are bound and therefore discrete, that the quarkonium term is used; for these energies, the energy levels of quarkonium and positronium are in exceptionally good correspondence. The task that Bell and Bertlmann gave themselves was to pick up the analogy and use nonrelativistic approximations to calculate the properties of the quarkonium states. The 'magic moments' were parameters they used in their calculations.

As well as doing good physics, the two men clearly struck up an exceptionally close relationship. Bertlmann of course had to return to Vienna at the conclusion of his fellowship, and he has been based there ever since, but, during the ten years or so that Bell had left to him, they were to meet many times, when Bertlmann made further visits to CERN, or when, for example, they visited the University of Marseille at the same time. It is clear that their friendship encompassed, but was not limited by, very great respect if not awe felt by the younger man for the older, and in return tremendous fatherly affection. Also central in the relationship, of course, were Mary Bell and Renate Bertlmann, an important Austrian artist.

It is in that context that Bell played what might be described as a wicked trick on Bertlmann! He agreed to give a talk on quantum matters at a meeting at the Foundation Hugot in June 1980, and the papers from the meeting were subsequently to be published in the *Journal de Physique*. In this talk Bell wished to contrast the standard quantum approach of, for example, Bohr to measurements on entangled systems—a measurement on either particle affects both—to the approach that he considered most natural—a measurement on either particle merely reveals properties of either particle that were already there.

In quantum theory terms, what he was advocating, of course, was hidden variables, but he wished for a more homely example from everyday life that would make the listener or reader feel how natural Bell's view was. As mentioned in Chapter 7, he had noticed an unusual quirk in Bertlmann's wardrobe—as a minor act of revolution, he always wore socks of different colours. When Bell thought about this, he realized that he immediately had the example he needed. Which colour sock Bertlmann will wear on any foot on a given day is unpredictable. But when you see that the first sock is pink, you know that the second sock will *not* be pink.

When you observe this, you do not, of course, assume, as Bohr does in the quantum case, that your observation has not only forced pinkness on the first sock, but also non-pinkness on the second sock that you have not even seen. Rather you take for granted that, when Bertlmann awoke and got dressed he decided to put a pink sock on the first foot, and was then obliged by his usual code to ensure that the second sock was not pink. Your observation then only confirms this for the first sock, and your deduction that the second sock is not pink is merely that—a simple logical argument to determine what was already the case before you looked at Bertlmann at all.

Bell must have been sorely tempted to build his paper around Bertlmann's idiosyncrasy, and he did not resist this temptation. His paper is titled 'Bertlmann's Socks and the Nature of Reality', and on the first page there is a sketch of Bertlmann himself, although he perhaps also looks like a typical French peasant. We see one sock labelled 'pink'; the other is not visible and is labelled 'not pink'. Bell begins his article as follows: 'The philosopher in the street, who has not suffered a course in quantum theory, is quite unimpressed by Einstein-Podolsky-Rosen correlations. He can point to many examples of similar correlations in everyday life. The case of Bertlmann's socks is often cited.'

The first paragraph continues with the kind of argument we have just sketched, and ends: 'There is no accounting for tastes, but apart from that there is no mystery here. And is not the EPR business just the same?' Bell decides that he must show why some people conclude that it *is* much more complicated, and goes through the standard Bohrian approach, ending in 'Paradox indeed!' But he stresses that the paradox is for these others, not for EPR. 'EPR did not use the word "paradox",' he stresses. 'They were with the man in the street in this business.' (In this way he quietly drops in the point that the term 'EPR paradox', so common for so many years, was inappropriate, misleading and extremely unhelpful for anybody genuinely trying to understand the arguments made.)

Socks are also brought into Bell's account of his theorem. Instead of a description in terms of electrons or photons, the equivalent is (the number of socks that could survive a washing cycle at 0° but not at 45°) + (the number that could survive at 45° but not at 90°) is not less than (the number that could survive at 0° but not at 90°). This is then translated into physics as a hidden variable type of statement. It is important to stress that, leaving aside this jokiness, the paper is a beautifully written account of the topic. Indeed it could well be described as Bell's most successful presentation; it starts with a simple explanation of the various issues and then moves on to a general treatment.

Let us, though, go to the University of Vienna some time in spring 1981. Bertlmann, having returned from CERN, is working at this computer when a colleague, the person who was responsible for receiving preprints sent from other universities, laughingly thrusts a typescript into his hand. 'Bertlmann,' he says, 'You're famous now!'

Bertlmann glanced at the preprint, and was both confused and shocked. It should be said that, as we have seen, Bell did not flaunt his quantum work in CERN, quite the reverse. Bertlmann might have had a general idea that Bell was involved in these arcane matters, but, like the overwhelming majority of physicists in this time before even the Aspect experiment, he knew nothing about them and frankly probably did not wish to know anything. When he saw Bell's paper, he may even have thought that the whole thing was nothing more than a joke at his expense.

He rushed to phone Bell up. Bell's reaction was one of amusement; he is rumoured to have told Bertlmann that he had made his socks as famous as Schrödinger's cat. However, there was a serious point to the exercise as well; he asked Bertlmann to read the paper and to tell Bell what he thought. Bertlmann has reported that the cartoon 'really changed his life'. He must have felt that he had little alternative but to become an expert in Bell's theorem and all that it entailed, and, as we have said, his work on kaons enabled him to combine his expertise in elementary particle physics with his interest and enthusiasm in the ideas and discoveries of Bell himself.

However, after that rather long detour, let us return to attempts to close the detector loophole, and indeed to the experiment that claimed to be the first to do so. This was carried out by Mary Rowe and her co-workers [11] in the group of David Wineland at the National Institute of Standards and Technology (NIST) in Boulder, Colorado. The experiment centres on the idea of the *ion trap*.

The ion trap has been used with very great success to study individual ions exposed to static or oscillating electric fields. The very existence of the ion trap served to confound the orthodox Copenhagen position. It was very much a core belief of Bohr and his colleagues that it was, and always would be, in principle impossible to study individual atomic systems, and of course in the 1920s and 1930s this agreed with the experimental position. Thus difficult questions about how individual systems would behave at a measurement could be shelved, and attention could be given to the expectation values of physical observables and the much more straightforward properties of ensembles.

Schrödinger, of course, was by no means a supporter of this orthodox position but he too wrote, as late as 1952: 'We never experiment with just *one* electron or atom... We are not *experimenting* with single particles, any more than we can raise Icthyosauria in the zoo.'

However, from the 1950s physicists were beginning to develop techniques to trap single ions. One might guess that it would be possible, at least in theory, to use a configuration of electric fields to produce a global minimum where the ion might sit. However, a result from the nineteenth century called Earnshaw's theorem makes it quite clear that this is not possible, and most ion traps today supplement a quadrupolar electric potential with a very strong magnetic field produced by a superconducting magnet or a radio-frequency potential. They also use, of course, an extremely high vacuum.

As experimental techniques have become more and more sophisticated, and, one might say, in total defiance of Bohr and Schrödinger, it has been possible to trap a single ion, or, in fact, even a single positron for a period of weeks. Some researchers have even become so fond of a particular ion that they have given it a pet name! For their production of ion traps of such power Hans Dehmelt and Wolfgang Paul were awarded the Nobel Prize for physics in 1989.

A development of enormous importance has been laser cooling. When first trapped, an ion has a temperature of a few thousand Kelvin. It is highly important to cool it, as this diminishes the possibility of it escaping, and also means that its properties may be investigated more easily in the absence of thermal motion. In laser cooling of ions (or atoms), light acts mechanically on the ion. A clever arrangement of laser beams results in the interaction always reducing the speed of the ion. By this means the temperature may be reduced to about 10^{-4} K. Further Nobel Prizes were awarded for this discovery in 1997 to Steven Chu, Claude Cohen-Tannoudji and William Phillips. (We already met Cohen-Tannoudji as one of the authors of the book from which Aspect learned his quantum theory.)

We will mention a very important use of laser cooling as the first stage in one of the most important achievements in physics in the past few years. This is the production of Bose–Einstein condensation. In Chapter 1, we met the idea of Bose–Einstein statistics. In quantum theory, these statistics are obeyed by particles with integral spin, which are called bosons, while particles with half-integral spin, such as electrons, obey Fermi–Dirac statistics and are called fermions.

An important property of Bose–Einstein statistics is that the particles are inclined to congregate in the same state. This is in complete contrast to Fermi–Dirac statistics where at most one particle can be in each state; this latter in turn is just a statement of the well-known Pauli exclusion principle, which is all-important in building up the atoms of the Periodic Table from electrons in the various electron states.

A strange feature of Bose–Einstein statistics is that, at the lowest temperatures, virtually all the particles in a system of bosons will be in the ground state. This was a prediction from the mid-1920s, and was given the name the Bose–Einstein condensation, but for many years it was thought to be completely impossible even to dream of confirming it experimentally. Then in 1997 it was achieved in dilute gases of alkali-metal atoms. As we have said, the first stage was achieved by laser cooling. Further cooling is then required, and this is achieved by forced evaporation; this is rather like the way in which a cup of tea cools as it evaporates, since the faster molecules escape preferentially. Yet more Nobel Prizes were awarded for this work, to Eric Cornell, Wolfgang Ketterle and Carl Wieman in 2001.

We shall meet the ion trap later in the book in connection with quantum computation, but for the moment we arc looking at Rowe's work. In her experiment, pairs of beryllium ions are initially confined in an ion trap. Two laser beams are then used to stimulate a transition that entangles the ions. The resulting wave-function is a superposition of two states; in the first, both ions are in what we may call level A, while in the second they are both in level B.

Which level any ion is in is determined by probing it with circularly polarized light from a laser beam acting as a detector. During this pulse, ions in level A scatter a large number of photons and about 64 may be detected by a photomultiplier, but photons in level B scatter very few photons. In this way statistics may be built up on the occupancies of each level, and using these statistics it is possible to state that Bell's inequalities are violated by about eight standard deviations. But the really important point is that a measurement result is achieved for every entangled pair of ions, and so no fair-sampling assumption is required, and the authors may justifiably claim that the detector loophole has been closed.

This paper was published in *Nature*, one of the very most important journals carrying papers across the whole of science, and in the same issue, Grangier summarized the progress that had been made and stressed that all the hard-won experience developed while studying

Bell's inequalities was exactly what was required for work on quantum information theory.

Another route to closing the detector loophole has been suggested by the group of Franco De Martini in Rome. They have made good progress using an extremely brilliant source of entanglement. Again they are aiming at detecting all entangled pairs.

It is confidently expected that Nobel Prizes will be awarded for the work on Bell's inequalities if and when they are conclusively shown to be violated. This will be said to have been achieved when an experiment is carried out with all loopholes closed simultaneously. This has not been done so far. The Innsbruck experiment that closed the locality loophole was actually rather poor for detection efficiency, while the Rowe experiment left the locality loophole wide open.

There have fairly recently been suggestions for experiments using homodyne detection and continuous variables that may be able to produce results that close all the loopholes at once. If so this would be the conclusion of a process of steadily greater technical expertise and sophistication over a period of about 40 years since the first experiments. Then there will be the fun of selecting those who may receive the call to Sweden. It is certainly to be hoped that those from the earlier generations are not left out.

References

1. H. Rauch, 'Towards More Quantum Complete Neutron Experiments', in: R. Bertlmann and A. Zeilinger (eds), *Quantum [Un]speakables; From Bell to Quantum Information* (Berlin: Springer-Verlag, 2002), pp. 351–73.
2. D. M. Greenberger, 'The History of the GHZ Paper', in: R. Bertlmann and A. Zeilinger (eds), *Quantum [Un]speakables; From Bell to Quantum Information* (Berlin: Springer-Verlag, 2002), pp. 281–6.
3. D. M. Greenberger, 'The Neutron Interferometer as a Device for Illustrating the Strange Behaviour of Quantum Systems', *Reviews of Modern Physics* **55** (1983), 875–905.
4. Y. Hasegawa, R. Loidl, G. Badurek, M. Baron and H. Rauch, 'Violation of a Bell-like Inequality in Single-Neutron Interferometry Experiments', *Journal of Modern Optics* **51** (2004), 967–72.
5. S. Basu, S. Bandyopadhyay, G. Kar and D. Home, 'Bell's Inequality for a Single Spin-½ Particle and Quantum Contextuality', *Physics Letters A* **279** (2001), 281–6.
6. P. G. Kwiat, K. Mattle, H. Weinfurter, A. Zeilinger, A. V. Sergienko and Y. H. Shih, 'New High-Intensity Source of Polarisation-Entangled Photon Pairs', *Physical Review Letters* **75** (1995), 4337–41.
7. G. Weihs, 'Bell's Theorem for Space-like Separation', in: R. Bertlmann and A. Zeilinger (eds), *Quantum [Un]speakables; From Bell to Quantum Information* (Berlin: Springer-Verlag, 2002), pp. 155–62.

8. N. Gisin, 'Sundays in a Quantum Engineer's Life', in: R. Bertlmann and A. Zeilinger (eds), *Quantum [Un]speakables; From Bell to Quantum Information* (Berlin: Springer-Verlag, 2002), pp. 199–207.
9. E. S. Fry and T. Walther, 'Atom Based Tests of the Bell Inequalities—The Legacy of John Bell Continues', in: R. Bertlmann and A. Zeilinger (eds), *Quantum [Un]speakables; From Bell to Quantum Information* (Berlin: Springer-Verlag, 2002), pp. 103–17.
10. R. A. Bertlmann, 'Magic Moments: A Collaboration with John Bell', in: R. Bertlmann and A. Zeilinger (eds), *Quantum [Un]speakables; From Bell to Quantum Information* (Berlin: Springer-Verlag, 2002), pp. 29–47.
11. M. A. Rowe, D. Klepinski, V. Meyer, C. A. Sackett, W. M. Itano, C. Monroe and D. J. Wineland, 'Experimental Violation of a Bell's Inequality with Efficient Detection', *Nature* **409** (2001), 791–4.

12
Bell's theorem without inequalities

GHZ: Greenberger, Horne, and Zeilinger

Until now all the arguments concerning local realism that we have met have involved two-particle inequalities, and many, many more such inequalities have been written down. (That is why we mostly talk of Bell's inequalities rather than Bell's inequality.) Experiments to check these inequalities were necessarily lengthy and not particularly straightforward, since a substantial experimental run was required, and the results had to be expressed statistically, as we have, of course, seen. Until the middle to late 1980s it was taken for granted that this was the way things were. But since then two important arguments have been put forward that demonstrate the possibility of testing Bell's theorem without the use of inequalities.

The first is always written as GHZ, which stands for Greenberger, Horne and Zeilinger. We left these three in the previous chapter around 1986, with the MIT neutron laboratory closed down, Zeilinger returned to Vienna and Greenberger also in Vienna for a semester plus a summer as a Fulbright Fellow. Over the past few years Greenberger [1] had thought about three-particle (and, in fact, four-particle) correlations quite deeply, and he had discussed them at times with both Horne and Zeilinger, but he had not done any systematic work on the problem.

One reason why he felt that three-particle correlations in particular might be of interest is that positronium annihilation to two photons was an accepted, though, as we have seen, very difficult, method of investigating Bell's theorem. However, positronium does occasionally decay to three rather than two photons, which makes the three-particle case possibly of experimental interest.

Anyway when Greenberger reached Vienna, he decided to try to create Bell-type inequalities for multi-particle states. It was a fairly wide-ranging problem, and to begin with he had little idea of how to proceed; it was all rather hit-and-miss. In fact he never came up with a systematic approach to follow, but gradually experience told him how to produce inequalities that were more and more restrictive. And each day he told Zeilinger about his more and more interesting results. He phoned Horne as well, though Horne was a little less appreciative. It must be remembered that he was an old-timer in the subject from the days of CHSH and CH, and at that time large numbers of papers with new inequalities, which were exciting, at least to the authors, were sent to various journals, and the great majority came to Clauser, Horne or Shimony, or perhaps Bell, to referee. Horne was unlikely to be entranced any longer by new inequalities.

And then one day Greenberger came up with an approach leading to a relation that was totally restrictive—it had no freedom at all. It was not an inequality—it was an equality! While all the inequalities had required many measurement results, the new equality could, in principle, decide between local realism and quantum theory with a single measurement. This came as a great shock, but discussion with Zeilinger helped clarify what was behind the totally unexpected result, and they agreed that it should be published. Though it had been Greenberger who wrote down the result, it clearly came from the intense collaboration between the three of them, so it would be published in a paper authored by Greenberger, Horne and Zeilinger—GHZ as it is always known.

This was 1986 but it was not to be published yet. For some reason, perhaps because they did not quite realize how important and interesting others would find the result, perhaps just because circumstances conspired to frustrate the writing, perhaps because it became part of a much bigger project, the planned paper would not see the light of day for some time! (In 2001 Greenberger wrote that: 'The state of the field in those days was such that it didn't bother us at all that this result and all the others we had sat around for several years and we hadn't published them.')

First, Zeilinger mentioned to Greenberger that he and Horne had done some related work and this could be included in the same paper. Then Greenberger left Vienna and continued his sabbatical travelling around Europe. After that he returned to New York and his teaching duties, and the writing was not begun. In fact it was not until 1988 when he returned to Europe for a semester, working on a Humboldt Fellowship at the Max Planck Institute for Quantum Optics at Garching, that he had sufficient time to start the planned article. And then it seemed that the project had grown so much that it was too large. Greenberger rang Zeilinger in Vienna to tell him that he had started writing and already had 70 typed pages. But he felt it was turning into a monograph and he didn't want to write a monograph! Yet again the work stopped.

This time though Greenberger did not keep the discovery a secret. He spoke about the theorem at various conferences, including a workshop on quantum problems at Erice. At Erice, David Mermin from Cornell, an eminent solid state physicist who had also made major contributions to the foundations of quantum theory, was in the audience. Greenberger started by announcing that he would show a version of Bell's theorem with all probabilities either 0 or 1. Mermin was totally convinced that this was impossible and, since he was tired, did not bother to listen.

But other people did. One was Michael Redhead, Professor of Philosophy at the University of Cambridge, England, and author of a famous book on quantum foundations from the philosophical point of view. He clearly found Greenberger's conference talk absorbing because, he and two colleagues, Jeremy Butterfield and Rob Clifton, wrote a substantial paper claiming to make the GHZ theorem more rigorous. (This paper [2] would eventually be published in 1991.)

Greenberger was somewhat shocked at receiving a preprint of this paper, and he knew that other groups were quoting the result also, so he phoned Horne and Zeilinger to say that people were beginning to use their result before they had actually published it! By now he was back in America, and he took the opportunity to give a talk at a symposium organized by Michael Kafatos at Fairfax University in Virginia, and to write a short paper [3] for the proceedings. It was publication of a sort, but clearly it would still be essential that a paper was published in a regular journal. Horne and Zeilinger in fact remained unaware that the theorem had been published even at this level.

Meanwhile, in yet another twist of fate, Mermin had also read the Cambridge paper. He felt somewhat irritated that mathematical complications and philosophical niceties were being produced even before the actual argument had been examined and either accepted or rejected. However, reading the preprint he became convinced that he should have made the effort to listen in Erice, and that GHZ had made an important discovery—a great simplification of Bell's ideas and theorem.

Mermin wrote an occasional column called 'Reference Frame' in the magazine *Physics Today*, which is not a regular journal, but a publication of the American Physical Society; it is sent to all members of the society, which means most physicists in the USA, and many beyond. It does not contain research articles, but news and contributed reviews on particular topics in physics. Mermin's columns are always titled 'What's wrong with this/these...?', and he picks out various aspects of science or the administration of science with which he disagrees. In June 1990, his column was titled 'What's wrong with these elements of reality?', and he explained the GHZ argument, giving full credit, of course, to G, H and Z. By this means very many physicists came to know of the work.

Still there was no formal publication. The three main men decided they must at last go ahead, but, a little wary of the philosophical slant of the Cambridge paper, decided they needed to recruit their own philosopher to the writing team, and of course the only possibility could be Shimony. Thus the first (though see the paragraph below for a minor clarification) publication of GHZ had four authors, listed in alphabetical order as Greenberger, Horne, Shimony and Zeilinger [4]. The paper was published in the *American Journal of Physics*; this is primarily a journal of teaching methods and items of physics particularly appropriate in the teaching context, but at the time it did publish many of the papers on quantum foundations.

There was one further complication. Mermin had become extremely interested in the GHZ theorem, and he had reached an understanding of its underlying meaning that was perhaps rather deeper than those who had discovered it. For example, Greenberger had been inclined to discuss it in terms of a sequence of two decays from an initial single particle. Mermin was able to show that this superstructure is unnecessary, and he was able to explain the situation in a more fundamental and general way. He published his own version, giving, of course, full credit to the original discoverers. His paper [5] was also published in the *American Journal of Physics*, and, with the very last twist of fate, it was published in the same volume, but actually a little earlier than the publication of GH(S)Z.

After that lengthy account of how the theory came into being, we now proceed (at last) to how it actually works. We shall be quite close to Mermin's presentation. It can be said that the argument is more technical than most of the rest of the book. It is actually just a little complicated rather than actually difficult to follow, and it should be possible for the reader to get at least a general idea of what is being achieved.

The argument is presented in terms of states of a spin-½ system, and the states we shall mostly make use of are $\left|s_z = \frac{1}{2}\right\rangle$ and $\left|s_z = -\frac{1}{2}\right\rangle$.

Let us start with $\left|s_z = \frac{1}{2}\right\rangle$. If a spin-½ system is in the state $\left|s_z = \frac{1}{2}\right\rangle$, then we know that a measurement of the observable s_z, the z-component of spin, is certain to give the value $+\hbar/2$. We can usefully express this result in different terms. We can define an operator $\hat{s}_z$ that measures the value of s_z, and using this operator, we can write $\hat{s}_z\left|s_z = \frac{1}{2}\right\rangle = +\hbar/2\left|s_z = \frac{1}{2}\right\rangle$.

Yet another way of discussing the same basic result is to say that, since operating on the state $\left|s_z = \frac{1}{2}\right\rangle$ with the operator $\hat{s}_z$ just gives us a constant, $+\hbar/2$, times the original state $\left|s_z = \frac{1}{2}\right\rangle$, we can say that $\left|s_z = \frac{1}{2}\right\rangle$ is an eigenstate of $\hat{s}_z$ with the eigenvalue $+\hbar/2$.

Now let us consider the state $\left|s_z = -\frac{1}{2}\right\rangle$. If the spin-½ system is in the state $\left|s_z = -\frac{1}{2}\right\rangle$, we know that a measurement of the observable s_z, the z-component of spin, is certain to give the value $-\hbar/2$. When we define the operator $\hat{s}_z$ that measures the value of s_z, we can write $\hat{s}_z\left|s_z = -\frac{1}{2}\right\rangle = -\hbar/2\left|s_z = -\frac{1}{2}\right\rangle$.

Again we can say that, since operating on the state $\left|s_z = -\frac{1}{2}\right\rangle$ with the operator $\hat{s}_z$ just gives us a constant, $-\hbar/2$, times the original state $\left|s_z = -\frac{1}{2}\right\rangle$, we can say that $\left|s_z = -\frac{1}{2}\right\rangle$ is an eigenstate of $\hat{s}_z$ with the eigenvalue $-\hbar/2$.

Just to stress the significance of eigenstates, let us consider, rather than one of the eigenstates of s_z, a general state of the system, which is $\left\{c_1\left|s_z=\frac{1}{2}\right\rangle+c_2\left|s_z=-\frac{1}{2}\right\rangle\right\}$, where c_1 and c_2 are general constants. Then if we operate on this general state with $\hat{s}_z$, we will get $\hbar/2\left\{\left|s_z=\frac{1}{2}\right\rangle-\left|s_z=-\frac{1}{2}\right\rangle\right\}$. In other words, this general state is *not* an eigenstate of $\hat{s}_z$ because when we operate on it with $\hat{s}_z$, we do not just get a constant times the original state. (The exceptions are if either c_1 or c_2 is zero, in which cases, of course, we have the eigenstates we have already discussed.) In turn this means that when we perform a measurement of s_z on a system in this state, we do *not* know what answer we will get; we may get either $\hbar/2$ or $-\hbar/2$ (except, of course, for the exceptional cases).

Analogous to $\hat{s}_z$, we have operators $\hat{s}_x$ and $\hat{s}_y$. These operators have their own eigenstates, which are different from those of $\hat{s}_z$. It is essential to realize that these operators, just because they *are* operators rather than constants, do not commute. In other words, for example, $\hat{s}_x\hat{s}_y$ is not equal to $\hat{s}_y\hat{s}_x$. The fact that $\hat{s}_x$ and $\hat{s}_y$ do not commute is intimately connected to the fact that they have different eigenstates. If they did commute they would have common eigenstates.

For the GHZ argument, it makes things easier, though it is certainly not essential, to work not with $\hat{s}_x$, $\hat{s}_y$ and $\hat{s}_z$, but with related quantities, $\hat{\sigma}_x$, $\hat{\sigma}_y$ and $\hat{\sigma}_z$, where $\hat{\sigma}_x$ is equal to $(2/\hbar)\,\hat{s}_x$ and similarly for $\hat{\sigma}_y$ and $\hat{\sigma}_z$. Then the relationships we have just discussed take the slightly simpler forms $\hat{\sigma}_z\left|s_z=\frac{1}{2}\right\rangle=\left|s_z=\frac{1}{2}\right\rangle$, and $\hat{\sigma}_z\left|s_z=-\frac{1}{2}\right\rangle=-\left|s_z=-\frac{1}{2}\right\rangle$. In other words the eigenvalues are +1 and −1.

We can now begin to explain the GHZ argument itself. It analyses three spin-½ particles, *a*, *b* and *c*, which interact and then fly off to detectors all in the horizontal plane. They are thus in an entangled state. Rather than just $\hat{\sigma}_z$, we have $\hat{\sigma}_z^a$, $\hat{\sigma}_z^b$ and $\hat{\sigma}_z^c$, and similarly for the *x*- and *y*-components. Operators for components of σ for different spins always commute, so that, for example, $\hat{\sigma}_x^a\hat{\sigma}_y^b=\hat{\sigma}_y^b\hat{\sigma}_x^a$, but those for the same spin do not do so, or in other words, $\hat{\sigma}_x^a\hat{\sigma}_y^a$ is not equal to $\hat{\sigma}_y^a\hat{\sigma}_x^a$.

GHZ now consider three compound operators, $\hat{\sigma}_x^a\hat{\sigma}_y^b\hat{\sigma}_y^c$, $\hat{\sigma}_y^a\hat{\sigma}_x^b\hat{\sigma}_y^c$ and $\hat{\sigma}_y^a\hat{\sigma}_y^b\hat{\sigma}_x^c$. Since each of these compound operators contains individual operators for all three spins, one would naturally expect that these compound operators would *not* commute. However, GHZ chose these operators specifically so that they do, in fact, commute. It may be said that the various aspects of the non-commutativity are arranged to cancel out.

As we noted above, the fact that the three commute implies that they will have simultaneous eigenstates. In fact there will be eight of these. For each, the eigenvalues for the three compound operators will be respectively ±1, ±1 and ±1, but we may restrict our argument to the case where all three are +1.

In our proof we are concentrating very much on operators and eigenvalues rather than the actual states involved. Nevertheless, even though we do not need to do so, it is interesting to write down the state that we are considering. In fact it is:

$$\left(1/\sqrt{2}\right)\left\{\left|s_z^a = \tfrac{1}{2},\ s_z^b = \tfrac{1}{2},\ s_z^c = \tfrac{1}{2}\right\rangle - \left|s_z^a = -\tfrac{1}{2},\ s_z^b = -\tfrac{1}{2},\ s_z^c = -\tfrac{1}{2}\right\rangle\right\}.$$

This state is called the GHZ state. Clearly it is highly entangled.

Now we come to the logical part of the argument. Since the three operators commute, we may say that we can simultaneously measure $\sigma_x^a\sigma_y^b\sigma_y^c$, $\sigma_y^a\sigma_x^b\sigma_y^c$ and $\sigma_y^a\sigma_y^b\sigma_x^c$, and in each case we will obtain the result unity. Now since each of the three compound operators is a product of three operators for different spins, these individual operators commute and we can simultaneously measure each individual component of σ for each spin.

This in turn implies that of the measurements on individual spins, for each of the three products there must be an even number of results of -1, and an odd number of results of $+1$. So if we know two of the results, we can predict with certainty the third. For example, looking at the first of the products, if σ_x^a and σ_y^b are both the same, either both $+1$ or both -1, then σ_y^c must be $+1$.

Thus there is a clear analogy with EPR. The three particles are entangled but have separated, and measurements on two of them give information on the third. If locality is assumed, then all three results must be pre-determined, or in other words there are hidden variables and quantum theory is not complete, as EPR argued.

Then each spin must have instructions as to what result will be obtained for a measurement of spin along the x-direction, *and* for a measurement along the y-direction. Each component of σ for each spin must be ± 1, and each of the products, $\sigma_x^a\sigma_y^b\sigma_y^c$, $\sigma_y^a\sigma_x^b\sigma_y^c$ and $\sigma_y^a\sigma_y^b\sigma_x^c$, must be $+1$.

It turns out that there are four distinct ways of achieving this. The first three are very obvious: (a) all results for any spin and setting are $+1$; (b) one spin gives the result $+1$ and the others -1 whatever the setting; (c) each spin gives $+1$ for the x-component, and -1 for the y-component. The fourth (d) is not quite so obvious but it can easily be checked: one spin gives $+1$ for the x-component and -1 for the y-component, while the other two spins do the opposite. Of course it is rather more difficult to show that these *are* the only solutions, but checking through the various possibilities shows that this is indeed the case.

Now let us consider the product of the three x-components, $\sigma_x^a\sigma_x^b\sigma_x^c$. It is easy to see from the previous paragraph that, for all four cases, (a) to (d), $\sigma_x^a\sigma_x^b\sigma_x^c$ is equal to $+1$. This may be said to be the final result of this EPR-type analysis for the GHZ state.

Now let us turn to the quantum approach. It comes to a totally opposite conclusion. It is indeed the case that the GHZ state is an eigenstate of the operator $\hat{\sigma}_x^a\hat{\sigma}_x^b\hat{\sigma}_x^c$, but the corresponding eigenvalue is not +1 as the EPR-type of analysis demands but −1. Clearly the GHZ argument is a total negation of the possibility of combining agreement with quantum theory and retention of local realism. It has, of course, been obtained with an exact relationship rather than with an inequality.

Related to this is a rather more subtle result. Let us take the product of the three compound operators that we considered above: $(\hat{\sigma}_x^a\hat{\sigma}_y^b\hat{\sigma}_y^c)(\hat{\sigma}_y^a\hat{\sigma}_x^b\hat{\sigma}_y^c)(\hat{\sigma}_y^a\hat{\sigma}_y^b\hat{\sigma}_x^c)$. Let us multiply this out, and first of all let us be extremely naïve and forget problems with commutation. This is another way of saying: follow an EPR-type of prescription. Then each y-component of $\hat{\sigma}$ enters squared, and we can then replace each square by unity, and just leave it out. When we follow this procedure, we end up with $\hat{\sigma}_x^a\hat{\sigma}_x^b\hat{\sigma}_x^c$. This is exactly what the EPR approach gave us before. Each of the three triple products has an eigenvalue of +1, and so does $\hat{\sigma}_x^a\hat{\sigma}_x^b\hat{\sigma}_x^c$. This all seems very good, and if it were true would be the triumph of local realism.

Unfortunately it is not true. If we evaluate $(\hat{\sigma}_x^a\hat{\sigma}_y^b\hat{\sigma}_y^c)(\hat{\sigma}_y^a\hat{\sigma}_x^b\hat{\sigma}_y^c)(\hat{\sigma}_y^a\hat{\sigma}_y^b\hat{\sigma}_x^c)$ properly according to the rules of quantum theory, or in other words taking account of non-commutativity, the result is not $\hat{\sigma}_x^a\hat{\sigma}_x^b\hat{\sigma}_x^c$ but $-\hat{\sigma}_x^a\hat{\sigma}_x^b\hat{\sigma}_x^c$, and this represents not the triumph but the total failure of local realism.

GHZ—experimental proof

Like earlier proofs of Bell's theorem, the GHZ argument showed that there was a mismatch between quantum theory and local realism. The next question was as always—Which is right? The GHZ argument had finally been published towards the end of the 1980s, and it took another decade for an experimental test to be made.

The experiment was performed by Dirk (Dik) Bouwmeester and co-workers [6] in Zeilinger's group in Innsbruck. (Bouwmeester is now at the University of California at Santa Barbara.) We have mentioned before that the techniques developed in the early tests of Bell's theorem were often pivotal when experiments demonstrating the central results of quantum information theory started to be performed. The present case, though, is an example of the reverse procedure. The experiment to examine the GHZ state was based on experiments to demonstrate quantum teleportation and entanglement swapping, which were carried out by the same group, and which we shall meet in Chapter 19.

The experiment, like many used in a large number of areas of physics, uses a type of Mach-Zehnder interferometer, an instrument invented

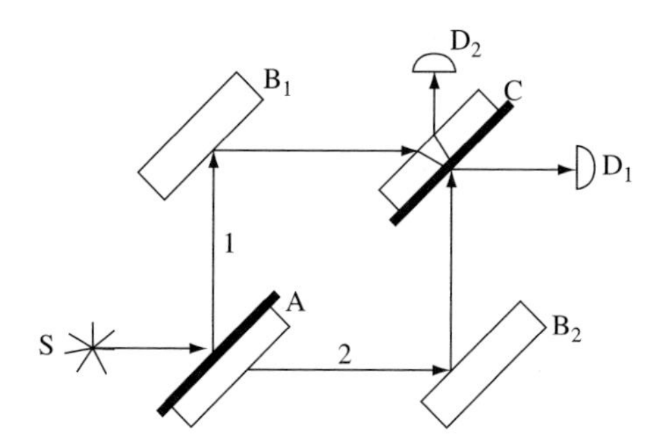

Fig. 12.1 The Mach–Zehnder interferometer. Light from source S has two possible paths through the apparatus, labelled 1 and 2. The diagram shows the possible paths to detectors D_1 and D_2, but if both the paths are exactly the same length, all photons are detected at D_1.

in the nineteenth century for optical measurements. The original from of the Mach-Zehnder interferometer is shown in Figure 12.1.

Light from source S is incident on a glass slab A, the top surface of which is half-silvered so that half the light incident on it is reflected and half transmitted. As we have already seen, this device is called a beam-splitter. Thus beams of light of the same intensity proceed along paths 1 and 2, and are reflected at mirrors at the lower side of glass slab B_1 and the upper half of glass slab B_2, eventually recombining in slab C, the lower surface of which is half-silvered. Two paths away from slab C are available, ending at detectors D_1 and D_2.

If the two paths between A and C are exactly the same length, all the light reaches detector D_1, none D_2. To understand this we need to know that the phase of a light wave changes by 180° or half a period when it is reflected at a medium of higher index of refraction than the medium the light is travelling in. However, there is no change of phase if the medium has a lower index of refraction than the medium the light is travelling in. Thus the phase changes in this way when a beam of light travelling in air is reflected at a mirror surface, but not if it is travelling in the mirror and is reflected at the same surface.

So let us consider the light travelling to D_1 via B_1. It will have suffered two phase changes of 180°, first at the upper surface of A, then at the lower surface of B_1. The light also travelling to D_1 but via B_2 will also have had two phase changes due to reflection at the upper surface of B_2 and the lower surface of C. Therefore the two beams of light will be in phase and there will be constructive interference.

However, let us now consider the light travelling to D_2. If it travels via B_1, it undergoes just two changes of phase; the reflection at the lower surface of C causes no change of phase since it is a reflection at a material of lower index of refraction. The light travelling via B_2 undergoes only one change of phase. Thus the light travelling along the two paths will be

180° out of phase, or, in fact totally out of phase, so the two beams will interfere destructively and therefore no light will be detected at D_2.

This is, of course, just the basic setup. The original purpose of the experiment was to insert in one of the beams a material whose optical properties may be investigated by analysing the change in detection rates at the two detectors. More recently, as the photon nature of light has become known and understood, the Mach-Zehnder interferometer has been used in a variety of experiments to elucidate and make use of the unexpected facets of its behaviour that result from this growth in understanding.

The central idea of the Bouwmeester experiment is to transform two pulses of polarization entangled photons into three entangled photons in the GHZ state and a fourth photon, known as the trigger photon. The GHZ entanglement is observed only under the condition that both the trigger photon and the entangled photons are actually detected, so we can say that the act of detection carries out two tasks—it projects the photons into a GHZ state and also performs a measurement of that state.

The apparatus is sketched in Figure 12.2. Pairs of polarization entangled photons are generated by down-conversion in the state $\left(1/\sqrt{2}\right)\left(\left|H_1, V_2\right\rangle - \left|V_1, H_2\right\rangle\right)$. Here H_1 means that the polarization of the first photon is in the horizontal direction and so on.

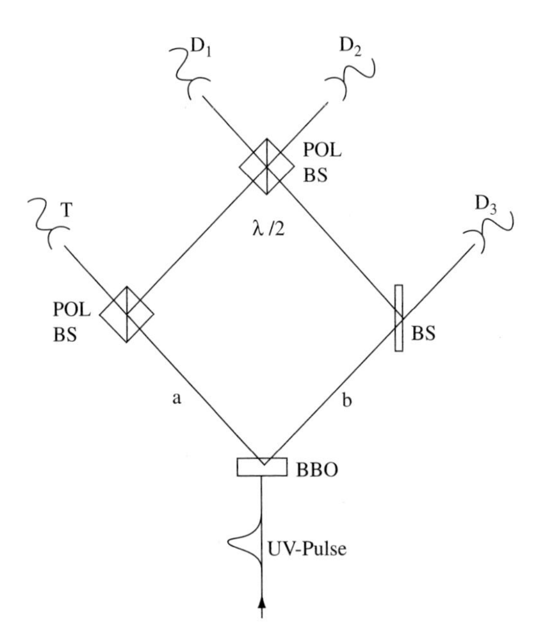

Fig. 12.2 The experiment that demonstrated the GHZ argument that was carried out by Dirk Bouwmeester and colleagues. With one photon registered at the trigger detector T, the three photons registered at D_1, D_2 and D_3 are in the GHZ state. [From Ref. [6]]

The path along arm *a* continues to a polarizing beam-splitter, where *V* photons are reflected and *H* photons are transmitted to a detector T, which lies behind an interference filter. Arm *b* continues towards a 50/50 polarization-independent beam-splitter. From each beam-splitter, one output is sent to a final polarizing beam-splitter, though in between the two polarizing beam-splitters, vertical polarization is rotated 45° by a $\lambda/2$ plate. The remaining three output arms continue through interference filters to single-photon detectors, D_1, D_2 and D_3. Taking into account all losses, the probability of collection and detection of a photon is about 10%.

Let us consider the case where all four photons are detected, one by each detector. The photon detected by detector T must be an *H* photon since it has been transmitted by the initial polarizing beam-splitter. Its partner must be a *V* photon, so it will travel along the *b* arm, and it will reach either D_2 or D_3, since if it reaches the final polarizing beam-splitter it must be reflected.

For the other pair, the photon travelling along arm *a* does not reach T, since, as we have said, we are looking at cases where each photon reaches a different detector. Therefore it must begin as *V* and is reflected at the first polarizing beam-splitter. The $\lambda/2$ plate means that it reaches the second polarizing beam-splitter as a linear combination of *V* and *H*, and it may finally be detected at D_2 as an *H*, or at D_1 as a *V*. Its partner is an *H* and will reach either D_1 or D_3.

Thus the possible states are $|H_1, H_2, V_3\rangle$ and $|V_1, V_2, H_3\rangle$, but at the moment they are, at least in principle, distinguishable. It might be possible to determine, for example, that T triggered simultaneously with either D_2 or D_3, and at a very slightly different time from the triggering of the other two detectors, indicating which was the partner of the photon triggering T, and hence which of the possible states we have constructed.

This is the purpose of the interference filters, which create an uncertainty in the timing of any detection longer than the duration of the ultraviolet pulse that causes the original down-conversion. Once it is no longer possible to distinguish between the two states, we must say that we do not have one or the other of the states, but a linear combination of both. Thus the state of the system is $\left(1/\sqrt{2}\right)\left(|H_1, H_2, V_3\rangle + |V_1, V_2, H_3\rangle\right)$, which is the GHZ state for this polarized photon system.

The efficiency for a single UV pulse to yield this fourfold coincidence is very low, around 10^{-10}, but with the rate of pulses around 7.6×10^7 s^{-1}, one event is registered every 150 s, just sufficient for the experiment to be performed.

Interaction-free measurement

We now want to move on to another method of demonstrating Bell's theorem without inequalities that was produced by Lucien Hardy [7] in

1992. This method is based on a thought-experiment of Avshalom Elitzur and Lev Vaidman [8], which itself was known for some time but was not published in a regular journal until 1993. It introduced what is known as *interaction-free measurement* and the thought-experiment is often called the 'bomb problem'.

In this problem, we must imagine that we have a stock of bombs, each of which operates via a sensor that explodes the bomb if any light at all is incident on it. However, in a consignment of bombs, it is known that some sensors are defective. For bombs with defective sensors, light may pass through them without exploding the bomb and without affecting the light in any way. The problem is to produce a stock of bombs in good working order.

Classically there is no solution. Shining light at the bombs will not solve the problem, since those bombs that do not explode will have been found to be defective, while those that were originally good will have been exploded and thus be good no longer! Quantum mechanically, though, we may take advantage of interaction-free measurement using a Mach-Zehnder interferometer.

We first return to Figure 12.1, and consider a non-transmitting object inserted into path 1. We consider a photon sent into the interferometer. There is then a probability of ½ of the photon reaching this object and being absorbed. However, there is also a probability of ½ of the photon travelling along path 2. In the latter case, there is no interference; the photon reaches beam-splitter C along a single path, and there will be a probability of ¼ of the photon leaving C along a horizontal route and being detected at D_1, but an equal probability of ¼ of it leaving in a vertical direction and being detected at D_2.

So now let us look at the bomb problem itself. The bomb is placed in path 1. A photon is sent through the interferometer. If the sensor is not working, both paths of the interferometer are available for the photon, and it is bound to be detected in D_1. However, if the bomb is working, there is a probability of ½ of the bomb blowing up. But there are also probabilities of ¼ of the photon being detected in D_1 or D_2.

So in this single run, if detector D_1 clicks, that tells us nothing; it could happen whether the bomb is good or defective. If the bomb explodes, that tells us that the bomb was good, but is obviously good no longer. However, if D_2 clicks, that tells us that the bomb is good and remains unexploded.

That, as we said, is for a single run. If D_1 clicks, we may send another photon through, and obviously there is an additional probability of determining that the bomb is and remains good. In fact we may speak of sending through photons until the bomb explodes or D_2 clicks. It is easy to calculate that the total probability of an explosion after some run is ⅔, while the probability of success, that is to say of obtaining a bomb to put on our stockpile, is ⅓.

The bomb has been studied without interacting with it, and it is natural to call the technique 'interaction-free measurement'. It is of considerable interest in its own right, but here we are interested in it mainly as a central component of Hardy's experiment.

Hardy's experiment

Lucien Hardy gained his PhD in 1992 working on quantum foundations at Durham University under the supervision of Euan Squires.

Born in 1933, Euan Squires himself was extremely interesting from the point of view of this book. He graduated from Manchester University in 1956, and at quite an early age became an authority on Regge pole theory, which was a hot topic in elementary particle physics. He spent a period working at Harwell, where he wrote some research papers with Bell, and, after other short spells in Cambridge, Berkeley and Edinburgh, became Professor of Applied Mathematics in Durham University in 1964 at the very early age of 31. Despite his chair being in maths, in spirit he was always a physicist, and he built up a large group of particle physicists in Durham.

Then his career took a change of path. In 1985 he wrote a popular book on quantum mechanics [9], and became aware, as Bell had remarked a quarter of a century earlier, and probably for the first time, that many foundational issues on quantum theory were unsolved and interesting. He studied them for the rest of his life, and commenced by publishing in 1986 a book [10] very different to the one he had written the year before, looking at the challenges at the heart of quantum theory rather than its successes. In 1990 he wrote a third book suggesting how these problems might be solved by explicitly including consciousness as the centre of quantum measurement [11].

He had an outward-going personality and brought many others into the fold of quantum foundations. He was the main spirit behind a series of annual conferences run in the UK from the mid-1990s that enabled those interested in the area to interact fruitfully. Squires was a good athlete, keen on all sports, but sadly died while playing cricket in 1996.

Following his PhD, Hardy himself had spells at Maynooth in Ireland, Innsbruck, Durham again and Rome (La Sapienza). He was at the Centre for Quantum Computation in Oxford from 1997 to 2002, when he moved to the Perimeter Institute in Waterloo, Canada, an institution set up to carry out research in the most important areas of theoretical physics.

In the work described in this section, which he carried out in 1992 while at Durham, he used two Mach-Zehnder interferometers as shown in Figure 12.3, one for electrons and one for positrons. The paths intersected so that there was a possibility of annihilation of electron and positron. Indeed, taking a particle approach, if the electron and the positron do meet then they are certain to annihilate each other.

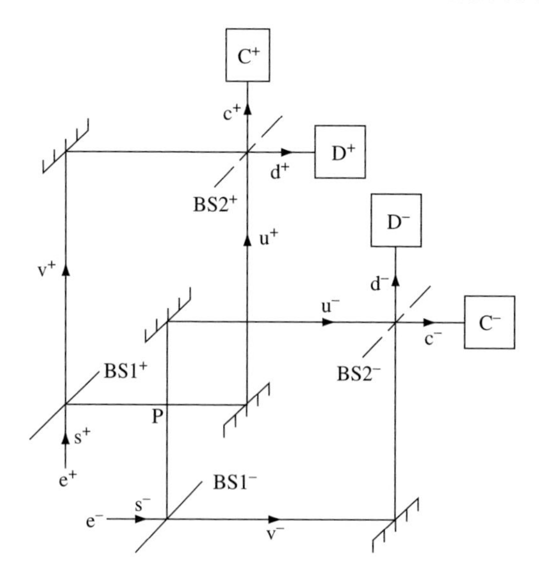

Fig. 12.3 Hardy's thought-experiment. Two Mach–Zehnder interferometers, one for positrons and one for electrons, are arranged so that if a positron takes the path u^+ and an electron takes the path u^-, they meet at P and annihilate each other. [From Ref. [7]]

Each interferometer has an input, $s^\pm$, two possible paths through the interferometer, $u^\pm$ and $v^\pm$, and two possible output modes, $c^\pm$ leading to detectors $C^\pm$, and $d^\pm$ leading to $D^\pm$. Each interferometer contains two beam-splitters, $BS1^\pm$ and $BS2^\pm$. The former, $BS1^\pm$, are always present, but either or both of $BS2^\pm$ may be removed. This means that there are four possibilities: (i) both are absent; (ii) $BS2^+$ is present and $BS2^-$ is absent; (iii) the reverse of (ii); and (iv) both are present.

If each interferometer were isolated, or alternatively if only electrons *or* positrons were sent into the system, and if both $BS2^+$ and $BS2^-$ are present, then as we saw earlier, electrons would leave along path c^+ and detector C^+ would click, or positrons would leave along path c^- and detector C^- would click. However, with the overlapping of paths, annihilation will occur in ¼ of the cases, and the detectors $D^\pm$ may click.

We may say that annihilation occurs if paths u^+ and u^- are followed, so we need to trace out the experimental possibilities for the other three cases: (a) paths u^+ and v^- are followed; (b) v^+ and u^-; and (c) v^+ and v^-.

The easiest case to consider is (i) because with both $BS2^+$ and $BS2^-$ absent, particles moving along $u^\pm$ register at $C^\pm$, and particles moving along $v^\pm$ register at $D^\pm$. Since we must consider the possibilities (a), (b) and (c), we find that clicks may occur at C^+ and D^-; or at C^- and D^+; or at D^+ and D^-, each with possibility ¼, the remaining ¼, of course, corresponding to annihilation. Clicking at C^+ and C^- should *not* occur.

Cases (ii), (iii) and (iv) are rather more complicated since we need to take into account interference at whichever second beam-splitters are present. In fact for (ii), the probability of C^+ and C^- clicking is ⅛; the probability of C^+ and D^- clicking is ½; the probability of C^- and D^+ clicking is ⅛; the probability of annihilation is ¼ as always, and D^+ and D^- do not click together. For (iii), we just interchange – and + in the results for (ii). For (iv), the probability of C^+ and C^- clicking is $\frac{9}{16}$; the probabilities of C^+ and D^-, C^- and D^+, and D^+ and D^- clicking are each $\frac{1}{16}$, with again annihilation taking up the remaining ¼.

We now try to build up a hidden variable model of the experiment by letting the state of the electron–positron pair, before any measurements are done, be described by a hidden variable λ, which takes a different value for each pair. We have a choice of two measurements to make on each particle, with the beam-splitter BS2 present or absent; we represent the two choices by 0 and ∞, respectively. Thus, just as an example, $C^+(\infty, \lambda) = 0$ would tell us that, with $BS2^+$ absent, the positron is not detected at C^+.

We wish our hidden variables to be local, and this implies two conditions. The result obtained for one type of particle depends neither on the measurement made on the other type of particle, or in other words whether the second beam-splitter is present for the other type of particle, nor on the result obtained on the other type of particle. (This refers to parameter independence and outcome independence, respectively.) This means we may write the overall result for the experiment as a product of the result for electrons and the result for positrons.

Thus from our results for (i) above, we may immediately write

$$C^+(\infty, \lambda)C^-(\infty, \lambda) = 0 \quad (12.1)$$

for all runs. Now from our summary of results for case (ii), we see that if the positron is detected at D^+, then the electron must be detected at C^-; similarly from (iii), if the electron is detected at D^-, then the positron must be detected at C^+. So we may write:

$$\text{If } D^+(0, \lambda) = 1, \text{ then } C^-(\infty, \lambda) = 1 \quad (12.2a)$$

$$\text{If } D^-(0, \lambda) = 1, \text{ then } C^+(\infty, \lambda) = 1. \quad (12.2b)$$

From the result for (iv), we may now say:

$$D^+(0, \lambda)D^-(0, \lambda) = 1 \text{ for } \tfrac{1}{16} \text{ of runs.} \quad (12.3)$$

Let us now consider one of these runs. Equations (12.2a) and (12.2b) then immediately tell us that $C^+(\infty, \lambda)C^-(\infty, \lambda) = 1$ for these runs, but this is immediately in total conflict with Eq. (12.1). So Hardy's argument has shown us that the results of quantum theory could not be reproduced using the simple local hidden variable approach.

Later Hardy was able to adapt the proof to rule out local realism for all entangled states, except, rather strangely, maximally entangled states such as the EPR state. Again the argument was without inequalities.

This theoretical argument was tested by de Martini and co-workers [12] in Rome (La Sapienza) with Hardy's support. Naturally they did not use beams of positrons, but instead used entangled photons produced by parametric down-conversion. The polarizations were rotated by amounts ψ_1 and ψ_2, and the photons were then injected into two different sides of a polarizing beam-splitter. On emergence, the polarization of each photon was rotated by further angles θ_1 and θ_2, and the photons then passed through further polarizing beam-splitters and then on to detectors.

The scheme was extremely flexible as all the four angles of reflection could be varied independently, so a complete picture of the behaviour of the system could be built up. Local realism was violated by much of the detail of the experimental results, which agreed well with the quantum predictions.

References

1. D. M. Greenberger, 'The History of the GHZ Paper', in: R. Bertlmann and A. Zeilinger (eds), *Quantum [Un]speakables; From Bell to Quantum Information* (Berlin: Springer-Verlag, 2002), pp. 281–6.
2. R. K. Clifton, M. L. G. Redhead and J. N. Butterfield, 'Generalization of the Greenberger-Horne-Zeilinger Algebraic Proof of Nonlocality', *Foundations of Physics* **21**(1991), 149–84.
3. D. M. Greenberger, M. A. Horne and A. Zeilinger, 'Going beyond Bell's Theorem', in: M. Kafatos (ed.), *Bell's Theorem, Quantum Theory and Conceptions of the Universe* (Dordrecht: Kluwer, 1989).
4. D. M. Greenberger, M. A. Horne, A. Shimony and A. Zeilinger, 'Bell's Theorem without Inequalities', *American Journal of Physics* **58** (1990), 1131–43.
5. N. D. Mermin, 'Quantum Mysteries Revealed', *American Journal of Physics* **58** (1990), 731–4.
6. D. Bouwmeester, J.-W. Pan, M. Daniell, H. Weinfurter and A. Zeilinger, 'Observation of Three-Photon Greenberger-Horne-Zeilinger Entanglement', *Physical Review Letters* **82** (1999), 1345–9.
7. L. Hardy, 'Quantum Mechanics, Local Realistic Theories and Lorentz-Invariant Realistic Theories', *Physical Review Letters* **68** (1992), 2981–4.
8. A. Elitzur and L. Vaidman, 'Quantum Mechanical Interaction-Free Measurements', *Foundations of Physics*, **23** (1993), 987–97.
9. E. Squires, *To Acknowledge the Wonder* (Bristol: IOP, 1985).
10. E. Squires, *Mystery of the Quantum World* (Bristol: IOP, 1986).
11. E. Squires, *Conscious Mind in the Physical World* (Bristol: IOP, 1990).
12. D. Boschi, S. Branca, F. De Martini and L. Hardy, 'Ladder Proof of Nonlocality without Inequalities: Theoretical and Experimental Results', *Physical Review Letters* **79** (1997), 2755–8.

13
The new age

Times have changed

We started Part II of the book decidedly in what we call the First Quantum Age. At this time, using quantum theory was taken for granted, but thinking *about* quantum theory was totally off limits. One was expected to believe that Bohr and von Neumann had managed to wrap that up decades before. We saw the stigma on those who dared to challenge this conventional wisdom.

While we have seen the hugely important work that had been responsible for the coming of the new age, and which dated from the 1960s and even earlier, we may say that the beginning of the New Quantum Age, defined as a period when new approaches were tolerated, or even mildly encouraged, may perhaps be dated from the early 1980s and Aspect's work. (As we said earlier, this does not imply that Aspect's work, though certainly extremely important, was necessarily more important than that of others, but it did rather capture the attention of the physics community.)

And happily today there is much more freedom for physicists to think about such issues, to carry out research, theoretical and experimental, on the foundations of quantum mechanics, and to express opinions on these matters without losing any credibility with the rest of the physics community. So far in this part of the book, we have studied almost entirely the work of Bell in this context, and the theory and experiments that flowed from it, but much else of interest has occurred.

Of course the most important and certainly the best-known aspect of this relaxation and new interest is that the study of quantum fundamentals has in a sense spawned the hugely important 'hot topic' of quantum information theory, and this is the subject of Part III of the book. However, there have also been new interpretations of quantum theory, and demonstrations of new and interesting, and in some cases useful, results of the quantum formalism, and we sketch these in this chapter.

Novel quantum interpretations

The Copenhagen interpretation held virtually total sway, of course, from 1928, when it was enunciated by Bohr until at the very least the 1970s and to a considerable extent much later. To say this implies not only that the vast majority of physicists accepted Copenhagen, which was obviously their right, but also that any novel ideas would be squashed without mercy, to do which should have been nobody's right.

In Chapter 6, we met two interpretations from the 1950s, both of which were naturally received with considerable hostility at the time, the Bohm theory and many worlds. We discussed the Bohm theory in much more detail in Chapter 8, stressing its great importance to Bell, first because it was a clear demonstration that hidden variable theories could be constructed, and secondly because its non-locality led him to his famous proof. The Bohm theory continues to be of interest but it is actively investigated by rather a small number of physicists.

Many worlds theories are discussed and used rather more, partly perhaps because they may be all things to all people! The most popular way of thinking about this type of theory is that at any measurement (or, for some authors, at any interaction) the world does literally split into a number of distinct worlds, each with its own properties and future behaviour. Probably the person whose work has most clearly espoused this view has been the eminent field theorist, Bryce DeWitt [1]. (In contrast, Everett's own writings [1] are capable of being read in several different ways.)

For some, this is as much as is required to play a particular conceptual role. An important example is cosmology, which cannot be handled by either Bohr's approach or that of von Neumann. Bohr speaks only in terms of a quantum experiment carried out by an external observer, while von Neumann also requires an external observer to collapse his wave-function. Since the entire Universe clearly has no external observer, many, perhaps even most, astrophysicists, including notably Stephen Hawking [2], speak in terms of a many worlds interpretation and, in a tour de force, Hawking and James Hartle [3], for example, have been able to discuss the wave-function of the entire Universe.

Staying with cosmology, we may mention the *cosmological anthropic principle* [4–6]. This is a name for the remarkable fact that, if many of the most fundamental constants of physics, in particular the charges and masses of elementary particles, had values only extremely slightly different from those they have in our universe, life in any form would not be possible. For example, we would have only hydrogen in the Universe; no nuclear fusion to form more massive atoms would have occurred.

There are two ways of coping with this rather dramatic piece of information. The first is to assume that the world was made specifically for life (not necessarily human life as we know it) to exist, in other words a very strong argument for design. John Polkinghorne [7], the elementary particle physicist and former Cambridge professor who became an Anglican priest, takes this position. The other solution is to say that there are many worlds, parallel or perhaps sequential, and that these fundamental constants may differ in the different worlds. If there are enough of these worlds, there will be one or a number that allow for the emergence of life, and it is obvious that we must live in one of these special worlds, and we will know nothing of the myriads of others in which life would be impossible.

Probably the most famous proponent of many worlds has been David Deutsch [8] (Figure 13.1), a polymath, whose famous book, *The Fabric of Reality*, gives his views across the whole of science. He describes what he calls his four principal strands of explanation of the Universe: epistemology, evolution, computation and quantum theory, and for each strand he has one 'hero', whose all-important views, Deutsch says, were ignored for far too long. For the first three strands, his heroes are Karl Popper, Richard Dawkins and Alan Turing, respectively. For quantum theory, his hero is Everett; Deutsch argues that a straightforward reading of the Schrödinger equation gives directly the Everett position, and that any other reading is deliberately contrived to give what is, wrongly in Deutsch's view, usually regarded as required by our common experience.

We shall meet Deutsch in Chapter 17 as the founder of quantum computation, and he regards quantum computation as completely inexplicable without the idea of many worlds backing it up. In one particular application of quantum computation, Shor's algorithm, which we shall

Fig. 13.1 David Deutsch [Corbis]

meet in Chapter 17, he claims that the computer resources used are around 10^{500} times those that seem to be available. For Deutsch, it is obvious that the algorithm is using the resources of 10^{500} universes.

However, it certainly should not be thought that to work in quantum computation implies a belief in many worlds, or indeed any great interest in interpretational aspects of quantum theory. Andrew Steane, for example, who is a well-known worker in quantum computation, has replied that a quantum computer requires only one universe, and that its power just comes from following the usual formalism of quantum theory.

Obviously some people find the idea of millions of universes unpalatable or perhaps just unscientific, and they may use the basic Everett formalism but understand it in different ways. One set of arguments [9, 10] may be called *many-views* or *many-minds* rather than many worlds. It accepts Everett's position that the evolving wave-function contains many branches, all with different experimental results, and different states of mind of the observer. Squires [9] has suggested that this may be quite reasonable. Being part of the wave-function, the observer cannot stand outside the world to observe it. The observer in any branch of the wave-function must be totally unaware of other branches, because, by making a conscious observation and obtaining a particular result, 'I' becomes that part of the wave-function corresponding to that particular result.

Another line of argument has been put forward mainly in Oxford, in particular by Deutsch [11], Simon Saunders [12] and David Wallace [13]. Wallace in particular argues that, defining the *ontology* of a theory as what the theory says actually exists, one may distinguish between the *fundamental ontology* and the *higher level or emergent ontology*. The fundamental ontology is what is written into the theory from the start, and, for quantum theory, this is the wave-function. The emergent ontology includes our many worlds. Thus, Wallace says, it is irrelevant and inappropriate to ask, as much of the discussion of many worlds does, where in the formalism of quantum theory one may find the many worlds. They are just not there!

Where then *can* we find them? We may start by recognizing that classical physics does not give a satisfactory account of physics. Indeed we may stop at that point, but if we do so, or in other words if we do not attempt to provide any interpretation of quantum theory, we shall certainly miss most of what the formalism should be telling us. To interpret quantum theory means just to render it understandable from a classical viewpoint. (Here there are certainly echoes of a Bohrian point of view.)

Thus to attempt to interpret quantum theory implies searching for patterns in the mathematics of the theory that relate to macroscopic objects—people, cats and so on. Such a process inevitably leads to a description in terms of many worlds. As Wallace says, this description is complete, since the entire wave-function can easily be obtained from it. The worst that may be said about it is that it remains a little arbitrary—one could, at least in principle, imagine obtaining different types of patterns—entirely different emergent ontologies.

Wallace sums up the present state of the many worlds interpretation as follows. He says that there have been three main classes of interpretation. The first, due mainly to de Witt, in which worlds are explicitly added to the formalism, he says, has no remaining defenders. The Everett approach, he says, was designed explicitly to avoid having to add anything to the formalism. The two surviving types, he says, are the two we have just discussed—many minds and the approach of Wallace himself.

We will here make one further point. This is that, in strong contrast to the 1950s, physicists and indeed philosophers are able to think and publish on such seemingly arcane matter as many worlds, and still remain in the mainstream of their disciplines. Presumably there will still be some who rather disapprove of such speculation, but there is definitely a greater willingness at least to tolerate even ideas of which one does disapprove, rather than regarding those who discuss them as being totally beyond the pale.

We have said that Bell was fascinated by the Bohm interpretation and promoted it whenever possible. Yet it must be said that towards the end of his life he became perhaps almost more enthusiastic about another approach, that of GianCarlo Ghirardi (Figure 13.2), Alberto Rimini and Tullio Weber, always known as GRW [14].

Fig. 13.2 GianCarlo Ghirardi [© Zentralbibliothek für Physik, Renate Bertlmann]

Most of the other interpretations of quantum theory we have looked at and shall look at—Bohm, many worlds, ensembles, consistent histories, knowledge—take the Schrödinger formalism for granted, and they use diverse conceptual arguments to explain or explain away the usual difficulties of quantum theory. In stark contrast, GRW add an extra term.

Bell had signalled his desire for such a theory in a paper he wrote in 1986, in which he agreed with Bohr that the very small and the very big must be described in totally different ways, in quantum and classical terms respectively, but suggested that the division should not be sharp, and certainly not 'shifty'. (The word 'shifty' refers to the fact that, according to complementarity, any division between quantum and classical is in principle in an arbitrary position.) Surely, he suggested, big and small should merge smoothly, and the merging should be described not by vague words but by precise mathematics.

The most likely way to achieve this was to add non-linearity to the Schrödinger equation. The Schrödinger equation is said to be linear in a mathematical sense—each term in the equation contains the wave-function to first power only. The result of this is that different solutions of the equation may be added together to give new solutions, giving the typical quantum behaviour of interference. Provided one steers clear of measurement, this mathematical behaviour is perhaps the main glory of quantum theory, but equally, when we do consider measurement, it leads to the measurement problem, the fact that in most circumstances quantum theory does not predict a unique result for any particular measurement.

Thus it seemed sensible to try out the possibility of adding a non-linear term to the Schrödinger equation. Clearly this would be a tricky business! While allowing the problem over measurement to be overcome, it must avoid altering to any extent any of the results of the standard theory that agree so well with experiment.

In addition there seems at first sight to be a fundamental argument against such a scheme. When deterministic non-linear corrections are added to quantum theory, Gisin and Polchinsky were able to show that faster-than-light signalling became allowed in an EPR-type experiment, and this was looked on as damning for such theories. Steven Weinberg, winner of the Nobel Prize for physics for his work on field theories and elementary particles, who was also very interested in fundamental ideas of physics, suggested that this fact, even more than the precise experimental verification of linearity, suggests that quantum mechanics is the way it is because any small change would lead to logical absurdities. However, what he neglected to take into account was that, as we said, the argument applied only to *deterministic* corrections, not to corrections that were stochastic (statistical). Stochastic corrections were the basis of the GRW theory.

This theory was published in 1986 and it was greeted by Bell with great acclamation at the Schrödinger Centenary Conference held in

London the following year. Indeed the actual GRW paper was rather lengthy and detailed, and Bell was able to re-express it in simpler terms, making the ideas behind the theory and also its considerable achievements rather clearer.

The intent of the GRW theory was as follows. For von Neumann, the system follows the Schrödinger equation apart from a collapse of wave-function at measurement, but as we have stressed, since the term 'measurement' is ill-defined, the von Neumann approach is conceptually unsatisfactory. In contrast GRW added a collapse process to the Schrödinger equation, but they describe it as a *spontaneous localization* process. At random intervals, the spread of the wave-function of a particle or composite system of particles is reduced to a very small range, d, not precisely zero because that would cause problems with the Heisenberg principle, but perhaps around 10^{-7} m. The intervals were random but nevertheless were described by an additional term in the Schrödinger equation, a stochastic term of course, so from this point of view, the scheme was more respectable mathematically that that of von Neumann.

Such an idea was not quite new. Philip Pearle had come up with several similar schemes, but GRW had a new idea, and it was this that delighted Bell. Their idea was that the rate of localization depended on the size of the composite particle involved. In fact the average time between localizations was given in the theory by τ/N, where N is the number of particles in the composite system, and τ is a new parameter, which GRW estimate could be about 10^{15} s, which is about 10^{8} years.

It is clear that this is a potentially winning scheme. For a single particle, where we do not wish to have collapse, we will have to wait 10^{8} years for it to occur. But for macroscopic objects, which we hope will behave classically, N may be 10^{20} for example, and then collapse will occur in about 10^{-5} s, exactly as required. As Bell says: 'Schrödinger's cat is not both dead and alive for more than a split second.'

Much detailed work has been carried out to improve the basic GRW idea. There have been attempts to make it relativistic. Ghirardi, Pearle and Rimini have produced a theory called *continuous state diffusion* (CSL), in which the spontaneous localization takes place continuously rather than randomly, and many other detailed accounts of steadily increasing generality and rigour have been written, by GRW themselves and their co-workers, and by Gisin and Lajos Diósi.

Two problems have caused considerable discussion. First, the localization produced by GRW in fact leaves functions with tails extending to infinity, so the associated particles cannot be regarded as truly classical. Second, an obscure feature of the localization is that it may produce situations where we know that n objects are in a box, but the probability of finding all n in the box is extremely small; this is known as the 'counting marbles problem' or, more eruditely, the 'enumeration anomaly'.

GRW have defended their approach against these complaints with passion and with considerable success.

Of considerable importance had been the degree to which the results of the GRW theory agree with those of standard quantum theory. Of course it is essential that, where these results are already confirmed by experiment, the results obtained from GRW do not differ essentially, and this is the case. But for GRW to move beyond being regarded as an interesting set of ideas, and to be taken fairly seriously as a genuine alternative to the standard theory, it is also essential that it makes some predictions clearly at variance with those of regular quantum theory that may then be checked experimentally. We may remember the comparison between Bohm theory, which made no predictions clearly different from those of quantum theory and so was rather generally disregarded, and Bell's work, which made clear-cut suggestions running contrary to the standard ones, and received a growing amount of attention.

For GRW, such predictions haves not been easy to obtain. The first task it to determine the acceptable ranges of τ and d, and then, on the basis of these values, one needs to look for results differing from those emerging from quantum theory. At that point one may compare with experiment. Pearle, in particular, has put in much effort in this task, but not so far with any genuine success. Where significant differences between the results of GRW (or, in fact, CSL) and standard quantum theory have emerged, they are still much too small to be tested experimentally.

Among other interpretations of quantum theory are the use of ensembles, which we discuss in the following chapter, and what we may call a knowledge interpretation, which we will meet at the beginning of Part III of this book. Here we discuss fairly briefly a few of the many other interpretations that have attracted attention.

First we discuss *consistent histories* and *decoherent histories*. Robert Griffiths [15] developed the idea of the consistent history, stressing the kind of grotesque wave-functions that arise in conventional quantum theory. For example, we can consider a particle reaching a beam-splitter, from which two paths emerge, each ending in a particle detector. Standard theory will say that, once the particle has passed the beam-splitter, the wave-function is straddled between two regions, each quite localized, but separated by a steadily increasing distance. Only when one of the detectors registers a particle does the wave-function collapse to the region of that particular detector.

A histories approach aims to build up a more complete and perhaps more 'sensible' account or 'history' of what has occurred. It will say that the particle was always on one particular path, the one leading to the detector that registers the particle. In general, a history may be described as a sequence of properties of the system holding at a sequence of times.

However, of course it is immediately obvious that this type of approach will not always succeed. Suppose, for example, that, as we saw for the neutron interferometer in Chapter 11, we manage to recombine the beams going along both paths so that interference occurs. In that case it would be quite incorrect to decide that the particle travelled along one path or the other; rather it must have 'sampled' both paths in order that we may have interference.

And indeed, for Griffiths, a consistent history may be described only in the first case where each path leads to its own detector, not in the second, where interference is possible, The term 'consistent history' essentially means that such a history does not interfere with any other, but evolves independently. Roland Omnès [16] introduced the term 'decoherent history', which is technically different from 'consistent history' but conveys much the same general idea. (The term 'decoherence' is discussed in some detail later in this chapter and is also important in Part III of this book.)

Another set of ideas on the same general lines was produced by Gell-Mann (Figure 13.3) and Hartle [17], and indeed they went rather further than Griffiths or Omnès by aiming to produce a complete cosmology for the universe. They explain how the observer, which they refer to as an IGUS, or *information-gathering-and-utilizing-system*, has evolved through history. They explain that the IGUS must concentrate on situations of decoherence because only in this way can useful predictions be made, and these are essential for survival.

Fig. 13.3 Murray Gell-Mann (second from left) with other Nobel Laureates from Caltech: (from left) Carl Anderson, Max Delbruck, Richard Feynman and George Beadle [courtesy of University of Houston]

Histories certainly provide an interesting window onto quantum theory, but it is not clear that they solve its central problems. While the individual proponents of the group of approaches certainly vary somewhat in their beliefs, it seems that most of them at least consider that the approach provides a rigorous and conceptually problem-free study of quantum theory that ends up by rediscovering Bohr's insights. It may be described as a reformulation of Copenhagen. While a histories approach does enable statements to be made that transcend what orthodox quantum theory would allow, as we saw for the beam-splitter case, it would seem that there is no claim that the measurement problem is actually solved.

Gisin and Ian Percival (Figure 13.4) [18] have presented an interesting approach to quantum theory which they call the *quantum state diffusion model*. They consider an individual quantum system interacting with an environment. Thus the Schrödinger equation is again modified by the addition of a term corresponding to this interaction with the environment. This extra term is stochastic or random because the actual interaction is itself random. This all sounds very much like the GRW approach, but conceptually the two are very different. As we have said, for Gisin and Percival the extra term added to the Schrödinger equation is an attempt to represent a physical interaction so we are not moving outside standard quantum theory; GRW very deliberately move away from standard quantum theory.

Gisin and Percival produced some very interesting results in graphical form. An interesting example is for a two-state system with the additional terms in the Schrödinger equation relating to

Fig. 13.4 Ian Percival [© Zentralbibliothek für Physik, Renate Bertlmann]

absorption and stimulated emission processes. The results show the system spending time in each state with transitions between them. These transitions are not instantaneous but typically take a few percent of the time that the system stays in one or other of the states. Another beautiful example is for a system that starts off with its wave-function a superposition of states with one, three, five, seven and nine photons, with the extra term now representing a measurement. In the measurement, there is a clear move to states with one *or* three *or* five *or* seven *or* nine photons, exactly the result that a measurement should produce.

A critic might argue that, interesting as the results may be, we are only getting out of the analysis exactly what we are putting in, or in other words the procedure is rather *ad hoc*. Nevertheless it does present a very convincing picture of some of the most central aspects of quantum theory in action.

We now move on to stochastic interpretations of quantum theory. These have a very long history going right back to Schrödinger in 1932, and they attempt to demonstrate a direct mathematical analogy between quantum theory and a classical stochastic theory. The most common classical theory to be used in this way is the theory of Brownian motion. This is the way by which molecules were (indirectly) observed for the first time; particles large enough to be seen under the microscope and suspended in a liquid are seen to move erratically under bombardment from the (invisible) molecules of the liquid. While many physicists and mathematicians contributed to the theory of the Brownian motion, the most important contribution was from Einstein in his *annus mirabilis* of 1905.

If quantum theory could essentially be put into a classical format, one must suspect that the specifically quantum problems must have been removed, and indeed the analogy was quite successful—up to a point. Actually some interesting physical insights were obtained, one of the most interesting concerning the quantum zero-point energy. Whereas in the standard approach as discussed in Chapter 2, this is a rather formal and mathematical aspect of the theory, in stochastic quantum theory it plays a central part as a real physical field. An atomic system exchanges energy with this field and so its own energy must fluctuate randomly. So the zero-point field causes the quantum fluctuations, but equally the fluctuations create the field in a self-consistent way.

Nevertheless there were fundamental mismatches between quantum and classical methods, as was pointed out by Schrödinger in his first work on the topic. In quantum mechanics the wave-function is (mathematically) complex, while in classical theories the analogous quantity, the subject of the equation, is (mathematically) real. In quantum theories the amplitude obeys the appropriate differential equation, and the probability density is the square of the amplitude, while in classical theories it is the probability itself that obeys the equation. And in quantum theory there must be quantization, which of course is not present in the classical case.

But the greatest difficulty for stochastic theories has come from Bell's theorem. The physicist who has made the greatest contribution to developing stochastic interpretations over several decades has been Edward Wilson [19], and for him, as for virtually all other contributors, the idea has been to create a local realistic theory. Indeed several of those whom we have called realists, for example Santos and Marshall, have constructed their own stochastic theories. But a stochastic theory is essentially a hidden variable theory and so must respect Bell's theorem.

Indeed Wilson had realized this independently when he applied his method to entangled systems. In 1985 he wrote: 'I have loved and nurtured... stochastic mechanics for 17 years, and it is painful to abandon it. But the whole point was to construct a physically realistic picture of microprocesses, and a theory that violates locality is untenable.'

Actually others may not feel quite so strongly about locality. Guido Bacciagaluppi [20] has more recently produced an up-to-date account of the properties of stochastic theories. However, it is understood that they cannot be regarded as a classical picture of quantum theory, but as a theory similar to that of Bohm, being broadly realist but non-local.

Roger Penrose (Figure 13.5) [21] has suggested that the problem involving the localization of energy in general relativity and the measurement problem in quantum theory may be linked so as to provide a solution to both. He notes that in a typical measurement process, before any collapse we may have a linear combination of two states of the detecting

Fig. 13.5 Roger Penrose [© Zentralbibliothek für Physik, Renate Bertlmann]

device in different positions, thus with different gravitational fields. His suggestion is that when two different space-time geometries are superposed, a spontaneous collapse of the wave-function occurs. One possibility of the scale of difference required for collapse is that the difference between the superposed geometries must be at least 10^{-35} m; this is the so-called Planck scale at which quantum gravity effect become significant.

Rough estimates suggest that the collapse time for a neutron or proton would be about 10^7 years, while for a water droplet of radius 10^{-6} m, the comparable time would be 0.05 s. Thus it may be said that, like GRW, Penrose's ideas seem to be able to cater both for the macroscopic and the microscopic cases. They are still, though, extremely speculative!

Two other approaches can claim to demonstrate collapse only in the limit of an infinite number of constituent particles in an apparatus. The *many Hilbert space approach* of S. Machida and Mikio Namiki [22] seeks to derive wave-function collapse from statistical fluctuations in the measuring apparatus. They argue that different incoming particles interact with a given detector in different macroscopic states, and so undergo random shifts of phase. Even for a detector of less than an infinite size, they argue that a 'partial collapse' occurs, but it is difficult to understand what this means in practical terms.

The Nobel Prize-winning physicist Ilya Prigogine [23], with several collaborators, has put forward an approach to quantum measurement involving irreversibility at a fundamental level. It seems, though, that transformation between reversibility and non-reversibility occurs more from philosophical principle than mathematical analysis, which perhaps makes the approach somewhat unconvincing.

What would probably please both Einstein and Bell about the present situation compared to that of fifty years ago is not *just* the freedom to present one's ideas, and the number of those doing so, but the fact that most of the interpretations are based on good old-fashioned physics, rather than the conceptual vagueness of the Copenhagen interpretation. Certainly the present situation is the far healthier one.

Environmental decoherence

The topic of this section, decoherence caused by interaction of the system with the environment, is definitely *not* an interpretation of quantum theory in the sense that, contrary to what is implied in many discussions, it is *not* an attempt to solve the measurement problem. Rather it is a successful attempt to explain the classicality of our common experience.

To explain this, we may go right back to one of Einstein's complaints about the Copenhagen approach, which was mentioned in a letter to Born in 1953. This is the suggestion that quantum theory tells us that the wave-function of even a macroscopic object would spread with time, and would dephase, so that the object itself would not remain localized. In other words it would lose its classical integrity.

Calculations in fact show that, if one considers a typical star, one of the examples suggested by Einstein, his argument actually does not hold. The wave-function of a star similar to the sun would dephase by only a very small amount even in the lifetime of the Universe. However, if we take a C_{60} molecule, often called a *buckyball*, we find that the spreading of the wave-function becomes appreciable; in fact the change in the width of the wave-function becomes larger than its original width, in a time around 10^{-3} s. Clearly Einstein had a good case. The buckyball, incidentally, may perhaps be defined as *mesoscopic*, intermediate between microscopic and macroscopic.

The reaction of the supporters of Copenhagen to such problems for truly macroscopic systems was either to declare that the system was macroscopic and so classical by definition or perhaps to talk in terms of persistent observation or measurement causing consistently repeated collapse, and therefore no sustained change in the wave-function. It might be added that mesoscopic systems were not discussed at the time as they were probably thought to be impossible to manufacture or study. Their very existence must cause drastic difficulties for Copenhagen, which allows only for microscopic and macroscopic systems, each with their own totally distinctive behaviour, and with nothing at all allowed in between.

The physicist most responsible for eventually addressing Einstein's problem was Hans-Dieter Zeh [24, 25], who was born in Germany in 1932. After completing his doctorate at the University of Heidelberg, he spent some time working in institutions in the United States before returning to Heidelberg to take up a lectureship in 1966. It was in late 1967 or 1968 that he wrote his first paper on environmental decoherence in quantum systems. We shall discuss the physics of this work shortly, but here we note the rather disastrous effect working on this topic had on his career. Yet again we note the stigma of wanting to work on the foundations of quantum theory.

Zeh himself called the period from the late 1960s to the early 1980s the 'dark ages of decoherence'. This first paper was rejected by several journals. *Nuovo Cimento* rejected it on the basis of a referee's report which said that: 'This paper is completely senseless. It is clear that the author has not fully understood the problem and the previous contributions in the field.' The editor of *Die Naturwissenschaften* commented much more mildly that some points in the paper were not clear for non-experts, but he still rejected the paper!

Even worse was to follow. Hans Jensen, Nobel Prizewinner for physics in 1963, was the most powerful member of the Physics Department in Heidelberg. Unable to understand Zeh's paper, he made what was, for Zeh at least, the crucial mistake of sending it to Léon Rosenfeld. Rosenfeld was the last person from whom Zeh could expect a fair response. He had been Bohr's assistant in the 1930s, and remained his most devoted follower, completely convinced that Bohr and complementarity had completely resolved all the problems of quantum theory for all time. After working in Bohr's institute for a considerable period, he occupied chairs in Utrecht and Manchester before returning to Copenhagen at NORDITA. For a period before and after Bohr's death in 1962, he could be described as 'Bohr's representative on earth' and perhaps as 'more Catholic than the Pope'!

He replied in totally derogatory terms, describing Zeh's work as 'a concentrate of wildest nonsense', and trusting that the paper had not been distributed around the world with Jensen's blessing. Jensen replied to Rosenfeld, attempting, as he saw it, to salvage Zeh's career, and an exchange of letters took place through the early part of 1968, but Rosenfeld's criticism was only slightly moderated. Jensen felt obliged to warn Zeh that any further research in this field would end his career, but Zeh, convinced that his career was almost certainly over in any case, continued to work in the same area.

Olival Freire [24] has quoted Zeh's views expressed in 1980 in a letter to John Wheeler:

> I have always felt bitter about the way Bohr's authority together with Pauli's sarcasm killed any discussion about the fundamental problems of the quantum...I expect that the Copenhagen interpretation will some time be called the greatest sophism in the history of science, but I would consider it a great injustice if—when some day a solution shall be found—some people claim that 'this is of course what Bohr always meant', only because he was sufficiently vague.

Yet all was not doom-and-gloom for Zeh, and once again the person to be thanked was Wigner. He was a member of the editorial board of the new journal *Foundations of Physics*, and recommended Zeh's paper for publication in this journal, where it appeared in the very first number [26]. He also invited him to talk at the important 1970 Varenna summer school, which has been mentioned several times already in this book. Indeed, although Wigner himself supported a different interpretation of quantum theory, one based on human consciousness, in his opening address at the School he described Zeh's ideas, buttressed by those of Everett, as an interesting interpretation of quantum theory, and one of six possible solutions to the problem.

At Varenna, Zeh also met Bell, who was also very supportive, although perhaps neither particularly liked the other's approach to quantum theory. In an interview with Freire, Zeh remarked that Bell was: 'very sympathetic, and he always asked the right questions. This was already a very great thing that there was somebody who was critical against the mainstream and put his finger on the right things. Only very few people did that.'

After all that—unfortunately mostly unsavoury—history, let us now turn to the physics of environmental decoherence. We first note that in quantum theory the term 'coherence' means that the wave-function is a sum of terms, which may relate to classically rather different situations. In some circumstances this may lead to the typical quantum behaviour of interference. An example might be a photon or a neutron in an interferometer. The photon or neutron may sample different paths, paths that are quite far apart in space, but may still be brought together and interfere.

However, what we may call *dephasing* or *decoherence* may readily take place. To keep things simple, let us imagine a system with two states—ϕ_1 and ϕ_2. Here it is probably easiest to think of these as two quantum states of an atom corresponding in most cases to two different energy levels. Let us imagine that at $t = 0$, the wave-function of a particular atom is ϕ_1. We now suppose, which will be a very common situation, that the atom is subject to what in quantum theory is called a small perturbation, probably an interaction with another atom. This may lead to the wave-function evolving into $a_1\phi_1 + a_2\phi_2$, where a_1 is initially 1 and gradually decreases in time, while a_2 is initially 0 and gradually increases in time.

Indeed there will be circumstances in which the wave-function may become precisely periodic: after a certain time τ has elapsed, the wave-function will have become equal to ϕ_2, after another period τ it will be back to ϕ_1 and so on. This is oscillatory behaviour at a particular frequency, and it is certainly the very epitome of coherent behaviour.

This is quite acceptable and interesting for microscopic systems, but for macroscopic systems, it gives rise to precisely the kind of difficulty Einstein objected to. We may suppose that the equivalent of ϕ_1 for this case is a perfectly proper classical state. ϕ_2 may also be such a state, but the superposition $a_1\phi_1 + a_2\phi_2$ will be expected to be non-classical—classically quite incongruous superpositions of totally different classical states.

It was Zeh who answered Einstein's objection by showing how macroscopic states lost this coherence, or in other words how we have *decoherence* for these states. Decoherence implies that the different states are essentially decoupled so that interference can no longer take place. From the 1980s, Erich Joos did important work in collaboration with Zeh [27], and Wojciech Zurek [28] also became very prominent working on decoherence.

Zeh showed that a macroscopic quantum system such as a measuring device can never be isolated from its *environment*. Such a system can never be treated as a closed system, but must be treated as an open system freely interacting with its environment. The term 'environment' naturally includes external influences but may also include internal factors, the behaviour of internal variables of the system that do not appear in its quantum description.

Briefly it may be mentioned that the effect of the environment cannot be neglected on the grounds that interaction between system and environment may be, in some sense, small. We are treating the environment as a perturbation, and there is a very elaborate theory called *perturbation theory* which tells us how to do this. Our system is macroscopic, and this implies that it will have a very large number of states so that the actual states will be extremely close together. Perturbation theory tells us that, when the states of the system are close together, even a small perturbation may have an extremely large effect. So we certainly cannot ignore the environment!

And the effect of the environment means that we must consider combined states of the system itself and the environment, $\phi_1\Phi_1$ and $\phi_2\Phi_2$, where Φ_1 and Φ_2 are states of the environment. Now we must imagine the environment to be extremely complicated, so that Φ_1 and Φ_2 are extremely complicated functions of many variables, and no simple perturbation may cause coherence between Φ_1 and Φ_2. This means that if initially the state of the environment is Φ_1, it will remain as Φ_1; there will be no decoherence. And of course if the states of system and environment is $\phi_1\Phi_1$ at $t = 0$, the presence of the Φ_1 prevents the system itself from acting in a coherent way. Environmental decoherence is indeed obtained.

Zurek describes the approach as creating 'a fixed domain of states in which classical systems can safely exist, but superpositions of which are extremely unstable…Events happen because the environment helps define a set of stable options that is rather small compared with the set of possibilities available in principle.'

It must be emphasized that decoherence is definitely *not* a solution to the measurement problem. For if the initial state of the system is, for example, the non-classical superposition $\left(1/\sqrt{2}\right)(\phi_1\Phi_1 + \phi_2\Phi_2)$, then it will not change; certainly it will not collapse down to one of the classical states, $\phi_1\ \Phi_1$ or $\phi_1\ \Phi_2$, which is what would be required for a collapse of the wave-function.

We may describe this in a different way. Let us think of an ensemble of systems with wave-function $\left(1/\sqrt{2}\right)(\phi_1\Phi_1 + \phi_2\Phi_2)$. For collapse to take place, the final state would have to be a mixture of states with wave-function $\phi_1\Phi_1$ and states with wave-function $\phi_2\Phi_2$. This is a change from a superposition to a mixture, but a fundamental law of quantum theory tells us that Schrödinger evolution can *never* transform a superposition into a mixture.

Zeh himself concocted a full interpretation of quantum theory, which included a solution of the measurement problem, but only by adding in an Everett-type scheme. A problem of the Everett scheme, which we have not mentioned previously, is that it does not *just* allow a set of classical states, each in its own world, to coexist, rather than just one classical state in one world as von Neumann would predict. It also seems to allow other types of situation, where different worlds see *non-classical* states.

The Zeh plus Everett combination says that first decoherence gives the required set of classical states, and then secondly Everett allows these to coexist. In fact we omitted to mention that Wallace's account of many worlds, which we sketched earlier in this chapter, also starts with a set of decoherent states; so the whole scheme again requires both decoherence and many worlds. And it is obvious, of course, that decoherence plays an important part in decoherent histories, which we also described earlier in this chapter.

The quantum Zeno effect [29]

When a radioactive system decays, provided the time of decay is not too short, the probability of survival of each system is given by the usual exponential function. We can say the probability of survival at time t is $\exp(-\alpha t)$, where α is a constant. Here α is related to the half-life, the parameter usually used to discuss the decay, though we shall not spell out the relationship here.

Now let us imagine we allow the system to decay for a time $t/2$, we then observe the system to see whether it has decayed, and then we allow it to decay for a further time $t/2$. At the end of all this, the probability of survival must be $\exp(-\alpha t/2)\exp(-\alpha t/2)$. (Essentially we have imposed a collapse at $t/2$.) But by the most basic property of exponential functions, this product is equal to $\exp(-\alpha t)$, so we can say that the evolution of the system is unaffected by the observation at $t/2$. We might think that this is just as well, because our mention of 'observing' the system was perhaps a little unconvincing—it is not quite clear how we might make the observation, so it is just as well that it does not alter the behaviour!

Such is actually essential behaviour for decay in a classical system. It corresponds to a constant rate of decay. If, as we might expect, it was also followed for a quantum system, there would be no surprises, and certainly no quantum Zeno effect. Yet stubbornly, as it were, quantum theory does not quite obey this rule. For the classical case we may note that, for small t, $\exp(-\alpha t)$ is very close to $1-\alpha t$, or in other words the probability of decay by time t is proportional to t, or we may say that the *rate* of decay is constant. But in the quantum case, and for very small times, the probability of decay is instead proportional to t^2, or we may say that the *rate* of decay is not constant but increases in proportion to t. (For the rest of this section, we shall assume that the times involved are short.)

So now let us allow the system to decay for time t. At the end of this time we can say that the probability of survival is $1-\beta t^2$, where β is another constant. On the other hand, we can let it decay for a time $t/2$, observe the system at this time thus collapsing the wave-function, and then let the system decay for a further $t/2$. The overall probability of survival now is $\lfloor 1 - \beta(t / 2)^2 \rfloor \lfloor 1 - \beta(t / 2)^2 \rfloor$ or $1 - \beta t^2 / 2$, where we omit the term in t^4, which is extremely small just because t is so short.

In other words, probability of decay of the system has been reduced by a factor of 2 by the intermediate measurement. In fact we may go further; if we perform two evenly spaced intermediate measurements rather than one, the probability of decay is reduced by a factor of 3, and if we perform n–1 evenly spaced measurements, thus dividing the total period into n sub-periods, the decay is reduced by a factor of n. It is fairly clear what must come next. If we allow n to increase without bound, or more technically we may say if we let it tend to infinity, so we can say that we have continuous observation, there is no decay at all. Not surprisingly this 'prediction' is sometimes known as the *watched-pot paradox*, because it is said that a watched pot never boils, and here we seem to be saying that observation stops the decay altogether.

The effect is perhaps more often called the *quantum Zeno paradox*. The word 'Zeno' refers to the ancient Greek philosopher. His original paradox was that he wondered how he could ever complete a walk of any distance at all. For first he must complete half the distance, then half the remainder, then half of that, and so on. There would be an infinite number of stages, so, he argued, the total time taken must be infinite. The standard reply today would say that an infinite series, or the sum of an infinite number of terms, may actually be finite. We note that there is at least a broad similarity of the quantum Zeno paradox of today, but in no way is the latter effect a modern version of the original paradox.

Why is the prediction we have just made so strange? It is because, as we hinted above, the idea of 'observation' is very difficult to pin down. It seems scarcely possible that just to put a detector somewhere in the vicinity of a decaying atom or a decaying system of atoms could itself affect their decay. Perhaps we would need to surround the decaying system totally with detectors. Even if that is felt to make sense, it still seems strange that the detectors could be controlling the situation, because, of course, if the full prediction of total freezing of the decay is correct, the detectors will never actually detect anything!

The quantum Zeno paradox is an example of a negative-result prediction, an experiment where the observer studies the system but does not actually interact with it. We are fairly used to the idea that in quantum theory an observer may interact with the system, and this interaction may drive the system into a new state. It is rather more strange that the observation may drive it into a new state by *not* interacting with it.

An example of a negative-result experiment is if it is known that an electron is in region 1 or region 2 of a box, but we don't know which. Quantum mechanically we must say that its wave-function is the sum of two wave-functions each for one of the regions. We now use a photon to observe whether it is in region 1, and, we may, say, find that it is not. We know now that its wave-function is localized in region 2. In other words the wave-function of the system has changed without us interacting with it. One of the strange facts about the quantum Zeno paradox is that the prediction is of a negative-result experiment of this type.

It is often said that the quantum Zeno paradox is now a standard and straightforward part of quantum theory, but this is not really true. The effect that is very much studied is a rather different one in which what was achieved by a measurement involving a macroscopic measuring apparatus is achieved by an interaction with another microscopic system. The behaviour of the system may be explained by following the Schrödinger equation for microscopic systems in a straightforward way. We prefer to call this type of experiment the quantum Zeno *effect* rather than paradox. There is certainly nothing paradoxical in it.

We can see the kind of experiment categorized as the quantum Zeno effect in the brilliant 1990 experiment of Wayne Itano and colleagues [30]. This experiment used trapped ions maintained at about a quarter of a degree above absolute zero by laser cooling. In the experiment, three levels of the ion are used. The main transition is between level 1, the ground state, and level 2, an excited state. This transition is driven by a field at an appropriate frequency. If the ion is in level 1 at $t = 0$, then a little time later the probability of it being in level 2 is proportional to t^2. (Note the significance of this t^2 factor, without which no effect of Zeno-type can take place.)

In the Itano experiment, this process is interrupted at times t / n, $2t / n$... by intense pulses of radiation causing transitions between level 1 and level 3, which is considerably higher in energy. Itano calls these pulses measurements because one of two things may happen. There *may* be a number of photons emitted, each of which having the frequency corresponding to the energy difference between levels 1 and 3. We can interpret this as the pulse finding the ion in level 1 and subsequently cycling between levels 1 and 3 and emitting photons of the corresponding energy. On the other hand, there *may* be no photons emitted. We can interpret this as telling us that the ion is in state 2, where it will remain.

Itano speaks of the 'measurement' as finding the ion either in level 1 or in level 2, and this he calls a collapse of wave-function. His main result is that the progress of the ion from level 1 to level 2 is inhibited by the 'measurements' exactly as the quantum Zeno argument predicts. Since, as we have said, there is no involvement of a macroscopic measuring device and no negative-result aspect to the experiment, and since everything may be explained by microscopic analysis, we must call it the quantum Zeno 'effect' not 'paradox'.

Since Itano's experiment, a great deal of work has been done on the quantum Zeno effect, and also it has been shown that, by suitable variation of the time between interactions, one may achieve a acceleration of decay rather than a slowing down. This may be called an inverse quantum Zeno effect. Another interesting feature is that the slowing down or freezing of a process in the quantum Zeno effect may be said to prevent decoherence in certain circumstances. Since decoherence is the factor that makes construction of a quantum computer extremely hard, it will not be surprising that the quantum Zeno effect may play a role in the construction of the first quantum computers.

Macroscopic quantum theory

Supporters of Bohr and Copenhagen had no concerns about handling macroscopic systems in quantum theory; indeed there were areas of quantum theory where some acknowledgement of macroscopic physics was possible or even essential. For example, a famous result called Ehrenfest's theorem is a clever way of showing that, in certain very limited circumstances and with some important assumptions, basic equations of quantum theory could be made to look like the fundamental equations of classical mechanics.

Paul Ehrenfest incidentally was a Dutch physicist, who was born in 1880 and made some moderately important contributions to physics, but was perhaps more important as a confidant of both Bohr and Einstein, and sometimes a bridge between them. Sadly, depressed by what he saw as his own lack of scientific achievement, and some depressing family circumstances, he committed suicide in 1933.

Bohr also took advantage of the classical mechanics in his famous *correspondence principle*, by which the laws of electromagnetism in macroscopic physics are assumed to apply in quantum physics, and in this way selection-rules for transitions between different atomic energy levels can be deduced.

However, the more general idea was that microscopic and macroscopic were to be treated separately as a matter of principle, the former by quantum theory dealing with wave-functions, and knowledge obtained only by measurement, the latter by classical Newtonian methods, with real particles existing independently of any observer.

There have been many claims to bridge this gap under the names of the classical limit of quantum theory, bridging the quantum–classical divide, or macroscopic quantum theory. In the Born–Einstein letters, Born attempted to answer Einstein's criticisms along the lines of our previous paragraph by claiming to show that a wave-packet localized in position space in a box-like potential moves exactly like a classical particle. But Einstein did not accept that Born was considering a sufficiently general case, and in any case, as Peter Holland [31] was to say: 'How can one pass from a theory in which the wave-function represents statistical knowledge of the state of a system to one in which matter has substance and form independently of our knowledge of it?'

One might ask—what would happen if one started with atomic systems or photons, which would clearly obey quantum theory and readily demonstrate interference, and gradually increased the size of the systems until one reached purely classical and presumably non-interfering macroscopic systems? Within the formal constraints of Copenhagen the question would appear to be unanswerable, but until recently the proposition would have seemed completely unthinkable, and Copenhagen would have declined to discuss the situation.

Experimentally things have changed a little in that Zeilinger and colleagues [32] have rather famously demonstrated with the C_{60} buckyball that we mentioned earlier in this chapter, which as we said could be defined as mesoscopic rather than microscopic. In this experiment, the molecules were prepared in an oven at 900–1000 K, and interference was demonstrated in a double-slit experiment with the slits separated by 100 nm. Actually further progress has been made since then to systems involving around 100 atoms.

On the theoretical side, most of the work has been carried out by Tony Leggett (Figure 13.6) [33]. Leggett won the Nobel Prize of physics in 2003 for his work on the properties of liquid helium, which loses all viscosity at sufficiently low temperatures and becomes *superfluid*, but he has also challenged orthodox views on quantum theory for over a quarter of a century or more. He has suggested that

genuine doubts remain about the ultimate meaning and indeed the self-consistency of the quantum mechanical formalism—doubts so severe that some physicists believe that in the end a much more intuitively acceptable world-picture will supersede the quantum theory, which will then be seen as merely a collection of recipes which happened to give the right answers under the experimental conditions available in the twentieth century.

Fig. 13.6 Anthony Leggett [courtesy of Professor Leggett]

Leggett has concentrated his attention on the problems of quantum theory when it encounters macroscopic bodies, and he questions whether the superposition principle can really be extrapolated to the macroscopic level. At some point between the atom and the human brain, he believes that the principle must break down. Indeed he suggests, probably quite rightly, that most physicists' feelings about the quantum theory of macroscopic bodies are somewhat ambivalent; indeed one might trigger two totally different answers by asking rather different questions.

On the one hand they might feel duty bound to say that macroscopic objects should obey the usual rules of quantum theory, and that complexity of itself should not change the state of affairs. Thus it should be possible to give meaning to a wave-function for a macroscopic system that is a superposition of wave-functions for states that are macroscopically distinct.

On the other hand, Leggett suggests that most physicists implicitly retain the idea of realism at the macroscopic level. This, he says, involves two assumptions. The first, *macroscopic realism* (MR), states that a macroscopic system with two or more states available to it will at all times *be* in one of these states (not just be *found* in one of the states if a measurement is performed). The second, *non-invasive measurability at the macroscopic level* (NIM), says that it is possible, in principle, to determine the state of a system with arbitrarily small perturbation on its subsequent dynamics. It should be noted that both these assumptions imply that conventional quantum theory does not apply to macroscopic systems.

It is not, in actual fact, at all easy to distinguish between these two opposing points of view. To test MR, or in effect to disprove it, we must demonstrate what Leggett calls quantum interference between macroscopically distinct states (QIMDS). He points out that, although several types of system, such as lasers, superconductors and superfluids, are often spoken of as demonstrating quantum theory at the macroscopic level, the different macroscopic states that are coupled in these types of system are not sufficiently distinct to be able to test QIMDS.

Rather the most used system for experiments of this type has been the SQUID, or **s**uperconducting **qu**antum **i**nterference **d**evice. The SQUID consists of a superconducting ring interrupted by a *Josephson junction*. A superconductor is a material whose electrical resistance has dropped to 0 at very low temperatures, and a Josephson junction uses the so-called *Josephson effect*, which had been predicted theoretically by Brian Josephson in 1962. Josephson showed that at a junction of two superconductors, a current will flow even if there is no drop in voltage; that when there is a voltage drop, the current should oscillate at a frequency related to the drop in voltage; and that there is a dependence on any magnetic field.

The Josephson effect has led to much important physics. It has become the best way of measuring the constant e/h, and has led to an international standard of voltage. SQUIDs themselves are used in low-temperature and ultrasensitive geophysical measurements, and in computers.

In the experiments on the SQUID that Leggett has championed, the current, a macroscopic variable, is controlled by a microscopic energy, while the temperature must be lower than 0.5 K. In these circumstances, the current obeys a Schrödinger-type of equation for a single isolated particle, and we can study how the potential energy changes as the current is varied. We may hope to see QIMDS, which would seem to be a clear indication of quantum behaviour and, we might feel, a mortal blow for MR.

Actually we need to be a little cautious in our pronouncements. We may remember Bell's theorem, where it could certainly be shown that quantum theory was incompatible with local realism, but it was then necessary to confirm experimentally that quantum theory was right. Here the observation of QIMDS would certainly indicate that we have a superposition (and we may note that a number of experiments, including the buckyball interference, push us in this direction), but the mere existence of a superposition of macroscopic states is not necessarily in conflict with MR if we are allowed to assume that quantum theory is not necessarily true.

To go further than this, an analysis similar to that of Bell is required, and this is exactly what was provided by Leggett and Anupam Garg. Their argument was analogous to that of Bell, but with the field or polarizer setting replaced by time, and with Bell's local realism replaced by a combination of MR and NIM. Much as CHSH did, they were able to study a quantity which, with MR and NIM assumed, must take values less than 2, while for the quantum case it can take values as high as $2\sqrt{2}$. Such experiments may soon be feasible.

Incidentally there have been arguments that in any clash between quantum theory and the combination of MR and NIM, it is NIM that is the culprit. It seems clear that quantum theory may be able to coexist with MR but not with NIM.

Leggett believes that any definitive result of such experiments would be extremely interesting. If it is found that macroscopic superpositions exist, that seems to show that the usual statement of the measurement problem is incorrect. This usual statement is that superpositions break down, or in other words collapse occurs at measurement, *because* measuring devices are macroscopic. But if we, in fact, allow superpositions of macroscopic states, clearly we must assume that collapse occurs for a totally different reason, of which we have no idea, if, of course, collapse occurs at all!

On the other hand, an absence of superposition could be even more exciting. It would indicate that an understanding of the microscopic building blocks, where superposition occurs, does not yield an understanding of the macroscopic whole, where superposition is absent.

Leggett actually admits that he would not expect to see superposition violated for SQUIDs; he thinks it is much more likely to occur at the level of complexity and organization required for biology or psychology. But, as he says, one must start somewhere! Violation might mean that a completely new theory would be required, probably moving further away from, rather than closer to, classical physics.

All this, together with the discussion of environmental decoherence earlier in the chapter, must surely convince the reader that there is enormously more to the quantum physics of macroscopic systems than was realized by advocates of the Copenhagen interpretation, which used, as we have implied, a little mathematics when it worked, and a lot of hand-waving when it did not.

References

1. B. de Witt and H. Graham (eds), *The Many-Worlds Interpretation of Quantum Mechanics* (Princeton: Princeton University Press, 1973). (This collection includes papers by Everett, de Witt and others.)
2. S. Hawking and R. Penrose, *The Nature of Space and Time* (Princeton: Princeton University Press, 1996).
3. J. B. Hartle and S. W. Hawking, 'Wavefunction of the Universe', *Physical Review D* **28** (1983), 2960–75.
4. J. D. Barrow and F. J. Tipler, *The Anthropic Cosmological Principle* (Oxford: Clarendon, 1986).
5. B. Carr (ed.), *Universe or Multiverse?* (Cambridge: Cambridge University Press, 2009).
6. J. R. Gribbin, *In Search of the Multiverse* (London: Penguin, 2010).
7. J. Polkinghorne, *Belief in God in an Age of Science* (New Haven, CT: Yale University Press, 1988).
8. D. Deutsch, *The Fabric of Reality* (London: Allen Lane, 1997).
9. E. J. Squires, 'Many Views of One World', *European Journal of Physics* **8** (1987), 171–3.
10. M. Lockwood, *Mind, Brain and the Quantum* (Oxford: Blackwell, 1989).
11. D. Deutsch, 'Quantum Theory of Probability and Decisions', *Proceedings of the Royal Society A* **455** (1999), 3129–37.
12. S. Saunders, 'Time, Decoherence and Quantum Mechanics', *Synthese* **107** (1996), 19–53.
13. D. Wallace, 'Worlds in the Everett Interpretation', *Studies in the History and Philosophy of Modern Physics* **33**, 637–61 (2002).
14. G.-C. Ghirardi, A. Rimini and T. Weber, 'Uniform Dynamics for Microscopic and Macroscopic Systems', *Physical Review D* **34** (1986), 470–91.
15. R. B. Griffiths, *Consistent Quantum Theory* (Cambridge: Cambridge University Press, 2003).

16. R. Omnès, *Quantum Philosophy: Understanding and Interpreting Contemporary Science* (Princeton: Princeton University Press, 1999).
17. M. Gell-Mann and J. B. Hartle, 'Classical Equations for Quantum Systems', *Physical Review D* **47** (1993), 3345–82.
18. I. C. Percival, *Quantum State Diffusion* (Cambridge: Cambridge University Press, 1998).
19. E. Nelson, *Quantum Fluctuations* (Princeton: Princeton University Press, 1985).
20. G. Bacciagaluppi, 'Nelson Mechanics Revisited', *Foundations of Physics Letters* **12** (1999), 1–16.
21. R. Penrose, *The Road to Reality* (London: Cape, 2004).
22. M. Namiki and S. Pascazio, 'Quantum Theory of Measurement Based on the Many-Hilbert-Space Approach', *Physics Reports* **231** (1993), 301–411.
23. I. Prigogine and I. Stengers, *Order out of Chaos* (New York: Bantam, 1984).
24. O. Freire, 'Quantum Dissidents: Research on the Foundations of Quantum Theory circa 1970', *Studies in History and Philosophy of Modern Physics* **40** (2009), 280–9.
25. K. Camlleri, 'A History of Entanglement: Decoherence and the Interpretation Problem', *Studies in History and Philosophy of Modern Physics* **40** (2009), 290–302.
26. H.-D. Zeh, 'On the Interpretation of Measurement in Quantum Theory', *Foundations of Physics* **1** (1970), 69–76.
27. D. Giulini, E. Joos, C. Kiefer, J. Kupsch, I.-O. Stamatescu and H.-D. Zeh, *Decoherence and the Appearance of a Classical World in Quantum Theory* (Berlin: Springer, 2003).
28. W. H. Zurek, 'Decoherence and the Transition from Quantum to Classical', *Physics Today* **44** (10) (1991), 36–44.
29. D. Home and M. A. B. Whitaker, 'A Conceptual Analysis of Quantum Zeno: Paradox, Measurement and Experiment', *Annals of Physics (New York)* **258** (1997), 237–85.
30. W. M. Itano, J. C. Bergquist, J. J. Bollinger and D. J. Wineland, 'Quantum Zeno Effect', *Physical Review A* **41** (1990), 2295–300.
31. P. Holland, *The Quantum Theory of Motion* (Cambridge: Cambridge University Press, 1993).
32. M. Arndt, K. Hornberger and A. Zeilinger, 'Probing the Limits of the Quantum World', *Physics World* **18** (3) (2005), 35–40.
33. A. J. Leggett, 'Schrödinger's Cat and Her Laboratory Cousins', *Contemporary Physics* **25** (1984), 583–98.

14
Bell's last thoughts

Bell's six possible worlds of quantum mechanics

Towards the end of his life Bell published two rather general papers on the foundations of quantum theory, and, as they are of considerable interest, we review them briefly in this chapter. The first, published in 1986, is titled 'Six Possible Worlds of Quantum Mechanics', and he used this paper to present his favourite interpretations and his bugbears. The latter are what he calls the 'romantic' worlds, which make good journalistic copy, but, in Bell's view, should not appeal to hard-nosed professional physicists. He approves much more of the 'unromantic' but fully professional alternatives.

His first world is the merely pragmatic one, which he suggests is the working philosophy of most of those, including himself, who use quantum theory practically. Upholders of this world recognize that when one probes the very large or the very small regions far removed from ordinary experiencc, we have no right to insist on retaining our usual concepts, or to be able to construct a picture of what is going on at the atomic level. Indeed we are lucky that quantum theory provides excellent rules for calculation. In this world, the 'shifty split' between classical observer and quantum observed can be dealt with by moving it sufficiently far into the observed part that moving it any further would make no difference.

Bell would say this is a good unromantic world for working with quantum theory, though not, of course, for thinking about it. The romantic counterpart of the pragmatic world is the world of complementarity, which Bell characterizes as stressing the bizarre. He feels particularly uncomfortable with Bohr's famous aphorism: 'The opposite of a deep truth is also a deep truth!'

Bell's second unromantic world is that of de Broglie and Bohm. In his opinion: 'The pilot wave picture undoubtedly shows the best craftsmanship among the pictures we have considered.' But he adds: 'Is that a virtue in our time?' The corresponding romantic world is that of many worlds, which he regards as 'extravagant, extravagantly vague, and almost silly', though he admits that it may have something useful to say about EPR, and that the existence of all possible worlds might make us more comfortable about the existence of our own world, which seems to be in some ways (i.e., the anthropic principle, discussed in the previous chapter) a highly improbable one.

The unromantic member of our last pair is the attempt to join up small and large in a rigorous way; in the following year he would meet and become enthralled with GRW, as discussed in the previous chapter. Its romantic partner is the world in which consciousness and the mind are used to collapse wave-functions and provide specific results for measurements. This is attributed to Wigner, and it is a little unfair for him to bear the brunt of criticism, because, as we have stressed, he did a great deal to further alternative views on quantum theory. Bell admits, in fact, that he is convinced that mind has a central place in the ultimate nature of reality, but is doubtful that contemporary physics is at the stage where the idea could be professionally useful.

Against 'measurement'

Bell continued what might be described as his account of his final views on quantum foundations in a paper presented at the Erice conference *62 Years of Uncertainty*, which took place in August 1989. This took the form of a polemical article titled 'Against "measurement"'. This could also be regarded as a follow-on to the paper on measurement in quantum theory that Bell and Nauenberg had published 23 years earlier, and which was discussed in Chapter 8. Indeed, though the tone of the later paper was more relaxed and also more open in its criticism, it might be said that the actual content of the two papers was not much different! The paper was published in *Physics World* as well as in the proceedings of the conference, and thus attracted quite a lot of attention. Sadly Bell was to die just after its publication.

In the paper, Bell admitted that quantum theory is fine FAPP [*for all practical purposes*, his famous new acronym], but he called for a exact formulation of the theory in mathematical terms, leaving no discretion for the theoretical physicist, in that the measuring apparatus should not be separated from the rest of the world, as though it were not made of atoms and not ruled by quantum mechanics.

It is here that Bell's criticism of the use of the word 'measurement' reaches its final form. Accounts of quantum theory suggest, he complains, that the theory is concerned exclusively with the 'results of measurement'. This, he says, is extremely disappointing. In the beginning natural philosophers tried to understand the world around them. It is true that they hit on the idea of contriving simple situations in which the number of factors involved is reduced to a minimum, and in this way experimental science was born. But, Bell insists, experiment must remain a tool and the aim must remain that of understanding the

Universe. To restrict quantum theory to laboratory operations is to betray the great enterprise.

What, he demands, qualifies a particular physical operation to be a 'measurement'? What, he inquires, qualifies a particular organism to be an 'observer': possibly merely to be living, or perhaps a PhD in physics is required? He brings up again the 'shifty split' between system and apparatus, and complains that the mere use of the word 'measurement' suggests some pre-existing property of the object, quite at variance with how quantum theory can actually be made to work.

Bell criticizes the fairly straightforward applications of collapse in the classic books of Dirac [1], and Landau and Lifshitz [2]. If anything, though, he is even more critical of the discussions of measurement in a paper by Nico van Kampen [3], and the very well-known textbook of his old friend from CERN, Kurt Gottfried [4]. He accuses Gottfried in particular of claiming a considerable amount of sophistication in his treatment, but of still, in effect, smuggling in the collapse idea.

Van Kampen uses the idea of collapse, but quite unusually believes that, rather than being a process in conflict with the Schrödinger equation, it is obtained in a perfectly straightforward way from the equation for system and apparatus together. Bell, though, considers his manouvre illegitimate. After the publication of Bell's paper, van Kampen published a reply and a more detailed account of his argument, but I would suggest that, rather than actually justifying collapse, his argument only produces the superposition of terms involving measuring and measured systems from which collapse should occur, but presents no rigorous argument for the collapse itself [5].

Gottfried's argument is much more sophisticated. The first point to be made is that he uses an ensemble approach. Since the early days of quantum theory, a reasonably common proposed solution to the conceptual problems of quantum theory has been to regard the Schrödinger equation as applying to an ensemble of systems rather than a single system. It was, in fact, Einstein's own solution restricted to dealing with quantum theory as it actually exists, though, of course, he really wished to replace the theory altogether as a by-product of his work on unified-field theories. The most well-known positive account of ensemble theories is that of Leslie Ballentine [6], but, with Dipankar Home, I [7] have argued that the technique works only by illegitimately replacing a superposition by a mixture.

In the work of Gottfried and quite a few others—in fact, Leggett, while hugely disapproving of it himself, is inclined to call it the most favoured approach of working physicists, and indeed practically 'orthodox'—the idea is made much more plausible by use of what is called the density-matrix. I won't go into too much detail about this, but I would say that it describes a physical system by a square array of numbers. The great advantage of using the density-matrix for Gottfried's argument is that it may describe both superpositions and mixtures, and 'all' that needs to be done in the argument is to change the density-matrix from superposition to mixture, by, it might be said, fair means or foul.

For Gottfried his means were perfectly fair, but for Bell they were definitely foul! In fact Bell described Gottfried's account of the interaction between measured and measuring systems as producing a 'butchered' density-matrix, in which terms off the main diagonal of the density-matrix are stripped off, but he considered that even this butchered density-matrix represents a superposition, rather than a mixture in which each system actually has a wave-function corresponding to a specific measurement result. Unless you were actually looking for probabilities, Bell says, you would be extremely unlikely to find them!

In June 1990 Bell and Gottfried both participated in a workshop on the foundations of quantum theory at Amherst College, and they continued to argue about these matters, though always with great respect for each other. In his initial reply to Bell's criticisms, Gottfried remarked that the fact of Bell devoting great efforts to discussing his ideas, although highly critically, was the highest compliment he had ever been paid.

His reply [8] was presented at a symposium on quantum physics held in memory of Bell at CERN in 1991. He admitted that some at least of Bell's charges were justified, though he also said that he was not squeamish about QED FAPP. In response to Bell's suggestion that a crisis in physics was imminent, he said that he did not agree, but even if there was, he felt that any replacement for quantum theory would be even further than quantum theory itself from our philosophical and conceptual prejudices, and that Bell would be left pining for the time when he had 'good old' quantum theory to kick around.

Gottfried retained a very great interest in Bell's work, and acted as joint editor for a new edition of many of his papers. In 1987 Cambridge University Press had collected and published Bell's papers on quantum theory up to that date, apart from a few that were, with small changes, to be published again elsewhere. The title of the book, *Speakable and Unspeakable in Quantum Mechanics*, related to a short paper with that title that Bell published in which he distinguished between the 'speakable' apparatus that we can talk about and the 'unspeakable' quantum system that we are not allowed to discuss. In 2004 a second edition was published, which included the very few extra papers he published before his death.

For the new collection, which was published in 1995, Gottfried edited the section on quantum foundations; Mary Bell, John's wife, the section on accelerators; and Martinus Veltman the section on particle physics. While the Cambridge book presented the papers in a uniform format, making the book perhaps more readable, the new collection, published by World Scientific, reprinted them straight from the original journals, perhaps bringing them to life a little more. Some might prefer one, some the other! To make things just a little more complicated, a little later the quantum section from the World Scientific book was published on its own. (References are given in Chapter 8.)

Over the next few years, Gottfried became somewhat dissatisfied with his initial response to Bell, and produced what he thought was a much more satisfactory argument [9], which was published in the leading science journal, *Nature*. It was also included in the second edition of the book initially criticized by Bell. (The co-author of this new edition [10] is Tung-Mow Yan.)

In this paper, Gottfried addresses Bell's complaint that quantum theory should be formulated exactly, with nothing left to the discretion of the theoretical physicist. Gottfried starts by remarking that, given Maxwell's equations and the equation for electromagnetic force, Newton could have understood the meaning of the electromagnetic field without having Maxwell standing at this shoulder with extra explanation. He then asks whether, in similar fashion, given the Schrödinger equation, Maxwell could have derived the statistical interpretation of quantum theory, the fundamental role of probability and the Born rule.

To answer this question, Gottfried says that, in the classical limit, quantum theory relates only to ensembles. He argues that, if he were provided with information about the behaviour of spin-½ particles, Maxwell could have come very close to the Heisenberg principle, and would have been able to deduce the statistical interpretation of quantum theory for discrete variables such as spin, though not for continuous variables such as position or momentum. However, I have suggested [11] that Gottfried's argument still involves an unjustified collapse, or in other words subtly but illegitimately replaces a superposition by a mixture.

In his paper, Bell states that the word 'measurement' is the worst on his list of 'bad words', but others also make him angry. He writes that: 'Here are some words which, however legitimate and necessary in application, have no place in a *formulation* with any pretension to physical precision; *system, apparatus, environment, microscopic, macroscopic, reversible, irreversible, observable, information, measurement*.' The first three imply, he comments, an artificial division of the world. The next four defy precise definition. On the word 'observable', Bell notes Einstein's remark that it is theory that decides what is 'observable', so 'observation' is a complicated and theory-laden notion that should not appear in a *formulation* of quantum theory.

That just leaves 'information'. Bell merely comments: '*Whose* information? Information about *what?*' At the time, these sardonic questions may have been felt by many to have justified his dismissal of the term. But over the next twenty years, the rise of quantum information theory has made a much deeper discussion absolutely essential, as we shall see in Part III.

References

1. P. A. M. Dirac, *The Principles of Quantum Mechanics*, 4th edn (Oxford: Clarendon, 1981).
2. L. D. Landau and E. M. Lifshitz, *Quantum Mechanics*, 2nd edn (Oxford: Pergamon, 1977).
3. N. G. van Kampen, 'Ten Theorems about Quantum-Mechanical Measurements', *Physica A* **153** (1988), 97–113.
4. K. Gottfried, *Quantum Mechanics* (New York: Benjamin, 1966).
5. M. A. B. Whitaker, 'On a Model of Wave-Function Collapse', *Physica A* **255** (1998), 455–66.
6. L. E. Ballentine, 'The Statistical Interpretation of Quantum Mechanics', *Reviews of Modern Physics* **42** (1970), 358–81.
7. D. Home and M. A. B. Whitaker, 'Ensemble Interpretations of Quantum Mechanics: A Modern Perspective', *Physics Reports* **210** (1992), 223–317.
8. K. Gottfried, 'Does Quantum Mechanics Carry the Seeds of Its Own Destruction?', *Physics World* **4** (10) (1991), 34–40.
9. K. Gottfried, 'Inferring the Statistical Interpretation of Quantum Mechanics from the Classical Limit', *Nature* **405** (2000), 533–6.
10. K. Gottfried and T.-M. Yan, *Quantum Mechanics: Fundamentals* (New York: Springer, 2003).
11. M. A. B. Whitaker, 'Can the Statistical Interpretation of Quantum Mechanics be Inferred from the Schrödinger Equation: Bell and Gottfried', *Foundations of Physics*, **38** (2008), 436–47.

PART III
AN INTRODUCTION TO QUANTUM INFORMATION THEORY

15
Knowledge, information, and (a little about) quantum information

Peierls, knowledge, and information

Among the people who published responses to Bell's 'Against "Measurement"' paper was Rudolf Peierls [1], who was certainly a very great physicist. He had been a student of Heisenberg, and spent a considerable amount of time working in Copenhagen with the group of physicists around Bohr. As such, he would always have been among those regarding complementarity as the final answer to the dilemmas of quantum theory. In 1986, in fact, he was to complain [2] that use of the term 'Copenhagen interpretation... makes it sound as if there were several interpretations of quantum mechanics. There is only one. There is only one way in which you can understand quantum mechanics, so when you refer to the Copenhagen interpretation of the mechanics what you really mean is quantum mechanics.'

Peierls (Figure 15.1) had been born in 1907, and he was studying in Cambridge on a Rockefeller Scholarship when Hitler came to power, so he stayed in England as a refugee. He became Professor of Physics at Birmingham University in 1937, moving to Oxford in 1963, and in both places he would have been regarded as being in charge of the leading department in Britain specializing in what we might call down-to-earth theoretical physics such as solid state and nuclear physics; Cambridge may be pre-eminent in the more esoteric and mathematical fields and in astrophysics.

Peierls himself was perhaps the last theoretical physicist who appreciated and could contribute massively to every area of physics: for example, solid state physics, nuclear physics, superconductivity and liquid helium, quantum field theory and statistical mechanics. In 1940 with Otto Robert Frisch, he crucially showed that, contrary to general belief at that time, an atomic bomb could be made from a comparatively small amount of uranium-235, and thus they kick-started the process of making the bomb, but after the war, concerned about the spread of nuclear weapons, he played a large role in the Pugwash movement, the physicists' movement to attempt to attain world peace.

Fig. 15.1 Ruudolf Peierls (left) with Francis Simon about 1951. Both Peierls and Simon came to Britain as refugees, both became Professors at Oxford University and both were knighted.

Peierls undoubtedly deserved a Nobel Prize; but perhaps the committee felt that he had a very large number of important achievements, but not the crucial one that might have sent him to Stockholm.

We met him in Chapter 8, where he welcomed Bell to work for a year in Birmingham; he was certainly enormously impressed by him, and helped him greatly in his career. As such a staunch supporter of Copenhagen, clearly he must have been rather irritated by Bell's dismissal of the work of Bohr and his chief colleagues. Nevertheless he came to appreciate Bell's theorem itself, and it is pleasant to note that Bell and Peierls were on very good personal terms for the remainder of Bell's life. Indeed at the conference dinner of the meeting in memory of Bell, which was held at CERN shortly after his death, Mary Bell, of course, was in the central position, and on one side of her was Rudolf Peierls.

It is certainly not surprising that Peierls wished to reply to Bell's views which were so much opposed to his own. Yet rather strangely the words with which Peierls actually expressed his views were not particularly in accord with those of Bohr and Heisenberg. Rather, perhaps picking up Bell's prohibition on use of the word 'knowledge', he based his approach exactly on that quantity. Mermin suggests, incidentally, that the words 'knowledge' and 'information' should be regarded as synonymous, and that is exactly what we shall do.

It is quite common for those who think just a little about quantum theory to deduce that the actual formalism is all about information. The reason why the wave-function is said to collapse, they suggest, is that the observer now knows the value, having just measured it, and the wave-function changes to correspond to this value. It is easy to believe that the actual state of the system has not had to change at all, just the information of the observer. Of course that implies that the system does have an 'actual state' that may even be broadly classical in nature, and this is totally against everything Bohr stands for. At first sight, though, this general type of approach may make things seem very easy!

And it may seem to be backed up, at least to some extent, by some of the giants of quantum theory. Von Neumann suggested that the collapse is caused by the 'abstract *ego*' of the observer. Wigner, as we have mentioned, argued that it was caused by consciousness. Both of these rather vague ideas were at least rather similar to putting it down to changes in knowledge/information.

Bell's brief barbs may pull us down to earth a little. 'Information about what?' makes the obvious suggestion that, if the information is about something, why do we not concentrate on the something it is about, rather than the information itself? The other question, 'Whose information?', leads to the fact that the analysis must be much more complicated then we have suggested so far, and this takes us on to Peierls' ideas.

Peierls agrees with Bell that a precise formulation of quantum theory is a must, and he also agrees that no textbook explains these things adequately. However, he says that he does *not* agree with Bell that it is all very difficult! He starts his reply by stating his view that the most fundamental statement of quantum mechanics is that the wave-function, or more generally the density-matrix, represents our knowledge of the system we are trying to describe. If the knowledge is complete, in the sense of being the maximum allowed by the uncertainty principle and the general laws of quantum theory, we use a wave-function; for less knowledge we use a density-matrix.

Uncontrolled disturbances may reduce our knowledge. Measurement may increase it, but if we start off with the wave-function case, so we have maximum allowed knowledge, and if we gain some new knowledge in a measurement, this must be compensated by losing some of the information already there. (In other words, if we know the value of the z-component of spin, and choose to measure the x-component, at the end we will indeed know the value of the x-component but will have lost all knowledge of the z-component.)

He remarks that when our knowledge changes, the density-matrix must change. This, he stresses, is not a physical process, and the change will definitely not follow the Schrödinger equation. Nevertheless our knowledge has changed, and this must be represented by a new density-matrix. For a measurement he remarks that the density-matrix has certainly changed by the time the measurement is complete, but it only reaches its final form, which corresponds to the result of the experiment, when we actually *know* the result.

He raises the slightly awkward question of how one might apply quantum theory to the early Universe, when there were no observers. His answer is that the observer does not need to be contemporary with the event. When we draw conclusions about the early Universe, we are, in this sense, observers.

On Bell's second question, 'Information about what?', we have already quoted Peierls as saying 'Knowledge of the system we are trying to describe'. This is a little vague, but it seems to indicate that there is a system, and the knowledge is about it. (This may seem obvious, but we meet a contrary suggestion shortly.)

On the first question, 'Whose information?', Peierls is quite forthcoming. Many people, he says, may have some information about the state of a system, but each person's information may differ, so they may all have their individual density-matrices. While these may be different, they must all be consistent; we cannot allow a situation where one observer has a density-matrix that yields a definite value of the z-component of spin, while the density-matrix of another observer gives a definite value of the x-component, because simultaneous knowledge of both is forbidden by the laws of quantum theory.

Putting this more mathematically, Peierls uses the idea of commutativity, which is actually central in quantum theory. If we write general matrices as A and B, it is an important rule of matrix algebra that, in the general case, AB is not equal to BA; we may say two matrices do not usually commute. However, Peierls says that, if A and B are the density-matrices for the same system as used by two different observers, then AB *must* equal BA, or in other words the matrices must commute.

Information, information, information

At the time, Peierls' position would not have been particularly popular. It seemed to lump together the information we could not have, such as simultaneous knowledge of two different components of spin, with knowledge that we could have but have just been too lazy to obtain. Yet times have definitely changed. With the coming of quantum information theory, information has clearly moved towards centre stage.

Mermin [3] was at the Amherst meeting and, he reports, at the time he very much agreed with Bell's remarks about knowledge/information. Yet as the 1990s wore on, he became involved in quantum information theory, and so he was very much exposed to the view that quantum theory was 'obviously' a theory about information pure and simple.

He has come to regard Bell's question, 'Information about what?', as not very interesting. He considers it 'a fundamentally metaphysical question that ought not to distract tough-minded physicists'. There is no way, he says, that one can decide whether information is about something objective or is just information about other information. Ultimately it is just a matter of taste! One might feel that, while Bell would very naturally be delighted that his work has led to something as stimulating as quantum information theory, this is the kind of remark that he would *not* appreciate. For him surely it was the 'something objective' that was the whole point of the exercise!

Mermin recognizes that Bell's other question, 'Whose information?', is extremely important, and has thought about it deeply, and in fact he has shown that Peierls' criterion of commutativity is not quite right and has made considerable efforts to construct a suitable replacement.

Many others have also stressed the pre-eminence of quantum information. For example, Zeilinger and Časlav Brukner [4] have argued that the basic concept of quantum theory is indeed quantum information, and that quantum physics is indirectly a science of reality, but fundamentally a science of knowledge. They show that their ideas lead to a natural explanation of complementarity and entanglement, and a derivation of the quantum evolution of a two-state system.

In a recent book, Vlatko Vedral [5] has also stressed the centrality of information in the Universe. The Universe and everything in it, he argues, can only be explained in terms of information, which is also the most profound concept of modern science, and the only entity on which we may base the most fundamental theories. Evolution is nothing but the inheritance of information with occasional change of its basic units, the genes. If we ourselves are real, we experience reality only through construction from information.

Interestingly both Zeilinger and Brukner, and Vedral claim that these ideas can provide answers to some of the deepest conceptual questions asked by the famous physicist John Wheeler (Figure 15.2). Wheeler was born in 1911, and made major contributions to many areas of physics. In particular, he wrote a very important paper with Bohr in 1939 shortly after the discovery of nuclear fission, in which many of the details of the process were worked out using the liquid-drop model. He was perhaps most famous for his work in general relativity; he was probably the leader in this area for several decades, and incidentally coined the term *black hole*. Wheeler worked at Princeton University from 1938 to 1976, and then at the University of Texas until 1986. He lived to the grand age of 96, dying in 2008.

He was always extremely interested in the most fundamental ideas. One of Wheeler's famous questions is 'Why the quantum?', and Zeilinger and Brukner suggest that the idea of the primacy of information may answer the question. Information is necessarily quantized in integral numbers of propositions, and if quantum theory is to be regarded as the description of information, it must also naturally be quantized.

Another saying of Wheeler was 'Law without law'. This refers to the apparent infinite regress, whereby laws of physics are 'explained'

Fig. 15.2 John Wheeler [from Denis Brian, *Genius Talk* (London: Plenum, 1995)]

only in terms of deeper or more fundamental or 'a priori' laws. Wheeler reasoned that our explanations would only be genuine if we need invoke no other laws of any type, and Vedral argues that this may be achieved in the acknowledgement of information as the fundamental quantity in the Universe. Quantum theory, he says, allows information to be created from emptiness. There is no requirement for prior information in order that information may exist. Thus the primacy of information is able to break Wheeler's logjam, and indeed Vedral insists that it is this fact that distinguishes information from any other concept that might conceivably unify our view of reality, such as matter or energy.

Lee Smolin [6] has expressed similar views in his attack on the famous or just possibly infamous string theory, not so much just as a proposed 'theory of everything', but more on the fact that it has dominated the physics departments all over the world for at least thirty years, without ever living up to its undoubted promise of solving many, if not all, of the fundamental problems of physics.

It is interesting that, in what he has called the five great problems in theoretical physics, Smolin includes resolution of the problems in the foundations of quantum mechanics, either by making sense of the theory as it stands or by inventing a new theory that does makes sense. Indeed he recognizes this problem as probably the most serious facing modern science. This suggestion is a salutary reminder that, for all the progress described in the previous part of the book, for all the proud list of proposed interpretations that was also drawn up, it may still be said that all the brilliant scientists involved have really just drawn (very necessary) attention to the problem, rather than actually solving it! Smolin almost echoes Bell and Nauenberg's remarks from 1966, as mentioned in Chapter 8, when he says that most physicists believe the problem either unimportant or solved; he insists that it is certainly not!

Smolin, who incidentally describes himself as a realist and a follower of Einstein in these matters, considers it unlikely that these foundational problems of quantum theory will be solved in isolation. Rather he believes the solution of several of these great problems will come together. The other four of his problems are: (i) to produce a theory of quantum gravity, that is to say, to combine quantum theory and general relativity to produce a single theory that can claim to be a complete theory of matter; (ii) to combine all the various particles and forces in today's physics to give a theory explaining them as manifestations of a fundamental unity; (iii) to explain how the values of the free constants in the standard model of particle physics (the masses and lifetimes of the various elementary particles) are chosen in nature; and (iv) to explain the existence and properties of dark matter and dark energy.

To give a little explanation of (iv), dark matter is a large fraction of the total mass of the Universe; we know it is there, but we cannot see it and we have no idea what form it takes! Dark energy is the energy that causes the expansion of the Universe rather unexpectedly to accelerate. Given the knowledge of the Universe we have obtained by a wide variety of experiments, both dark matter and dark energy are required by the laws of physics, and Smolin allows, as an alternative solution to (iv), that we may learn to explain how the laws of physics, in particular the laws of gravity, may be changed so that we may be excused the embarrassment of the existence of dark matter and dark energy!

Smolin regards the use of quantum information as one of the possible alternatives to string theory as an attempt to solve these problems. One possible way forward is to give up the attempt to apply quantum theory to the Universe as a whole, and instead, to define various sub-systems, and to regard quantum theory as the record of quantum information that one sub-system may have about another as a result of their mutual interaction. In this way, ideas from the study of quantum information may demonstrate how elementary particles may emerge from quantum space-time, much as Vedral suggested, and in this way the basis of our scientific view of the Universe is entirely changed.

Seth Lloyd [7] in particular has argued that the Universe should be regarded as a quantum computer. Many years ago, by setting up simple rules on the computational development of initially simple computational systems, it was shown that what one might call, in a very general way, 'universes' could develop, and these 'universes' could be of any degree of complexity.

It was a fairly natural conjecture in the 1960s that our actual Universe might be no more than an example of such a 'universe', or in other words, that our Universe might in fact itself just be a digital computer. One of the people who studied this idea was Edward Fredkin, then at MIT, who we shall meet in the following chapter as a pioneer of quantum computation. However, these early ideas were expressed fairly naturally in terms of a *classical* computer. As we shall see in the next two chapters, a classical computer cannot simulate a quantum mechanical system, so it is fairly clear that the Universe, being quantum mechanical, cannot be a classical computer.

However, we shall also see that a quantum computer may simulate another quantum system, and so it seems natural to contemplate the possibility that the Universe is just a quantum computer. The history of the Universe, Lloyd suggests, is just an ongoing quantum computation. To the question—'What does the Universe compute?', Lloyd gives the answer—'It computes its own behaviour.' At first the patterns it produces are simple, but as it processes more and more information, it produces more intricate and complex patterns—on the physical side, giving rise to galaxies, stars and planets, while on the human side, producing life, language, human beings, society and culture.

An introduction to classical information and computation

The word 'classical' is, almost by definition, only used in retrospect. When Newton, Maxwell and many others developed what we refer to as classical physics, they, of course thought of themselves as doing physics (or, perhaps one should better say, science). This is, of course, obvious, but the point being made is that, when the pioneers of the study of information and computation performed their excellent work, they had another reason for not entering into a distinction between classical and quantum, even though this was the 1930s and 1940s, so technically they might have considered the matter—quantum theory was well established by this time.

The reason why this distinction did not occur to them is that they had no understanding that they were doing physics at all. Even today we perhaps feel instinctively that the information processed by a computer is abstract, just like pure mathematics. By all means, it is probably written down on paper, but it could in principle be presented on a whole range of other media, or even remembered in one's own memory. We might well expect that in a university faculty or building, the subjects of mathematics and computer science might be close together, and a little away from the mainstream sciences such as chemistry and physics.

And indeed this was exactly how the major work on information and computation was presented—as a series of developments in pure mathematics. The initial development of information theory [8] was carried out by Claude Shannon [9] in the 1940s. Shannon (Figure 15.3) worked for what is usually called the Bell Laboratories, or what was formerly

Fig. 15.3 Claude Shannon [©IEEE; from *Claude Elwood Shannon: Collected Papers*, ed. N. J. A. Sloane and A. D. Wyner (New York: IEEE Press, 1993)]

called in full Bell Telephone Laboratories. For many decades the Bell Labs employed many of the world's best scientists and engineers, who were given great freedom to work in areas that were at most tenuously connected to the actual Bell business, named after Alexander Graham Bell, which was, of course, telephone communication.

Over the years, a number of Bell scientists were awarded the Nobel Prize including Clinton Davisson, for demonstrating the wave nature of matter; John Bardeen, Walter Brattain and William Shockley, for the development of the transistor; Philip Anderson, for many important discoveries in condensed matter physics; Arno Penzias and Robert Wilson, for discovering cosmic microwave background radiation, the most direct proof of the Big Bang; Steven Chu, for the invention of laser cooling; and as recently as 2009, Willard Boyle and George Smith, for the invention of an imaging semiconductor circuit, the CCD (charge-coupled device). Despite this late flurry, which recognized work actually done around forty years ago, it would have to be admitted that, as a result of commercial pressures, the glory days of the Bell Labs have been over for at least a decade. It was recently reported that only four scientists remain in basic research, and even since then the laboratories have pulled out of fundamental research into materials and superconductors, to concentrate on more immediately marketable areas such as networking, high-speed electronics, nanotechnology and software.

To return to Shannon, his work was of tremendous importance to pure science, but it was also, of course, well within the central remit of the Bell Company. Shannon was born in Petoskey, Michigan, in 1916, and graduated from the University of Michigan in 1936 with two bachelor's degrees, one in mathematics and the other in electrical engineering. His MSc thesis has been called the most important master's thesis of the twentieth century; it applied rigorous methods of mathematical logic to telephone routing switches, and his idea is the basis of all electronic digital computers. During the Second World War, Shannon joined Bell Labs specifically to work on cryptography, and, rather strangely, fire-control systems. During 1943, Alan Turing visited Bell Labs to share with the US Navy some of the methods used at Bletchley Park, and much of Shannon's important work was to run in parallel with that of Turing. In 1945 he wrote a secret memorandum on cryptography, a declassified version of which was eventually published in 1949.

His all-important work on communications also dated from the war years, though it too was published only after the end of the war in 1948. His task was to discover how to make maximum use of a telephone cable, to discover how many calls could be sent down a wire simultaneously without any of them losing important detail. Today we would undoubtedly write the end of the last sentence as: without any of them losing *information*, but the technical use of that term was only invented by Shannon in the course of this work. Of course the more calls that a telephone line could support, the greater the profits for the Bell company.

Shannon first produced a robust expression for the amount of information contained in a sequence of numbers. Broadly we may say that the more unlikely an event is, the higher the related information. This seems very reasonable; if my team has lost every match this season, and I hear that it has lost another one, that was a highly probable if unfortunate event, and it changes my mindset only marginally. On the other hand, if they have, against all probability, actually won, then that is a matter of great interest to me, and it certainly makes sense to say that the news carries much more information.

With this all-important definition, Shannon was then able to describe how to process a number of telephone calls with greatest efficiency, first considering systems without noise. For calls conveying little information, a great deal of compression will be possible using suitable transmission codes. Just to give a flavour of this, let us imagine a climate where nearly every day is sunny, just an odd one cloudy. We may write the annual sequence of weather as sssssssssssscssssssssssss...., using 365 characters. However, we may easily think of ways of expressing the same data much more efficiently, for example by just stating the numbers of the days that were cloudy.

However, if the sequence is much less repetitive, for example ssccscssscсssсccc, so it carries much more information, it will be much less easy to present the data more simply by compression, and the general mathematical structure of this compressibility of data is known as Shannon's noiseless channel coding theorem.

Of course in real systems there will always be noise. Thus a channel of information will always produce some errors when a signal is sent along it. Error codes enabling the receiver of the message to detect, locate and finally correct the errors, and Shannon's noisy channel coding theorem tells us that, however noisy the channel, it is possible to produce a satisfactory error code that will leave an arbitrarily small remaining error probability. Naturally the noisier the channel, the more elaborate the error code must be. We shall see in Chapter 17 that error correction is even more important when working with quantum information than as it is in this classical case.

Incidentally, given that Shannon's interest was so strongly directed to practical ends, the reader may be suspicious of our characterization of his work as mathematics. It was published in the *Bell Systems Technical Journal*, which, while it conveyed papers of the highest quality and originality even apart from those of Shannon, could still be regarded as a superlative trade journal. It might be asked—was Shannon's work not engineering? We noted earlier that he had degrees in both mathematics and engineering, and the answer might be that, while its immediate application was indeed highly technical, Shannon would certainly have thought of his ideas as abstract and mathematical in nature. It would certainly never have occurred to him that the fact his signals travelled on telephone lines meant that the lines themselves were of any significance whatsoever.

Shannon himself moved to MIT in 1956 and spent the remainder of his career there. He died in 2001.

We now move on to the classical theory of computation. The famous name here is that of Alan Turing [10]. Turing (Figure 15.4) is most famous, of course, for his work at Bletchley Park during the Second World War, in which he was responsible for breaking the code used by the German Enigma cipher machine, and, as we have just seen, he interacted positively with Shannon while on a visit to the United States in 1943. However, he made exceptionally important contributions through the whole early history of computation.

The central ideas in the classical theory of computation are the 'Turing machine' and the 'universal Turing machine'. A Turing machine is a conceptual device that performs computations, or in other words it takes an input sequence and transforms it to an output result. It acts on a tape on which there is a sequence of binary symbols, 0's and 1's; this is our input. The machine itself commences in one of a finite number of internal states. If the machine is in state G, and it reads a character s on the tape, it may do any of three actions. It may replace s by another symbol s'; it may change its own state to G'; it may move one step in direction d, left or right. Thus we may specify the machine by a finite set of transition rules: $(s, G) \rightarrow (s', G', d)$.

The machine proceeds step by step according in each case to the appropriate rule, and the computation consists of the input sequence on

Fig. 15.4 Alan Turing [from Andrew Hodges, *Alan Turing: The Enigma* (London: Burnett Books, 1983)]

the tape being transformed to an output sequence. Of course, one may immediately ask whether, in fact, the machine eventually stops—actually 'halt' is the term always used, or whether it will continue operating indefinitely. This, as we shall see, is a crucial question.

Turing then described a machine that can simulate the action of any other Turing machine. Essentially it takes over all the rules in our set of transition rules for any Turing machine. Then in 1936, virtually simultaneously, Turing and two other mathematicians, Alonzo Church and Emil Post, each working independently and indeed in very different conceptual frameworks, stated versions of what is known as the Church–Turing hypothesis, which is that the universal Turing machine can compute any function 'naturally regarded as computable'.

Two comments are required about the Church–Turing hypothesis. The first is that it is indeed and remains a hypothesis; there is no proof of it, but no counter-example has been found either, even though there have been attempts to find one for several decades. The second is that the statement of the hypothesis seems exceptionally vague. What does 'normally regarded as' actually mean? We may first say that each of the three was saying that the Turing machine (or Church and Post would have used the corresponding entity in their own systems) does indeed act as a computer as we would normally define one. In fact it can achieve any task that we would expect it to, acting as a computer. It can change any initial sequence to any output sequence.

In fact any means of computation at all may be shown to be computationally equivalent to a universal Turing machine. We may say that computation is *universal*. This implies that any computer, no matter what its form of design or construction, no matter what its apparent power, can perform the same range of tasks, though naturally some computers will be much faster than others.

The original point of the Church–Turing–Post work was not centred on computation, but on a fundamental point in mathematical logic. The *Entscheidungsproblem*, which is included in the titles of the papers of both Church and Turing, was posed by the great mathematician David Hilbert, who called it the fundamental theorem of mathematical logic. If it were answered positively, it would mean that any mathematical problem could be solved by an algorithm, a clearly stated sequence of mathematical operations.

However, the work of the three mathematicians in 1936 showed that this was not true. Turing showed that the so-called halting problem was insoluble. It was not possible to construct a computable function that will tell us whether a Turing machine with a particular input will eventually halt, or will continue operating indefinitely.

This was another blow for Hilbert to add to the Gödel theorem of 1932. This work itself emerged from Hilbert's famous list of 23 challenges to mathematicians issued in 1900. The particular challenge was to provide a complete and consistent set of inferences for mathematical proofs. Unfortunately, Gödel showed that it could not be done. In fact, he showed that mathematics is incomplete; for any formal system containing arithmetic, there must be true statements of the system that cannot be recognized within it. The Church–Turing–Post result was, from the mathematical point of view, another indication that mathematics was much more subtle than one might have thought; even the seemingly straightforward conjectures might be untrue.

However, from the point of view of computation, the work of Turing and the others was highly positive. It seemed that they had constructed a general, powerful and seemingly final theory of computation. Turing used aspects of this theory in his work at Bletchley Park, though the machines he constructed there were limited in being designed specifically to carry out one particular task. The achievement of the programmable computer, with data and programs stored in exactly the same way, had to wait until the end of the war, when both Turing and von Neumann came up with the idea.

In any case, by the 1950s, it seemed that information theory and computation theory were in an extremely satisfactory state. It seemed that the basic ideas had been clearly and convincingly stated; all that was required was to apply them in different and more complicated cases. Yet at least in retrospect we can see that unnecessary and extremely restrictive assumptions were being made. It was tacitly taken for granted that information and computation were abstract or mathematical in nature.

It was obvious that in fact they had to be manifested physically. Information could be represented as physical marks on paper, fluctuations in air pressure or electronic signals. Computation could be carried out by an abacus, a computer or the firing of neurons in our brains. However, the physical representation was assumed to be unimportant, and so its form did not need to be discussed. Inevitably it was taken for granted that classical notions would be quite satisfactory, not, as we said before, because those involved thought explicitly that classical physics was correct or good enough, but because they did not think they were doing physics at all! It has been said that all Turing had to understand was paper, but unfortunately he didn't—he thought it was classical! And Deutsch has said that 'computable' was taken to mean computable by mathematicians, again limiting consideration to classical methods.

It took a long time for it to be realized that these unspoken assumptions were actually wrong. The main credit for this should go to Rolf Landauer of IBM, who later became one of the pioneers of quantum computation, though Landauer himself says that the idea originally came from Leo Szilard (Figure 15.5). From the 1960s on, Landauer [11] was saying: 'Information is physical.' He has written that:

> Information is inevitably tied to a physical representation. It can be engraved on stone tablets, denoted by a spin up or spin down, a charge present or absent, a hole punched in a card, or many other physical phenomena. It is not just an abstract entity; it does not exist except through physical embodiment. It is, therefore, tied to the laws of physics.

Fig. 15.5 Leo Szilard in 1953 [courtesy of Karl Maramorosch, from I. Hargittai, *The Martians of Science* (Oxford: Oxford University Press, 2006)]

Clearly if information is physical, we must be careful to use the correct physics. We must recognize that physics is actually quantum mechanical!

In 1985, David Deutsch [12] applied the same argument to computation. He wrote that: 'The theory of computation has traditionally been studied almost entirely in the abstract, as a topic in pure mathematics. That is to miss the point of it. Computers are physical objects, and computations are physical processes. What computers can or cannot compute is determined by the laws of physics alone, and not by pure mathematics.'

Once that is stated, it is clear that the study of quantum computation is essential. And we must not make the mistake of thinking that classical and quantum computers are in a sense equals. We must remember that classical mechanics is always just an approximation, often admittedly an excellent one, to quantum mechanics. Similarly classical computation must in principle always be an approximation to the actual theory of computation, which is quantum computation, though again the approximation may be a very good one.

Some elements of classical computers

Here we explain a few of the very most basic elements of classical computers. The first, very simple but very important, is that the basic unit of information is the bit, which takes either of the values 0 or 1. Here we shall write the bits as (0) and (1).

A register contains a number of bits. For example the number 14 is 1110 in binary, so we may represent it in a register as (1110). It is easy to see that an n-state register can represent 2^n different numbers.

The other main component that we shall examine is the logic gates, which perform operations on the bits. These gates may then be combined into networks to perform the actual computation. Let us start with gates acting on a single bit. There are two possibilities of interest, the identity and the *NOT* gate, and the truth tables for these gates are shown in Figure 15.6. We may summarize the results for the identity gate as $(1) \rightarrow (1)$, $(0) \rightarrow (0)$, and for the *NOT* gate as $(1) \rightarrow (0)$, $(0) \rightarrow (1)$.

A	*A*	NOT *A*
0	0	1
1	1	0

Fig. 15.6 Truth table for single-bit operations. The first column gives the initial bit. The second and third columns show the result of applying the identity and the *NOT* operators.

We now move on to the standard two-bit gates of classical computation, the truth table of which is shown in Figure 15.7. Here the gate *A AND B* produces the output 1 only if both *A* and *B* are 1 in the input. The gate *A OR B* gives 1 if either *A* or *B* is 1, or both. *NAND* and *NOR* mean just *NOT AND* and *NOT OR*, so for cases where *AND* or *OR* gives 1, *NAND* or *NOR* gives 0, and vice versa. *XOR* is the exclusive *OR*, and gives 1 if either *A* or *B* gives 1, but not both.

It will be noted that all the single-bit gates are reversible, in the sense that if we were given the output we can trivially work out the input. Another way of putting this is that it is trivial to undo the action of the gate. For the identity there is no action to undo, while, for the *NOT* gate, we just apply another *NOT* gate.

However, the two-bit gates of Figure 15.7 are all irreversible in the sense that, given the output, we clearly cannot, in all cases, work out the input. Of course if the output of *AND* is 1 or that of *OR* is 0, or that of *NAND* is 0, or that of *NOR* is 1, we *do* know the input, but in each case if we get the other possible answer in the output, we do not know it. The same is true for any output for *XOR*. This feature is unavoidable, of course, because there are two outputs but only one input. We are just here mirroring common arithmetic, where, for example we may easily calculate $9 + 8 = 17$, but, if we are told that the result is 17, it is clearly impossible to find out the two numbers that were added together.

When we turn to quantum computers, we are keen to use reversible gates, and, for a two-bit gate, we clearly need two inputs as well as two outputs. The gate we will look at here is called the controlled-*NOT* or *CNOT* gate. A control input in any gate remains unchanged during the operation, but does determine the behaviour of another input. For the *CNOT* gate, there is a control input, *C*, and the other input, *A*, changes its value if *C* is 1, but keeps the same value if *C* is 0. The truth table for the CNOT gate is shown in Figure 15.8, where the outputs are *A'* and *C'*.

It is easy to see that the *CNOT* gate is reversible. Just as, for any pair of values of *C* and *A*, we know the value of *C'* and *A'*, equally, for any pair of values of *C'* and *A'*, we know the values of *C* and *A*. Alternatively

A	*B*	A AND *B*	A OR *B*	A NAND *B*	A NOR *B*	A XOR *B*
0	0	0	0	1	1	0
0	1	0	1	1	0	1
1	0	0	1	1	0	1
1	1	1	1	0	0	0

Fig. 15.7 Truth table for double-bit gates. The first two columns give the input bits, *A* and *B*, while the remaining columns give the output obtained by use of the various gates.

C	A	C'	A'
0	0	0	0
0	1	0	1
1	0	1	1
1	1	1	0

Fig. 15.8 The truth table for the *CNOT* gate. C is a control, and it is unchanged by the gate; A changes if, but only if, C is equal to 1.

we may easily see that, if we apply the *CNOT* gate to the output, we just recover the input state. In other words we may say that the *CNOT* gate is its own inverse.

References

1. R. Peierls, 'In Defence of "Measurement"', *Physics World* **3** (10) (1991), 19–20.
2. R. Peierls, in: P. C. W. Davies and J. R. Brown (eds), *The Ghost in the Atom* (Cambridge: Cambridge University Press, 1986), pp. 70–82.
3. N. D. Mermin, 'Whose Knowledge?', in: R. Bertlmann and A. Zeilinger (eds), *Quantum [Un]speakables; From Bell to Quantum Information* (Berlin: Springer-Verlag, 2002), pp. 271–80.
4. Č. Brukner and A. Zeilinger, 'Quantum Physics as a Science of Information', in: A. Elitzur, S. Dolev and N. Kolenda (eds) *Quo Vadis Quantum Mechanics* (Berlin: Springer-Verlag, 2005).
5. V. Vedral, *Decoding Reality: The Universe as Quantum Information* (Oxford: Oxford University Press, 2010).
6. L. Smolin, *The Trouble with Physics: The Rise of String Theory, the Fall of a Science and What Comes Next* (London: Penguin, 2008).
7. S. Lloyd, *Programming the Universe: A Quantum Computer Scientist Takes on the Cosmos* (New York: Knopf, 2006).
8. R. W. Hamming, *Coding and Information Theory*, 2nd edn (Englewood Cliffs, NJ: Prentice-Hall, 1986).
9. C. E. Shannon, 'A Mathematical Theory of Communication', *Bell Systems Technical Journal* **27** (1948), 379–423, 623–56.
10. A. Hodges, *Alan Turing: The Enigma* (London: Random House, 1992).
11. R. Landauer, 'Information is Physical', *Physics Today* **44** (5) (1991), 23–9.
12. D. Deutsch, 'Quantum Theory, the Church-Turing Principle and the Universal Quantum Computer', *Proceedings of the Royal Society A* **400** (1985), 97–117.

16
Feynman and the prehistory of quantum computation

Feynman and miniaturization

The date of the founding of quantum computation is undoubtedly 1985, with the seminal work of David Deutsch. Yet there was interesting work done, principally by Richard Feynman and colleagues, for at least fifteen years before Deutsch's work, and indeed one can go back a further 10 years to find the real start of Feynman's own contributions. (We may also admit that the field did not actually make much progress for nearly a decade after Deutsch's work, and we shall discuss the reasons for this in the following chapter.)

Julian Brown [1] has suggested that the Feynman period can be best described borrowing terminology from the history of quantum theory. The period following Planck's work and up to 1925 is often called the period of the old quantum theory. During this period, with the work, after Planck, of Einstein, Bohr, de Broglie, Compton and others, we can say that many important ideas, insights and results were obtained, but there was no rigorous foundation to the work, and no understanding of how all the nuggets of information fitted together. Only with the new quantum theory of Heisenberg and Schrödinger in 1925–6 did we get the rigorous foundation to the subject, from which all the *ad hoc* results of the old quantum theory could be obtained in a straightforward way. (We do not use this terminology in this book because the terms 'old quantum theory' and 'new quantum theory' do not fit in well with our 'First Quantum Age' and 'New Quantum Age'!)

Brown though suggests that in the period before Deutsch's work, many ideas of interest were produced, but there was no overall structure to the subject, and no genuine understanding of the basic ideas. That did not come until the work of Deutsch.

Feynman (Figure 16.1) [2] himself was probably the best-known physicist of the second half of the twentieth century. Already at Los Alamos, in the creation of the atomic bomb, he was prominent among the galaxy of important physicists. Then after the war he was one of those who came up with the generally accepted solution of the problems of quantum electrodynamics (QED).

After the problems of quantum theory were unravelled, it was very natural to attempt to bring the photons into the theory, in other words to develop a quantum theory of the electromagnetic field—QED. During the 1930s, many of the physicists who had made major contributions to the development of quantum theory, such as Dirac, Pauli, Heisenberg and Jordan, naturally moved onto QED, but unfortunately, though, they met great difficulties. Specifically when one went to high energies, it seemed that infinite quantities inevitably emerged from the theory.

Only after the war did three younger physicists, Feynman himself; the American Julian Schwinger; and the Japanese Sin-Itiro Tomonaga, present solutions to the problem. In fact they removed the infinities. Essentially all produced the same mathematical technique, called *renormalization*, but they all described their work in vastly differing mathematical terms. At first some of their older colleagues rejected their work; Dirac, for example, thought that renormalization was just an arbitrary and inconsistent trick. But within a short period it was accepted as the way forwards, and indeed as broadly providing a template to attempt to describe the other fields of physics.

For his version of the work, Feynman invented a rather amazing technique called the Feynman diagram. Rather than using acres of algebra, such as that produced by Schwinger and Tomonaga, this was a visual means of representing a possible way in which a particular outcome of an interaction involving elementary particles could be obtained. For any outcome there will probably be several diagrams representing different ways of obtaining it, and all these diagrams must be summed quantum mechanically to give a result for the probability of the process occurring that might be compared with the experimental value.

Though the idea is often used in elementary discussions to provide merely a picture, it should be realized that the method is actually designed to provide fully quantitative answers, which are often obtained

Fig. 16.1 Richard Feynman [courtesy of California Institute of Technology]

with less work than that needed for performing the same calculation using more traditional methods. For over sixty years, Feynman diagrams have been central in all theoretical work involving elementary particles.

Feynman contributed to many areas of physics. In particular he was instrumental in the arguments that there was experimental evidence for the existence of quarks, and he made very many important contributions to the physics of elementary particles. His many incisive and crucial observations in a wide range of areas of physics, together with his vision to invent and promote totally new fields, built up his supreme reputation within the physics community. He was also, unlike many great researchers, extremely interested in passing on his knowledge and wisdom, and his course on fundamental physics, which was published in 1965 as *The Feynman Lectures on Physics* [3], inspired generations of students with his ability to explain well-worn topics in new and thought-provoking ways.

He had a charismatic personality and an unrivalled gift for self-publicity, and became a figure of great reputation and considerable influence completely outside the university and the world of physics, right up to his death in 1988.

In this chapter, we note the fact that, from early on, he was a great publicist for the view that the quantum nature of computers deserved study. Yet it must be admitted that, in this book, he cannot be treated as an untarnished hero. In Chapter 9, we met his highly negative response to Clauser's ideas on testing quantum theory. Also Feynman was inclined to pick up a suggestion, or even an account of a complete theory, from some probably junior worker, and to think about this with great concentration, coming up with his own approach to the subject, but in the course of this totally forgetting that he had borrowed the idea from somebody else. We shall see that both Paul Benioff and Bell himself were treated this way in the course of our narrative.

Let us anyway start our account of Feynman's contributions by remarking on an after-dinner speech given at a meeting of the American Physical Society in December 1959. The title of the talk was 'There's Plenty of Room at the Bottom', and it was an inspiring and wide-ranging call to miniaturize. Feynman said that he hoped the field of study he was outlining could be as important and useful as low-temperature physics, high-pressure physics or the physics of high vacuum.

He suggested the following objectives: writing and reading the *Encyclopedia Britannica* on the head of a pin; putting the contents of all the books in the world in a cube roughly 10^{-2} cm across; and improving the electron microscope so that biologists could solve problems by direct inspection of the structure of interest. He actually offered two \$1000 prizes, the first for production of an operating electric motor in a cube of size roughly 0.005 cm, the second for reducing the area of the page of a book by a linear factor of 25,000, so that it could be read by an electron microscope. He was to pay up on both these prizes, the first as early as 1960, but the second not till 1986.

His paper may be regarded as the first hint of the field of nanotechnology, an extremely hot topic today in terms of scientific interest and technological potential. Feynman discussed the construction of layered materials atom by atom, selecting the nature of each individual atom, much as is done today in the production of computer chips. He also suggested the possibility of making the miniature machines by working downwards by successive factors of perhaps 4. Or as he puts it: 'training an ant to train a mite to do [what you want]'.

Feynman's comments on computing in this early paper were preliminary but exceptionally interesting. First he noted that, if the density of storage of information were sufficiently high, we should not need to erase it. As we shall see, later in this chapter, this can have profound implications for reversibility, which is very significant for quantum computation. He went on to say that computers, which at the time still occupied extremely large rooms, should be made progressively smaller. Of course this has very much been the theme of the subsequent fifty years, as we shall see in the next chapter, but Feynman's demand was that one should aim at constructing a computer with components of little more than atomic size, with the obvious demand that quantum mechanics would be required in their manufacture and mode of working.

Other suggestions made by Feynman were that computers could be made so that they could make judgements on how to do their tasks, rather than having to be told in detail how to do every step, and also that computers should be able to be trained to recognize voices. The latter has certainly been achieved to a great extent since 1959.

Feynman and quantum simulators

We now move forwards to the 1980s. Feynman had, according to his great friend and an important pioneer of computer science, Marvin Minsky, become enamoured with every aspect of computation, and, from 1981 to 1986 he gave an interdisciplinary course at Caltech called 'The Physics of Computation'. This included many aspects of computer science, including the organization and theory of computers, coding and information theory, and the physics of computers, but it also included discussion of quantum computers and especially reversible computation, which is important for quantum computation.

Since Feynman's death, many of his lectures on quite a wide variety of topics have been published, and an edited version of these lectures on computation [4], gleaned from tapes and notebooks, was prepared by Tony Hey and Robin Allen in 1996. A companion volume [5] was also published, which included contributions from many of Feynman's collaborators at the time, discussing their work with Feynman in the 1980s and also bringing the various stories up to date. These associates included John Hopfield, Carver Mead and Gerard Sussman, all of whom had participated in the original course, and also Landauer, Minsky, Charles Bennett, Paul Benioff, Norman Margolus, Ed Fredkin and Tom Toffoli, most of whom will play significant roles in our story as Feynman learned from them, inspired them and in many cases built on foundations they had laid. Feynman's own important papers, which are described in this section, are also included in these books.

A very important event was a conference organized by Fredkin, Landauer and Toffoli at MIT in 1981 on the Physics of Computation. At the time this was a practically unknown field, and the organizers were delighted when Feynman agreed to give the keynote lecture. Landauer remarked that Feynman's mere participation in the conference, and his willingness to give an occasional lecture in the area, helped to emphasize that it was an interesting topic for study. His keynote lecture was one of two very important papers he was to write on the subject; it was probably the first to discuss using quantum ideas in computation and to use the term 'quantum computer'.

Actually Benioff had a little earlier studied whether quantum systems could be used for efficient simulation of classical computers, and was able to show that, rather contrary to what was generally thought at the time, the uncertainty principle would not prevent this being achieved. Feynman was undoubtedly stimulated by Benioff's ideas, but regrettably did not manage to give Benioff the credit he deserved.

Feynman asked the reverse question: Can a classical computer simulate either a classical system or a quantum system exactly? Of course one could always replace the physics by a series of numerical algorithms, and solve these as exactly as one might require, but what Feynman meant was *exact* simulation. He also required that the number of computer elements required should only increase in proportion to the size of the system being simulated, not exponentially.

He was first able to show that a classical computer may certainly simulate a classical system, but if the system is quantum mechanical, then the amount of work does indeed increase exponentially. In fact if the simulation requires calculation of the probability of each of R particles being at each of N points, the amount of memory is proportional to N^R, obviously increasing explosively with R.

The next question was whether a classical computer might simulate a quantum system probabilistically. The answer was again negative, and the reasoning is interesting—it relies on Bell's theorem. If one could replace the quantum system by hidden variables, then essentially it would be treated as classical, and the result from the previous paragraph would allow simulation. However, of course this is exactly what Bell's theorem rules out. It is a little sad that, at this point in his argument, Feynman, who, it will be remembered, had dismissed Bell's work when

Clauser had wished to discuss it with him a decade earlier, now presented the complete Bell argument—without any mention of Bell at all!

Feynman did, though, suggest that quantum systems could be simulated by other quantum systems, and he talked of a 'quantum computer', explicitly meaning a quantum system acting as a computer and going beyond a Turing machine. He discussed whether there might be a new type of universality; some types of quantum system might simulate each other, and there might be quantum systems that could simulate anything, including the (quantum) physical world, and thus be called universal quantum simulators.

This was clearly a very important paper. It showed that there were tasks that a classical computer could not perform, and explicitly that the problem was that the amount of work increased exponentially with the size of the task, and this idea has been extremely important in modern developments. Yet the paper was restricted to simulation. Feynman did not realize that a quantum computer would be able to perform a much greater range of tasks, and in these tasks would also be able to outshine classical computers dramatically.

Reversibility in physics and in computation

In 1985, Feynman wrote another very important and influential paper on quantum theory and computing, which was considerably more detailed than his 1981 paper. Much of the paper relied quite heavily on work done on reversibility and irreversibility in computers that had been carried out over the previous twenty years or so by Feynman himself and also by Landauer, Bennett, Margolus, Toffoli and Fredkin.

The question of reversibility and irreversibility in computation was intimately connected with the same question in physics, which was a very important part of the nineteenth-century theories of thermodynamics and statistical mechanics. The science of thermodynamics [6] was articulated in particular by Rudolph Clausius and Lord Kelvin in the 1850s, though they made use of extremely important contributions of James Joule and Sadi Carnot.

One of the most important aspects of thermodynamics was that heat, rather than being a substance called *caloric*, as had been an almost universal belief until the work of Joule, was a form of energy. The First Law of Thermodynamics then established what everybody knows today as the law of conservation of energy. Initially it just said that the total amount of heat and work (or mechanical energy) must be constant. Subsequently other forms of energy—light energy, sound energy, electromagnetic energy and so on—had to be brought into the sum.

The Second Law was somewhat more subtle, and there have been several different statements of it, or approaches to it. The earlier statements decreed that certain physical processes, which were definitely allowed by the First Law, nevertheless did not take place. These statements were nothing more than generalizations of experience, but experience so embedded in the study of natural phenomena that they could not easily be questioned.

Kelvin's statement referred to the fact that the aim of nineteenth-century heavy engineering was to use heat energy obtained from burning coal or oil to run machinery; in other words to take heat from a high-temperature source and transform it all to work, as would certainly be permitted by the First Law. The Second Law says that this cannot happen; there must be a sink of energy at a lower temperature into which some of the energy obtained from the coal or oil must be deposited.

What is the justification for this? Without the law, it would seem possible to sail across the ocean by extracting energy from the sea water and thus cooling it down slightly. Common experience says this does not happen because there is not a reservoir cooler than the sea into which energy may be deposited. Incidentally it should be stressed that the Second Law applies to cyclic processes, as in operating machinery, not necessarily to an individual stroke of an engine.

Clausius' statement is probably a more obvious distillation of everyday truth. It says that heat will not move of its own accord from a cooler to a hotter object. ('Heat won't go from a cooler to a hotter; you can try it if you like but you'd really better nott-er,' in the words of the song.) Of course if we provide energy we can make the heat go 'uphill'. Thus a fridge will not work transferring heat from the food inside to the relatively warm outside world if we do not plug it in; if we do plug it in, we thus provide energy to make this happen, and this is, of course, reflected in our electricity bill.

The Second Law says that there is a natural direction to physical processes, which is often called an *arrow of time* [7]. The process will go readily in one direction, but it will not of its own accord go in the opposite direction. For example, heat flows from hot to cold, but not back from cold to hot. We may say that physics is, in this way, *irreversible*.

Let us look at some further examples of this. First, let us imagine a chamber of gas divided into two compartments of equal volume by a removable partition. At the beginning of the process, let us imagine that one of the compartments is filled with gas, while there is a vacuum in the other compartment. Now we remove the partition. If we come back to the system after a little time, we will find that gas occupies both compartments; the pressure, density, and temperature will be the same throughout the whole gas. (We may say that the gas is by then in *equilibrium* or *thermodynamic* equilibrium.) Thus the process proceeds readily in this direction. However, if we start off with the gas present through both compartments, we will never, the Second Law says, following common experience, find a situation where all the gas is in one or other compartment. The process we discussed is clearly irreversible.

As another example, suppose we drop an ice-cube into water at 50ºC. Clearly the ice-cube will melt, the temperature of the water dropping a little to say 49ºC. But if we start off with water at 49ºC, we will certainly never find that an ice-cube has formed and is sitting the top of the liquid. Again the process is irreversible.

What about reversible processes—is it ever possible to imagine one? For a start we would have to avoid any process involving a flow of heat, or in other words we must never put two objects in contact that are at different temperatures. We must avoid friction or viscosity, and any other similar type of dissipation of energy. We must carry out the process exceptionally slowly—the term usually used is *quasistatically*, so that we are always exceptionally close to an equilibrium state. Even then the idea of a reversible process is always an idealization, to which we may, if we are lucky, come extremely close.

The usual example in the textbook is a gas cylinder with a piston moved outwards so as to increase the volume of the gas. The piston must be moved exceptionally slowly, and the cylinder should be thermally isolated so that heat can neither enter nor leave the system. The process can be called *adiabatic*, essentially meaning there is no flow of heat. The process can then be reversed, the piston being moved equally slowly inwards so as to return the system to its initial state. This shows that the first stage of the process was clearly, to a very good approximation, reversible.

Another way of stating the Second Law, also due to Clausius, is in terms of a quantity called the *entropy* of the system. For nearly everybody, this quantity is very difficult to get a really good feel for. Technically it is defined so that, if an object at temperature T receives an amount of heat Q, its entropy goes up by an amount Q/T. (To makes things simple, we are assuming that Q is so small that T does not change appreciably in the process.) Similarly if an object at temperature T emits an amount of heat Q, its entropy decreases by Q/T.

The importance of the concept of entropy is that in any irreversible process the total entropy goes up, while in a reversible process it stays the same. Since in the Universe overall there are many, many irreversible processes, and indeed, as we have said, reversible processes are always an idealization, it is clear that the total entropy of the Universe must increase. This is another statement of the Second Law of Thermodynamics, and it demonstrates a very clear arrow of time.

As we have said, the concept of entropy in thermodynamics is not an easy one; it becomes more straightforward in the context of statistical mechanics. However, it is instructive to follow the changes of entropy in a passage of heat Q from an object at a higher temperature T_1 to another object at a lower temperature T_2. The first object will lose entropy Q/T_1, while the second will gain entropy Q/T_2. Thus the total change in entropy in the process is $Q/T_2 - Q/T_1$, which is positive since T_1 is greater than T_2. Thus entropy increases as it must. (Again we are assuming that Q is so small that T_1 and T_2 do not change appreciably in the process.) Of course if the Clausius statement of the Second Law were disobeyed so that heat passes from a colder object to a hotter object, then this statement involving entropy would be disobeyed as well.

So far we have been using the ideas of thermodynamics, which is, as we have said, a distillation of common experience into concrete laws of nature. It uses only macroscopic concepts, making no use of atomic models of nature. The advantage of this is that its statements are eternal (provided, of course, common experience does not change!), and so do not depend on one's microscopic model of nature, which may, at least in theory, change from one year to the next.

The discipline that tackles essentially the same set of problems from a viewpoint definitely embracing the microscopic is called *statistical mechanics* [8]. The most significant contributors to this discipline from the point of view of this book were Maxwell and Ludwig Boltzmann. Boltzmann was, for the last 40 years or so of the nineteenth century, the most powerful advocate of atomic ideas. He met fierce opposition from philosophers and physicists adamantly opposed to the actual existence of atoms, although atomic ideas in chemistry were, of course, well established by this time. It has often been supposed that Boltzmann's suicide in 1906 was at least in part due to this bitterness of this debate, and ironically it came just as atomic ideas were becoming established. It must, though, be recorded that Boltzmann had suffered throughout his life from intense depressions.

One of the main areas where statistical mechanics complemented thermodynamics highly effectively was in assisting the understanding of entropy. To explain this, we shall introduce the idea of the *microstate*. As an example of its use, we will return to the chamber of gas with two compartments, and suppose we are putting each of N molecules into one or the other compartment, where N is a very large number.

We first consider the situation where all N are on one side or other, and we call this a *macrostate* of the system. Now we may ask how many microstates does this correspond to?—in other words, how many ways can I arrange the actual molecules so that we have this particular macrostate? The answer, which we shall call W, is obviously just 1—we really have no flexibility at all.

Now we ask—how many different ways can we arrange the molecules so that there is just one molecule (any molecule) on the left-hand side, and all the others on the right-hand side? Now the answer, the new value of W, is clearly N. And to cut a long story short, as the number on the left-hand-side increases to $N/2$, W will get larger and larger, and it will be its maximum value for that particular case. As the number on the left-hand side increases further beyond $N/2$, W will, of course, drop.

Now suppose we put the molecules into the system at random, either compartment being equally likely for every molecule. It is clear that there will be a far greater probability of the molecules being roughly equally divided between the two compartments than all of them being on one side or the other. In other words we are almost certainly going to reach a situation with a high value of W. In yet more words, W is just a measure of the probability of a particular macrostate being obtained. Equally since the macrostate with $W = 1$ is in a sense totally ordered, while those with much higher values of W are much less ordered, we may say that W is a measure of the *disorder* of the particular macrostate.

Finally we imagine that the molecules are deliberately inserted into the system in a state with a low value of W, perhaps the state with all the molecules in one particular compartment so that $W = 1$. Now if the molecules are free to travel through the whole volume, it is clear that in time the number in this compartment will fall, the number in the other compartment will rise, and, in what we may call the equilibrium situation, there will be roughly equal numbers on each side. (It is an equilibrium state from the macroscopic point of view because the *numbers* on each side will vary only little, but of course not from a microscopic point of view because individual molecules will move freely from one compartment to the other.)

Thus in this natural process, W will steadily increase, just as we have said the entropy S does, and it may seem likely that the two are related. The link was made by Boltzmann, and the relevant equation is $S = k \log W$. Since as W increases, $\log W$ increases as well; this means that S also increases with W. This supremely important equation was inscribed on Boltzmann's grave. (The log, by the way, is to base e, the exponential number.)

Another result of this analysis is that we may say that S is a measure of the disorder of the system. This implies that, as time advances, systems will, if left to their own devices, get more and more disordered, and this is again just common experience. In irreversible processes the entropy increases, and so does the disorder; in reversible or adiabatic processes, both remain unchanged. (Another word for an adiabatic change is just as 'isentropic' change—a change at constant entropy.)

We must now speak of the k in the above equation, which is always called Boltzmann's constant. To explain the significance of this exceptionally important constant, we must examine the idea of the degree of freedom of a system. Essentially a degree of freedom of a microscopic system is an independent way of being able to store energy. Thus a molecule of a monatomic gas, which is just a single atom, has three degrees of freedom, each corresponding to kinetic energy for motion in one of the three spatial directions. A molecule with more than one atom has further degrees of freedom corresponding to rotation and vibration of the molecule. A simple harmonic oscillator in one dimension has two degrees of freedom, one for kinetic energy, the other for potential energy and so on.

And the very important *Law of Equipartition of Energy*, which is a centrepiece of classical physics, states that, at temperature T, the average amount of energy for each degree of freedom is just $\frac{1}{2}kT$, where k is Boltzmann's constant. Boltzmann's constant will play an interesting role as we move on to consider the energy involved in computation.

We first, though, look at the troubling problem of Maxwell's demon. This will be hugely relevant to the study of reversible computation because the problems very much complemented one another, and to a large extent they were resolved by the same people.

Maxwell's demon was an irritating, enticing and, for many decades from 1871, an unsolved problem for physicists and philosophers of all types. To introduce it we may imagine that, in the divided chamber we have discussed so much, the partition is complete except for a small hole where the demon stands guard. In Maxwell's original version, he lets through only fast molecules travelling from right to left, slow molecules moving from left to right, so that the temperature will increase on the left, decrease on the right, obviously violating the Second Law. Alternatively, he may let any molecule travel from right to left, and prevent any molecules attempting to travel from left to right. Thus the pressure will increase on the left and decrease on the right. In both cases entropy decreases, order increases.

This little problem has stimulated a vast amount of discussion. Well over 550 papers have been published, and some of the most important have been collected in a volume edited by Harvey Leff and Andrew Rex [9]. A first edition of this book was published in 1990. By the time a second edition was published in 2003, and because in the intermediate period the connection between Maxwell's demon and quantum computing had been exhaustively explored, the title was adjusted to include the word 'quantum', and, just because of space limitations, many of the papers selected for the first edition had to be dropped in favour of those expounding the new insights.

One of the most celebrated of the older discussions of Maxwell's demon was written by Leo Szilard in 1929. Szilard's important insight was to concentrate on how the demon obtained the information to allow him to know whether to let any particular molecule through the gap. The word 'information' has been dropped into the previous sentence fairly casually, but in fact Szilard's paper was exceptionally important because it presented the first substantial discussion of the concept in a physics journal; it was this paper that Landauer credits as first recognizing that 'information is physical'. Szilard in fact announced that performing a measurement requires the expenditure of a certain amount of energy, $kT \log_e 2$. 'Expenditure' of energy is essentially dissipation of energy, so will correspond to an increase in entropy, and so irreversibility. For four decades or more, this was looked on as a crucial result, but unfortunately it was not actually correct.

There was a strong influence from the study of quantum computation in finally getting the argument correct, as we shall see in the following section. The work of Landauer, specifically on quantum computation, suggested that there was actually no need for any expenditure of energy in carrying out the measurement. And it was Charles Bennett (Figure 16.2) [10] who took the final step, though it should be mentioned that Oliver Penrose [11], Roger's brother, also came up with the same argument, and Leff and Rex call their resolution of the problem the Landauer–Penrose–Bennett solution.

Bennett had reached effectively the same solution for reversible computation [12]. He will appear in later chapters of the book, and is undoubtedly one of our heroes. The important point in the solution is that energy is dissipated and there is an increase in entropy, not, as Szilard had suggested, when the demon makes a measurement, but when it discards the information that it has gained in the measurement. In fact without the necessity for discarding the information, the change in entropy of the process would be negative; this, of course, would violate the Second Law, and we might say that it is essentially the original argument of the demon. It is the increase in entropy due to discarding the information that makes the overall change in entropy zero, thus saving the Second Law and making the whole process in principle reversible.

We might wonder whether the demon actually needs to discard the information. Could he not retain it and thus defeat the Second Law?

Fig. 16.2 Charles Bennett [© Zentralbibliothek für Physik, Renate Bertlmann]

Feynman's suggestion will be remembered; if the memory of the demon is large enough, it might have no need to discard information. In fact it may also be remembered that we stressed that the law applies to cyclic processes; if we wish to discuss the matter rigorously, we must return the demon to his initial state—he must discard information gained in the process.

Feynman and reversible computation

Much the same sequence of events occurred in the story of reversible computation. Just as Szilard suggested that a measurement required the expenditure of $kT \log_e 2$ of energy, in 1949 von Neumann argued that in a computation exactly the same amount was the minimum required to manipulate one bit of information. Here a manipulation may be, for example, a decision to choose one of two paths. This seemed very reasonable, as the information had to stand out against background noise that would increase with temperature. Von Neumann's result soon became generally accepted, although it was not actually published until 1996, when, after his death, some of his general writings were edited and published.

In 1961, Landauer analysed the way in which computer operations were actually performed, and he came to the conclusion that von Neumann was quite wrong. He claimed that if a manipulation were carried out slowly and steadily, as for our adiabatic or isentropic change in thermodynamics, one can avoid dissipation of energy altogether so there is no need for expenditure of energy. However, when information is erased or discarded, energy must be dissipated. This would include re-setting a register to an initial value, or writing over the value in a register. The amount of energy then dissipated is just (yet again!) $kT \log_e 2$. This is often called the Landauer constant, and the general idea is known as Landauer's principle.

What is it about erasure that requires dissipation of energy? A physical argument is as follows. In erasure, the two states of a particular bit, 0 and 1, must both end up as 0, but the means of achieving this must be independent of whether the initial state is 0 or 1. It is simply not possible to produce a physical model that can achieve this without dissipation of energy. We may reach the same conclusion from a more abstract argument. If two different initial states are transformed to the same final state by the forwards process, clearly there can be no backwards process that would restore the particular initial state and make the forwards process reversible. The reverse process clearly would not 'know' which of the initial states should be recovered.

While these insights of Landauer and Bennett were obviously extremely significant in increasing our understanding of reversibility, it was not obvious that they would actually be of any use in constructing a reversible computer. For after all, like the demon, surely the computer would simply have to discard information whenever it was re-set to commence a new computation, and when it did so energy would be dissipated and any possibility of reversibility lost. The only alternative appeared to be for the computer to refrain from discarding information and to 'choke under its own garbage'.

Yet again it was Bennett who came up with the solution to this seemingly intractable problem. He found that it was possible to carry out any computation in three stages. In the first, the required answer is obtained, but other information that is not required is also generated. In the second stage, the answer is copied reversibly onto a spare blank tape. In the third stage, the first stage is essentially carried out in reverse; all the surplus information generated in the first stage disappears, automatically removed by the program, not by an erasure. At the end of the three stages, the computer has been returned to its initial state and no energy has been dissipated, but the answer is on the spare tape.

This work was carried out in the early 1970s, and at that time it seemed of great conceptual interest but not of practical use. Production of heat had been a problem for the earliest valve-based computers. Once computers replaced valves, though, and despite the fact that computers of the time did in fact use between 10^6kT and $10^{12}kT$ of energy for each elementary manipulation of information, most of it expended in keeping different parts of circuits at fixed voltages, heating was not a problem. And of course these energies were, in any case, vastly greater than the Landauer limit.

Yet, as we shall see in the following chapter, over the years since then, computers have become steadily smaller, meaning that, to avoid the computer becoming excessively hot, achieving reversibility has become an important goal even for classical computers. But in quantum computation, achieving reversibility has a double importance; first, useful quantum computers will inevitably have components packed close together, but second, since quantum theory is reversible, quantum computation should also use reversible methods.

Feynman took a lot of convincing that the ideas of Landauer and Bennett were correct, and he thought about the possibility from a number of different points of view, always coming up with the same answer—reversible computation was in theory possible! But could it be achieved in practice? Feynman set himself and others the task of determining what components would be required to construct a reversible computer. Fredkin was also particularly keen to achieve this, but from rather a different point of view. As was mentioned in the previous chapter, he wished to develop the idea that the Universe was just a computer program, and, for this to be the case, it must be possible for computers, like the laws of physics, to be reversible.

As we explained in the previous chapter, two-bit computer gates with only a single output, such as *AND* and *OR*, cannot be reversible. For a reversible gate, we need at least as many outputs as inputs. We looked at the *CNOT* gate, which is a reversible gate, and is extremely important in quantum computation. For reversible classical computation, the most important gate that Feynman and Fredkin found was the Fredkin gate. Classically it may be said to be universal in the sense that any classical reversible logic can be constructed using Fredkin gates.

The Fredkin gate has three inputs and three outputs, like the *CNOT* gate including one control *C*. The other two inputs swap values if and only if *C* is equal to 1. The truth table is shown in Figure 16.3. Like the *CNOT* gate, the Fredkin gate is clearly reversible, because, just as each input leads to a unique output, each output is the result of a unique input. Another way of seeing this is to note that, if the output is passed through another Fredkin gate, we regain the input state; the Fredkin gate, like the *CNOT* gate, is its own inverse.

Another useful gate discovered in this period was the Toffoli gate. This may be called the controlled-controlled-*NOT* gate or the C^2NOT gate because, as seen in the truth table in Figure 16.4, there are two control inputs C_1 and C_2, and the other input *A* swaps between 0 and 1 only if *both* the control bits are 1. Like the Fredkin gate, the Toffoli gate is clearly reversible and also clearly its own inverse.

C	A	B	C'	A'	B'
0	0	0	0	0	0
0	0	1	0	0	1
0	1	0	0	1	0
0	1	1	0	1	1
1	0	0	1	0	0
1	0	1	1	1	0
1	1	0	1	0	1
1	1	1	1	1	1

Fig. 16.3 Truth table for the Fredkin gate. *C* is a control; *A* and *B* interchange values if, but only if, *C* is equal to 1.

C_1	C_2	A	C_1'	C_2'	A'
0	0	0	0	0	0
0	0	1	0	0	1
0	1	0	0	1	0
0	1	1	0	1	1
1	0	0	1	0	0
1	0	1	1	0	1
1	1	0	1	1	1
1	1	1	1	1	0

Fig. 16.4 Truth table for the Toffolli gate. C_1 and C_2 are both controls; *A* changes its value if, but only if, both C_1 and C_2 are equal to 1.

Unlike quantum computation, reversible classical computation requires a three-bit gate, and either the Fredkin gate or the Toffoli gate is suitable; the Toffoli gate is much discussed, but the Fredkin gate has an advantage for some purposes as we shall shortly see.

Feynman was obviously impressed by the Landauer–Bennett argument and the truth tables for reversible gates, but felt it was all still a little abstract, and a practical model would be useful. And with some help from Toffoli, Fredkin now designed a so-called billiard-ball computer. In this scheme, billiard balls were rolled along a smooth table; suitable collisions and reflections by 'mirrors' were arranged, and timings were such that the pattern of balls emerging from the 'computer' gave exactly the information corresponding to a particular gate for a particular input, with the gates being as complicated as the Fredkin gate. The latter was chosen rather than the Toffoli gate, as the two bottom lines in the truth table for the Toffoli gate do not conserve 1's, which is highly inconvenient for a billiard-ball computer where the 1's are actually billiard balls!

Of course the computer could not be expected to work very effectively in practice. Energy would be lost through friction and in collisions, which would inevitably not be perfectly elastic. Timings and directions of travel would obviously degenerate, and the functioning of the computer would deteriorate throughout any computation. But the important point was that the work showed that the actual laws of physics did not rule out reversible computation. And indeed Fredkin invented, independently of Bennett, the method of running the program in reverse to remove 'garbage' reversibility, which was particularly important for a billiard-ball computer, for which leaving garbage is effectively losing billiard balls.

In the second of his important papers, Feynman explained much of the contents of this section, and sketched how the various gates might be put together to construct a quantum computer. It is interesting to ask how near he was to reaching the idea of genuinely quantum computation, as we shall describe it in the next chapter. Landauer suggests that, if only he had put together the contents of his two papers, he would have anticipated Deutsch, and argues that that it is actually remarkable that he did not do so. However, Hopfield believes that Feynman had no idea that quantum computers could in principle be better than classical computers. His important argument was that the scale and speed of computation should not be limited by the classical world, and he really did not realize that quantum computers required a totally different theoretical basis from that of classical computers. That insight would indeed have to wait for Deutsch.

References

1. J. R. Brown, *Quest for the Quantum Computer* (Riverside, NY: Simon and Schuster, 2001). [hardback: *Minds, Machines and the Multiverse: The Quest for the Quantum Computer* (2000)]

2. J. Gleick, *Genius: The Life and Science of Richard Feynman* (New York: Pantheon, 1992).
3. R. P. Feynman, R. B. Leighton and M. Sands, *The Feynman Lectures on Physics*, 3 vols (Reading, MA: Addison-Wesley, 1965).
4. R. P. Feynman, *Feynman Lectures on Computation*, ed. A. J. G. Hey and R. W. Allen (Reading, MA: Addison-Wesley, 1996).
5. A. J. G. Hey (ed.), *Feynman and Computation* (Reading, MA: Perseus, 1999).
6. R. Flood, M. McCartney and A. Whitaker (eds.), *Kelvin: Life, Labours and Legacy* (Oxford: Oxford University Press, 2008).
7. P. V. Coveney and R. Highfield, *The Arrow of Time* (London: Allen, 1990).
8. B. Mahon, *The Man who Changed Everything: The Life of James Clerk Maxwell* (Chichester: Wiley, 2003).
9. H. S. Leff and A. F. Rex, *Maxwell's Demon: Entropy, Information and Computing* (Bristol: Adam Hilger, 1990);*Maxwell's Demon 2: Entropy, Classical and Quantum Information, Computing* (Bristol: Institute of Physics, 2003).
10. C. H. Bennett, 'Demons, Engines and the Second Law', *Scientific American* **257** (November 1985), 88–96.
11. O. Penrose, *Foundations of Statistical Mechanics* (Oxford: Pergamon, 1970).
12. C. H. Bennett, 'Logical Reversibility of Computation', *IBM Journal of Research and Development* **32** (1988), 16–23. Also in Ref. [9] (2003).

17
Quantum computation

Moore's law

We shall start this chapter with an argument that suggests that, like it or not, we must grapple with the ideas of quantum computation. (Later in this chapter we shall see that there are good reasons why we *should* like it!)

As early as 1964, Gordon Moore, the co-founder of Intel, picked up the point that computing power was increasing with enormous rapidity. By 1964, in fact, computers were still enormous machines occupying very large suites of rooms and requiring constant attention to such matters as temperature and humidity. A large university or business might possess one computer, not particularly powerful by today's standards, and all the users would submit their programs on cards or paper tape. But even that was an enormous advance on a decade earlier, when it was confidently predicted that the market for computers in the whole world might be counted on the fingers of one hand. Computers at that earlier time were valve-based, massive, exceptionally expensive and requiring practically constant attention by experts.

So in 1964 Moore announced what has since become known as 'Moore's law', that computing power, to be specific the number of transistors on a silicon chip, was doubling about every eighteen months. This implied that the number of electrons in a transistor was similarly dropping, and of course it also implied that, as computer components become closer and closer together, signals have to travel a much shorter distance between components, and the speed of the computer and thus its power will also increase dramatically.

In the years since, Moore's prediction had certainly been upheld. The number of transistors per chip has increased from about a thousand in the mid-1960s to about a billion by around 2010. A schoolchild today has a computing power that would outclass a mainframe computer of a few decades ago. And the number of electrons per transistor has plummeted from millions to hundreds. The latter is the important aspect from the present point of view. If Moore's law is to continue to hold, it seems that, perhaps quite soon, transistors will require to be made out of a very small number of electrons—perhaps three or four, and we will be well within the quantum regime. In other words we will need quantum computers.

David Deutsch and quantum computation

The great achievement of Deutsch [1] was to recognize that quantum computation needed its own theoretical base. He therefore rewrote the Church-Turing argument we met in Chapter 15, at each point replacing the classical physics that Turing had implicitly assumed by quantum physics. Deutsch was thus able to deduce the possibility of a universal quantum computer that could compute anything that any other quantum device could compute. This was analogous to the universal Turing machine that could compute anything any other Turing machine, essentially any classical computer, could compute.

I have said 'analogous to', but that cannot be the full statement. In physics we do not have classical mechanics and quantum mechanics, as it were, on an equal footing. The Universe *is* quantum mechanical, and classical physics is never more than an approximation to quantum mechanics, sometimes, of course, an extremely good one. Similarly the quantum theory of computation is *the* theory of computation, and the Church-Turing scheme is, at best, a set of rules that are obeyed in certain circumstances, or an approximation to the new Deutsch scheme.

Thus, as Deutsch has said, the classical theory of computation, which had been unchallenged for more than fifty years, was now logically, if not practically, obsolete. The challenge was now obviously to explore the properties of quantum computers, to determine whether they could perform tasks that classical computers could not perform or could not perform efficiently, and, if it was felt worthwhile, to design and to make them.

This was all clearly extremely important from a conceptual point of view, but it was not immediately obvious that it was very interesting practically. For around a decade after Deutsch's work, to be frank, it was far from clear that quantum computers could perform tasks outside the repertoire of classical computers, and, since it was obvious that constructing a quantum computer would be extremely difficult, there seemed little point in pursuing the vision.

To explain why this seemed to be the obvious conclusion, we need to spell out some of the basic features of any quantum computer [2–4]. We shall indeed see that while, at first sight, the idea of the quantum computer seemed full of promise, at second sight this apparent promise rather evaporated. (Fortunately there was to be a third sight, but this would not be obvious for ten years or so.)

Perhaps the most fundamental difference between classical and quantum computing is that the bit must be replaced by the *qubit*. It will be remembered that the bit, or the basic unit of classical computing, takes one of the values 0 or 1. The qubit, christened by Ben Schumacher, is the basic unit of quantum computation. It also has two states, which, since they are quantum mechanical states, we may write as $|0\rangle$ and $|1\rangle$. In quantum computing these may be the two states of a spin-½ system, with the z-component of the spin being equal to $\hbar/2$ and $-\hbar/2$, respectively, or horizontal and vertical polarization of a photon, or the two lowest energy levels of an atomic system, or two states of any other system that may be convenient.

However, because of the superposition principle in quantum theory, the concept of the qubit is enormously richer than that of the bit. While the bit can *only* take one of the values 0 and 1, the qubit may be in a superposition of $|0\rangle$ and $|1\rangle$. In other words it may be in a state $c_0|0\rangle + c_1|1\rangle$, where c_0 and c_1 may take any values at all provided the state is normalized., If c_0 and c_1 are (mathematically) real, normalization requires that $c_0^2 + c_1^2 = 1$. If either is (mathematically) complex, and we shall meet examples like this shortly, the condition is $|c_0|^2 + |c_1|^2 = 1$, where $|c_0|$ is the modulus of c_0.

It is interesting, by the way, that this difference between the bit and the qubit means that the clear distinction between digital and analog classical computers is completely absent in the quantum case. In classical computation, a digital computer handles discrete quantities, while an analog computer uses continuous quantities, like, for example, a voltage. However, in the quantum case, the distinction between $|0\rangle$ and $|1\rangle$ is discrete, but the quantum superposition behaviour is essentially analog, since c_0 and c_1 change continuously. So there are no quantum digital computers and no quantum analog computers, just quantum computers.

We shall now move onto a number of qubits collected in a register. Let us consider, for example, a system of 4 qubits represented in a four-state quantum register. Since each state of the register may be $|0\rangle$ or $|1\rangle$, or in other words take one of two values, the number of possible states of the overall system is just 2^4 or 16, and they range from $|0000\rangle$, through $|0001\rangle$ and $|0010\rangle$, right up to $|1111\rangle$. For a system of n qubits and so an n-state quantum register, the total number of states that can be represented is 2^n.

At first sight it might indeed seem that we have a great advantage over the classical case. Suppose we want to perform a particular classical computation for each of the 2^n states of an n-state classical system. Clearly we must do 2^n computations. Now translate this to the quantum case. Rather than insisting that our initial state is just one of the 2^n possibilities, we may ensure that it is a superposition of *all* the possibilities. Mathematically we may write it as $\sum_x c_x|x\rangle$, where $|x\rangle$ is a general member of our set of 2^n states, and we ensure that all the cx 's are non-zero.

Then let us suppose that the output corresponding to a particular $|x\rangle$ is $|f(x)\rangle$. Then the output for our general input state of $\sum_x c_x |x\rangle$ must be $\sum_x c_x |x\rangle\, |f(x)\rangle$. Here we have deliberately retained the $|x\rangle$ as well as the $|f(x)\rangle$ to ensure reversibility as explained previously. The important point is that it seems we have done all the 2^n calculations in one fell swoop, whereas classically we would need to do each individually. Thus at first sight it seems that quantum computation should be enormously powerful, being hugely faster than classical computation. This is what Deutsch has described as 'quantum parallelism'.

Second sight, as we have said, must bring us down to earth. For it is one thing to say that the quantum computer has calculated all the 2^n values; it is quite a different matter for us to find out what they are. We must obtain a result, or, in quantum mechanical terms, we must perform a measurement on the output from the computer. But we know that when we do this, we will obtain just one of the answers that the computer has provided; all the others will be lost.

The frustrating conclusion is that, for all the apparent potential of the quantum computer, it is far from obvious that it will be of any use at all! As we shall see in the following section, Deutsch was able to take a very tentative step towards demonstrating that there is indeed a use for quantum computation, but it would be a further ten years before the argument was made totally convincingly.

Here, having described qubits and quantum registers, we move on to the other main components of the quantum computer, the quantum logic gates. These gates are assembled into quantum networks to construct the actual quantum computer.

We shall use quantum gates that are reversible, and some of our gates may be based on particular classical gates that are themselves reversible. However, just as the qubit has much richer possibilities than the bit, so there are many more possibilities for quantum logic gates than for reversible classical gates.

For single-qubit gates, we have, of course, the identity and *NOT* gates, but several other gates are much more important. An important type of gate merely changes the relative phase between the two qubits. The $\pi/8$ gate performs the operation

$$|0\rangle \to |0\rangle; \qquad |1\rangle \to \frac{1+i}{\sqrt{2}}|1\rangle,$$

while the phase gate does the following:

$$|0\rangle \to |0\rangle; \qquad |1\rangle \to i|1\rangle$$

In both cases the i does exactly what we said; it changes the phase of the $|1\rangle$.

The third gate we introduce is called the Hadamard gate, and the output of this gate consists of linear combinations of the input states. Its effect is

$$|0\rangle \rightarrow \left(1/\sqrt{2}\right)\left\{|0\rangle + |1\rangle\right\}, \quad |1\rangle \rightarrow \left(1/\sqrt{2}\right)\left\{|0\rangle - |1\rangle\right\}.$$

Superposition is one of the two main properties of the quantum computer that, we shall find, do make it superior to the classical computer, at least for some problems. Thus we shall not be surprised to find that the Hadamard gate plays a crucial role in the performance of quantum computation.

Indeed the Hadamard gate might be regarded as the simplest example of a means of performing what we shall call a Fourier transform. To give a very rough idea of the importance of such methods in quantum computation, we shall look briefly at Fourier methods in classical physics, in particular in music. Let us say that we would like to obtain some appreciation of a piece of music by measuring a few parameters. A brute force approach would be to pick a random time and measure the amplitude of the music at that time, or at a few such times. Clearly that would give practically no understanding of the music as a whole. We could say that this is analogous to the method we have described so far of allowing the quantum computer to produce many outputs, but then, by performing a measurement, getting a single answer, which is probably totally unrepresentative.

A more sophisticated approach to the analysis of the music would be to use Fourier methods, in other words perhaps to find the amplitude of the music, not at a fixed time, but at a few strategically chosen frequencies. This may enable us to develop a rather greater understanding of the structure of the musical piece.

And this in turn may give at least *some* understanding of the way in which the successful quantum computing algorithms may give us exceptionally valuable information. We may say that, rather than just taking the whole of a single answer obtained by a brute force measurement, we sample part of all of them, in a way cleverly designed to give us the result we actually require.

We have said that superposition is one of the two central tools of the quantum computer; the other is entanglement, and we shall now see that the *CNOT* gate achieves this. The operation of this gate was described in Chapter 15, and we may summarize it as follows:

$$|00\rangle \rightarrow |00\rangle; \quad |01\rangle \rightarrow |01\rangle; \quad |10\rangle \rightarrow |11\rangle; \quad |11\rangle \rightarrow |10\rangle.$$

In other words, the second qubit swaps between $|0\rangle$ and $|1\rangle$ if but only if the first qubit, which is the control qubit, is in state $|1\rangle$. The control qubit itself remains unchanged.

Let us now apply this *CNOT* gate to a state in which the first qubit is in state $\left(1/\sqrt{2}\right)\left\{|0\rangle+|1\rangle\right\}$, and the second qubit is in state $|0\rangle$. We may write the combined state as $\left(1/\sqrt{2}\right)\left\{|00\rangle+|10\rangle\right\}$, and this state is, of course, unentangled, since the state of each qubit may be stated separately, just as we did indeed initially state them.

However, if we apply the *CNOT* gate to this state, we obtain $\left(1/\sqrt{2}\right)\left\{|00\rangle+|11\rangle\right\}$, and this clearly *is* entangled, since it is not now possible to give the state of each qubit separately; both qubits are always in the same state, but this may be $|0\rangle$ or $|1\rangle$. Thus it is clear that the *CNOT* gate does cause entanglement, a crucial task in quantum computation.

In fact the $\pi/8$, phase, Hadamard and *CNOT* gates are a universal set of gates in quantum computation, in the sense that any reversible operation may be approximated to arbitrary accuracy by this set of gates.

This is in contrast to classical computation, where a three-bit gate is required in any universal set of gates; this may be either the Fredkin or the Toffoli gate. Actually it is possible to use either of these gates in quantum computation, but in quantum computation their use as such is unnecessary, because it is possible to construct these three-qubit gates, or in other words perform all the tasks of these gates, using gates with one or two qubits. This is not possible in classical computation.

It is highly beneficial for the possibility of constructing quantum computers that only a small number of different gates is required, and particularly that they are restricted to single-qubit gates and a double qubit-gate. Quantum computers are exceptionally vulnerable to many types of error, but particularly to those caused by interaction with the environment, which has a very high probability of causing decoherence. Decoherence is the degradation of the desired superposition to the very much undesired mixture.

The *CNOT* gate will obviously be considerably more difficult to construct than any single-qubit gate, because it must be extremely difficult to achieve the required interaction between two qubits, while entirely avoiding any interaction beyond these qubits. Obviously if a three-qubit gate were required, the difficulty would be increased yet further.

In the remainder of this chapter and the next, we shall sketch the types of problems that cannot be solved by classical computers but are soluble with quantum computers, and mention a little about how quantum computers may actually be constructed. First, though, we shall look at a further part of Deutsch's early work—the so-called Deutsch algorithm.

The Deutsch algorithm

Deutsch's paper did not at once attract a great deal of interest. One of the reasons may have been that, rather ironically because Deutsch was stressing that his work was physical in nature, it still seemed to a casual reader rather mathematical or even philosophical in nature. It certainly did require an interest in, and some knowledge of, both quantum theory and computation, and, apart from Feynman and some of his colleagues, there were very few scientists with this wide knowledge.

Probably even more important, though, was the fact that it was far from clear that quantum computation could be of any actual use. We have already noted that, though the quantum computer appears to be doing a vast amount of work, we can only take advantage of a minute fraction of this work.

In his original paper, Deutsch did come up with one application of quantum computation—the Deutsch algorithm. It was tremendously important from a conceptual point of view, because it was a clear violation of the original Church-Turing argument that no computer could perform better than a universal classical computer. For the Deutsch algorithm, a quantum computer could beat a classical computer by a factor of 2. Also Deutsch's argument was extremely inventive—simple in principle but subtle in detail.

Yet the general importance and considerable interest of Deutsch's algorithm could not disguise the fact that, though Deutsch tried hard, it was difficult for him to convince anybody that the algorithm was of any use in practice. In particular, although, as we have said, it could produce results twice as fast as any classical computer, it was only successful in half the runs.

It seemed quite plausible to generalize this fact to a rule that what quantum computers might give you in one way, they would take away in another—although this suggestion actually was quite wrong. Artur Ekert (Figure 17.1) and his group in Oxford were later able to improve Deutsch's original algorithm so that it gave a result for every run, rather than just one run in two, but this improvement was not produced until 1998, well after, with the production of the Shor and Grover algorithms, it was already absolutely clear that quantum computers had a great deal to offer.

Deutsch's algorithm gave a solution to a precisely posed question. He considered evaluation of a function of x, $f(x)$, which it is assumed is complicated and thus requires a lengthy time to compute. Both x and $f(x)$ take one of two values, 0 and 1. Thus there are four possibilities: (a) both $f(0)$ and $f(1)$ are 0; (b) both are 1; (c) $f(0)$ is 0 and $f(1)$ is 1; or (d) $f(0)$ is 1 and $f(1)$ is 0. The question to be answered is: do we have one of cases (a) and (b) where $f(0)$ and $f(1)$ are the same, or one of cases (c) and (d) where they are different?

To solver this problem classically we would need to do two runs, computing first $f(0)$, and then $f(1)$. The answer is then obvious. What Deutsch was able to show was that, with his algorithm one could obtain the answer with only one run, though unfortunately only half the runs would be successful. In the other half of the runs, the programs will

Fig. 17.1 Artur Ekert [© Zentralbibliothek für Physik, Renate Bertlmann]

report that it has failed, and whether a particular run ends in success or failure is a matter of chance.

Deutsch attempted to build up interest in his algorithm by showing that it might solve a real problem. Let us suppose that we have 24 hours to decide whether to make an investment in the stock market, and suppose again that we will invest only if two indicators, which will be our $f(0)$ and $f(1)$, agree. However, an individual computation of f takes 24 hours. Classically our two computations will take 48 hours in total, so they will provide our answer too late for us to discover whether the two indicators agree, and whether they do so, to invest. With quantum computation, though, we will only require a single run, and we will have time to make the investment decision. Of course we will only get the results on one day in every two, but we will still be able, we hope, to make a killing on these days if the indicators are right.

The story is amusing, though clearly not convincing enough to encourage us to put major resources into developing a quantum computer, even one with the small number of gates that would be required. We would probably prefer to buy two classical computers! However, we shall describe in some detail how Deutsch's algorithm works, because in this simplest possible application of quantum computation, it is possible to see clearly the tasks performed by the various quantum logic gates. Even those readers who do not wish to trace the algorithm through in detail may like to read it through, and get the general idea that it is a clever application of the principles of quantum theory, in particular the use of superposition and entanglement, and it applies the various gates cleverly; it is sufficiently long to be interesting, but each stage is actually reasonably straightforward.

We shall call the central processing unit of the circuit *U*; it has two inputs, and because it is reversible, two outputs. *U* has two tasks. The first is to calculate a value of $f(x)$. We send in an $|x\rangle$ as qubit 1, so if x is 0, *U* calculates $f(0)$; if x is 1, it calculates $f(1)$. Note that this is our lengthy calculation, taking 24 hours; it is a calculation we only have time to perform once, not twice. So far we do not seem to have advanced beyond the classical computational approach, since we have implied we will calculate one or the other of $f(0)$ and $f(1)$. We shall see shortly how we can do better.

However, we first explain the other task of *U*, which can be called an f-controlled *NOT* gate. The input to *U* for qubit 2 is $|0\rangle$, but the second function of *U* is to change it to $|1\rangle$ if but only if $f(x)$ is 1. So let us look at the various cases. If x is 0, then qubit 1 will remain as $|0\rangle$; qubit 2 will remain as $|0\rangle$ if $f(0)$ is 0, but will change to 1 if $f(0)$ is 1. In either case the combined state of the system will be $|0, f(0)\rangle$.

If, however, x is 1, then qubit 1 will remain as $|1\rangle$; qubit 2 will remain as $|0\rangle$ if $f(1)$ is 0, but will change to $|1\rangle$ if $f(1)$ is 1. In either case the combined state of the system will be $|1, f(1)\rangle$. To sum up both cases, we may say that this second task of *U* is to write the value of $f(x)$ onto qubit 2.

However, now we come to how we make an advance on the classical computation. Rather than sending in either $|0\rangle$ or $|1\rangle$ to *U*, we send in a linear combination of both. Our input to the whole circuit is $|0\rangle$, but immediately this is sent through a Hadamard gate, and the output $\left(1/\sqrt{2}\right)\left(|0\rangle+|1\rangle\right)$ is sent into *U* as qubit 1. We can see the power of the Hadamard gate in allowing us to sample a number of possibilities rather than having just one of the possibilities on a take-it-or-leave-it basis.

Since an input of $|0\rangle$ to *U* yields an output of $|0, f(0)\rangle$, and an input of $|1\rangle$ yields $|1, f(1)\rangle$, the laws of quantum theory tell us that an input of $\left(1/\sqrt{2}\right)\left(|0\rangle+|1\rangle\right)$ must yield $\left(1/\sqrt{2}\right)\left(|0, f(0)\rangle+|1, f(1)\rangle\right)$.

The interesting point now is that the output is of a very different nature depending on whether $f(0) = f(1)$. If they are equal, the output is unentangled. For case (a) from our initial list, it is $\left(1/\sqrt{2}\right)\left(|00\rangle+|10\rangle\right)$. This is clearly not entangled, because qubit 2 is definitely $|0\rangle$, while qubit 1 is definitely $\left(1/\sqrt{2}\right)\left(|0\rangle+|1\rangle\right)$ Similarly for case (b), the output $\left(1/\sqrt{2}\right)\left(|01\rangle+|11\rangle\right)$ is unentangled because qubit 2 is now definitely $|1\rangle$, and qubit 1 again $\left(1/\sqrt{2}\right)\left(|0\rangle+|1\rangle\right)$.

However, for cases (c) and (d), where $f(0)$ is not equal to $f(1)$, the output is entangled. For case (c) it is $\left(1/\sqrt{2}\right)\left(|00\rangle+|11\rangle\right)$, while for case (d) it is $\left(1/\sqrt{2}\right)\left(|01\rangle+|10\rangle\right)$. In each case, the results obtained in measurements on the two qubits are not independent. For (c), for example, if qubit 1 were found to be $|0\rangle$, then so also would be qubit 2, while if qubit 1 were found to be $|1\rangle$, then again so would qubit 2.

We see here how the use of a controlled gate may lead to entanglement. The gate used for the rest of the algorithm will be the Hadamard gate, and its task here may be described as causing interference between the various paths through the circuit.

First we send qubit 2 through a Hadamard gate. For cases (a) and (b) the state remains unentangled. In the case of (a), after passage through the gate, each qubit is in the state $\left(1/\sqrt{2}\right)\left(|0\rangle+|1\rangle\right)$. It is not necessary, but if we wish to, we can write the overall state as $\frac{1}{2}\left(|00\rangle+|01\rangle+|10\rangle+|11\rangle\right)$. Similarly, for case (b), after passage through the gate, qubit 1 is still in the state $\left(1/\sqrt{2}\right)\left(|0\rangle+|1\rangle\right)$, while qubit 2 is in the state $\left(1/\sqrt{2}\right)\left(|0\rangle-|1\rangle\right)$. If we wish we can write the overall state as $\frac{1}{2}\left(|00\rangle-|01\rangle+|10\rangle-|11\rangle\right)$.

For cases (c) and (d), the overall state remains entangled. Following the prescription for the Hadamard gate, we find that, for case (c), it becomes $\frac{1}{2}\left(|00\rangle+|01\rangle+|10\rangle-|11\rangle\right)$, while for case (d) it becomes $\frac{1}{2}\left(|00\rangle-|01\rangle+|10\rangle+|11\rangle\right)$.

We now wish to measure the state of qubit 2. In each case, of the four terms in the overall state, for two of them the state of the second qubit is $|1\rangle$ and for the other two it is $|0\rangle$, so it is clear that, again in each case, there must be a probability of ½ of obtaining each result. However, we are interested in how the state of qubit 1 depends on the result we obtain in measuring qubit 2. For cases (a) and (b), because the system is unentangled, the answer is completely straightforward. The state of qubit 1 is entirely independent of whatever happens with qubit 2, and so in each case qubit 1 will be left in the state $\left(1/\sqrt{2}\right)\left(|0\rangle+|1\rangle\right)$.

Things are not quite so straightforward for the entangled cases (c) and (d), where the state that qubit 1 is left in must depend on the measurement result for qubit 2. First suppose that qubit 2 is found to be in state $|0\rangle$. By the projection postulate, the final state of the system will contain only those terms where qubit 2 is in this state, and, for both cases (c) and (d), these are $|00\rangle+|10\rangle$ inside the bracket. The factor of ½ is the product of two factors, each of $1/\sqrt{2}$, one belonging to each qubit, we can say that that final state of qubit 1 will be $\left(1/\sqrt{2}\right)\left(|0\rangle+|1\rangle\right)$.

However, with equal probability of ½, the measurement on qubit 2 may find it in state $|1\rangle$. In this case, to find how qubit 1 will be left, we must extract the terms in the overall state for which qubit 2 is in this state. For (c) these are $|01\rangle-|11\rangle$, while for (d) they are $-|01\rangle+|11\rangle$. However, these differ only by an overall change of sign, and in quantum theory this corresponds just to an overall phase difference and is physically irrelevant. We shall therefore ignore it. Therefore we may write the final state of qubit 1 as $\left(1/\sqrt{2}\right)\left(|0\rangle - |1\rangle\right)$, again splitting the factor of ½ between the two qubits.

We may therefore sum up the situation as follows. If the measurement on qubit 2 gives the result $|0\rangle$, which occurs with a probability of ½, then in all cases qubit 1 is left in the state $\left(1/\sqrt{2}\right)\left(|0\rangle + |1\rangle\right)$. Since all cases are the same, clearly we cannot hope to distinguish between them and the method fails.

However, if the measurement on qubit 2 gives the result $|1\rangle$, which also occurs with probability ½, then, for cases (a) and (b), for which $f(0) = f(1)$, qubit 1 will be left in the state $\left(1/\sqrt{2}\right)\left(|0\rangle + |1\rangle\right)$, but for cases (c) and (d), for which f(0) is not equal to $f(1)$, qubit 1 will be left in the state $\left(1/\sqrt{2}\right)\left(|0\rangle - |1\rangle\right)$.

Clearly in the latter case we may hope to distinguish between the cases, and we may do so by yet another use of a Hadamard gate. Cases (a) and (b) will give a final state of $|0\rangle$ for qubit 1, while cases (c) and (d) will give a final state of $|1\rangle$. This is exactly what we hoped to find!

However, at double the speed of a classical computer, but only successful in half the cases, Deutsch's algorithm was obviously of great conceptual importance, but not at all powerful! Thus it was extremely interesting when in 1992, Deutsch, together with Richard Jozsa, extended the algorithm considerably to give the Deutsch-Jozsa algorithm. In this case, the input consists of n qubits, each with a value f of either 0 or 1. So the number of input functions, by which is meant the number of different sets of values of f for all n qubits, is 2^n. The stipulation is that either all the 2^n values are the same or they are balanced, in the sense that f takes its possible values of 0 and 1 on exactly the same number of occasions. The problem is to find whether the set of values are the same or balanced.

To solve the problem classically, all the 2^n values must be calculated, but the Deutsch–Jozsa algorithm for a quantum computer was able to solve the problem with a single run. There is now a serious gain for the quantum computer. The number 2^n increases exponentially with n, so Deutsch and Jozsa can claim an exponential gain on moving from classical to quantum computation. Theoretically this is extremely important, and it is the type of claim we shall be meeting again later in this chapter.

However, it must still be admitted that, with its rather artificial requirement for f to be either constant or balanced, the Deutsch-Jozsa algorithm is even less realistic than the Deutsch algorithm. Until 1994, it was still generally assumed that quantum computation would never be able to improve on classical computation for any serious problem. It was in 1994 that this was all to change with the coming of Shor's algorithm.

Shor's algorithm

We shall now briefly return to the earlier parts of the book. Part I summarized the achievement of the First Quantum Age, but also the limitations that it placed on free thought. Most of Part II dealt with the struggle of a gallant few, EPR, Schrödinger, Bohm, Bell, the gallant experimentalists and theoreticians who took up the challenge of testing Bell's inequalities, and others, to challenge conventional wisdom and take on the powerful men of Copenhagen.

While it was gradually admitted that Bell's work was a genuine and important contribution to our understanding of quantum theory, and even though some of the ideas reached the textbooks, in the 1970s and 1980s the field of foundational studies in quantum theory was still largely looked on as esoteric and very much a minority interest. The cases of Clauser and Zeh in particular show that the 'stigma' died very slowly. Even those few who were interested in the topic, or even worked on it, would hardly have thought that its status was likely to change in the foreseeable future, or that it would have practical application, or indeed that money would soon be flowing in.

Even among those who were indeed working in the area tended to use language that would be seen by outsiders as rather negative and troubled. Entanglement was definitely a great worry for Schrödinger, loss of realism, determinism and locality seemed to be a cause for dissatisfaction and concern, and while the intellectual challenges were certainly huge, the possibility of genuine progress seemed low.

Neither Feynman's nor Deutsch's important work on quantum computation seemed fruitful enough to change the common view. But quite amazingly, and indeed quite quickly around 1994, things changed dramatically. Suddenly entanglement, instead of being a worry, something to explain away, became an 'information resource', in the words of Steane [5] in an excellent review of quantum information theory that appeared in 1998. The phrases with a negative connotation, such as 'non-determinism', 'loss of realism' and 'non-locality' were quietly sidelined, and replaced by the kinds of topic that could be followed using this new approach to quantum theory—'quantum computing', 'quantum cryptography' and 'quantum teleportation'.

Within a period of months, and with the aid of a few crucial discoveries, quantum information theory had become the hottest of hot topics. National security demanded that governments put their money into quantum computation and quantum cryptography, and indeed money flowed in. The most prestigious physics journals contained numerous articles both on quantum information itself and on the study of quantum foundations, which was seen to have spawned the younger topic and was continuing to stimulate it.

While the all-important proposal for constructing a controlled-*NOT* gate for the ion-trap quantum computer played a part in all this, it was undoubtedly the demonstration by Peter Shor [6] in 1994 itself that a quantum computer could solve a problem that was not only insoluble by classical means, but was also of great importance for national and commercial security, that was largely responsible for this dramatic shift in opinion.

The problem was of factorizing large numbers, and I shall first explain why it is so interesting scientifically and so important practically [7]. We shall start by considering multiplication, in fact multiplying two prime numbers together. This is a straightforward process, with which we are all familiar. If both numbers have n digits, the number of operations in the method or algorithm we all learned at school needs roughly n^2 operations. Actually there is a more efficient algorithm available, which requires rather fewer operations, but in any case the task is easy. We ourselves can perform the multiplications for smaller numbers, while a computer can easily do the job for numbers with say 500 digits.

The crucial point for larger numbers is that the number of operations increases as a polynomial in n. Thus the increase in the computational resources as n increases is moderate. Problems like this are said to belong in complexity class P, standing for polynomial, and are regarded as tractable computationally.

However, if we are presented with a number that we are told is a product of two primes and asked to find them, the problem is enormously harder. For example, it is easy to show that 173 × 233 is 39,843, but, given, say, 40,309, it is enormously tedious to find its factors. Essentially we must start at the lowest prime numbers and work our way up, checking each to see whether it is a factor. We may need to go as far as the square root of 40,309, as one of the factors must be smaller than that.

Factorization of a number with n digits requires about $10^{n/2}$ operations, and the crucial difference from multiplication is that here the dependence on n is exponential, which means that the number of operations increases dramatically, far more speedily than any polynomial dependence. Problems of this type are called NP for non-deterministic polynomial. This means that they can notionally be solved polynomially, but only on a hypothetical computer able to make correct choices at every branching point in the program. In practice they cannot be solved polynomially.

Problems of NP class are referred to as intractable computationally. Factorizing a number of 250 digits might take the world's most powerful computer of today about million years. We perhaps wish to be more careful, as, according to Moore's law, computers are likely to increase dramatically in power, and there have indeed been cases where prizes offered by magazines for factorizing large numbers have been won by a large number of computer users acting cooperatively.

However, if we go up to 300 digits, it seems that numbers should be impregnable; to factorize them would take all the computers in the world working together a time greater than the lifetime of the Universe! But of course this statement is essentially restricted to classical computers; as we shall see, for quantum computers the situation is entirely different.

The inability of the most powerful (classical) computer to factorize sufficiently large numbers is at the heart of today's security industry. It is this that makes the famous RSA algorithm, developed in 1977 by Ronald Rivest, Adi Shamir and Len Adleman, secure.

Cryptography, or the study of codes, is always discussed in terms of Alice and Bob being able to communicate securely without Eve, the eavesdropper, being able to ascertain the contents of their message. And the enormous advantage of the RSA algorithm over previous techniques is that it requires only the use of a public key; it does not require the sharing of a private key.

To explain the difference between private and public keys, we shall first look at the method of enabling secure communication by means of the private key; we may call this the one-time pad. This will be particularly important when we reach the topic of quantum cryptography in Chapter 19, because a more accurate name for quantum cryptography would just be quantum (private) key distribution.

The use of the private key is one of most secure ways of passing confidential information between Alice and Bob. The method was actually developed and its total security established by Shannon, the founder of information theory, whom we met in Chapter 15. Alice and Bob are each in possession of a 'pad' of keys. Alice then uses the first key on the pad to encrypt her first message using a standard algorithm. The encrypted message is sent to Bob, who can decrypt it using the same key. If Eve intercepts the message, she will not be able to decrypt it since she does not have the key.

It must be stressed that it is essential in the use of this technique of cryptography that a particular key is *not* reused. Each key must be used only once, hence the term 'one-time pad'. The reason is that although, as we have said, from a single encrypted message Eve can deduce absolutely nothing about the message itself, with two encrypted messages she is able to make sophisticated guesses about the key and thus about the messages themselves. It is a historical fact that several of the most famous spies in history have been caught by using a one-time pad more than once!

We have said that the method is highly secure, so what are its potential difficulties? First there is the question of length. Each key will have a certain number of binary digits. In principle it would seem possible for Eve to check all possibilities until she finds one that produces a readable decrypted method. Thus keys must be made sufficiently long that the number of possibilities is too large. In practice the number chosen is 10^{18}, which is essentially the decision of the US Government. Naturally there is a certain tension between maintaining security of commercial transactions, and the government being able to decipher the communications of criminals and terrorists. It is assumed that the figure of 10^{18} should be beyond the power of commercial organizations, but not that of government with its superior computational resources.

The other difficulty is the highly practical one of distributing keys to Alice and Bob. It is easy, of course, if they may get together every so often to exchange keys, or, failing that, if couriers may be used, but in the latter case one needs to be sure that the couriers are trustworthy, and that their travel around the world will not be prevented by Eve. By the 1970s, the amount of material being transported by government, military and commercial organizations was practically overwhelming.

Since then another dimension has been added to the problem by the requirement for Internet security. More and more commercial transactions, and also exchanges of non-commercial material still requiring security, such as medical records, take place on the Internet. In exchanges between individuals and banks or commercial organizations, use of a one-time pad would not normally be possible. (It should be recognized, though, that banks may provide customers with calculators, primed to respond to the customer's password with a succession of numbers, each of which may be used in turn as part of the customer's logging-in.) A method of ensuring safe communication without use of a private key is clearly required, and this will almost certainly turn on the impossibility of factorizing large numbers.

In the RSA method, Alice picks two large prime numbers, p and q, and calculates the product of these numbers, pq, which we call N. She also picks another number, e, and N and e constitute her public key, which is announced openly for the benefit of anybody who wishes to send her coded messages. The numbers p and q themselves are kept strictly secret.

We now suppose that Bob wants to send her a message. He first changes the words into binary digits, probably using the ASCII (the American Standard Code for Information Interchange). The resulting binary number is then changed to a decimal number M.

Bob now calculates C, the encrypted message to be sent to Alice, using the formula $C=M^e \pmod N$. Here we are using so-called modular arithmetic. To obtain $x \pmod y$, we take the remainder when x is divide by y. Thus for example, when 51 is divided by 6, the remainder is 3, so 51 (mod 6) is 3. The importance of this step is that if Bob knows M, it is straightforward for him to calculate C, but the reverse does not apply. The nature of modular arithmetic is such that, if Eve was to discover the value of C, then, since the numbers used in practical applications of the scheme will be very large, it is effectively impossible for her to deduce the value of M, the unencrypted message. Technically her problem is spoken of as finding a discrete logarithm, which, like factorizing of large numbers, was intractable for classical computers.

Alice, though, knows the value of p and q, the factors of N, and she can use some very old, but still powerful, mathematics. She finds another quantity d from the relation

$$de = 1\left(\mathrm{mod}\frac{p-1}{q-1}\right).$$

We may call d her *private key* or *decryption key*, and using it she may reverse Bob's encryption with the equation $M = C^d$ (mod N). So she how has the message that Bob sent, but it is essential to note the only reason that Eve cannot also obtain the message is that she cannot factorize N; if she could, she could do exactly the same as Alice. It is exactly this inability that quantum computation in the form of Shor's algorithm threatens to nullify.

So let us turn to Shor's work. Shor was yet another participant in our story who was working at Bell Labs at the time. His work was a theoretical tour de force, tackling together the problems of factorization, period finding and obtaining discrete logarithms, all related, and all intractable using classical computers. By calling the problems intractable, we mean that the computer resources required increase exponentially with the scale of the problem. We have already had other examples of such problems—Feynman's attempt to simulate quantum systems with classical computers, and the problem solved by the Deutsch-Jozsa algorithm.

Shor was able to devise a method by which all these problems could be solved in polynomial time. It is not possible to give specific times, because much depends on details of the implementation. However, Richard Hughes [8] has estimated that, even with arrangements that may be far from optimal, factorizing a number of ℓ bits might require about 5ℓ qubits, and would take a time proportional to ℓ^3—note the all-important polynomial rather than exponential dependence on ℓ.

A 512-bit number, which would be about 155 decimal digits, which is about the largest number factorizable today, might take about 30 seconds to factorize using Shor's algorithm. Those whose security relies on schemes such as RSA will be even more concerned by the fact that, for a 1024-bit number, today regarded as completely impregnable, would take only 2^3 or 8 times as long to factorize—about 4 minutes.

Before looking at some aspects of the algorithm itself, I shall make two general comments. The first is that Shor's achievement is made even more remarkable than already implied by the fact that it came at a time when it was assumed by most people that such schemes could not exist—that quantum computers could not do anything very remarkable. To cause such a rapid change in general belief was breathtaking.

The second question is whether we should be worried that at some times in the future today's secrets may be revealed. Our answer will depend on the type of secret. When you readily submit PIN numbers and credit card numbers to a computer banking system, to be used in RSA-type procedures, you certainly don't want the system to break down today, so that all your details become known to all and sundry. However, you will not be at all bothered by the fact that the ability of a quantum computer to factorize large numbers may mean that, in ten or fifty years time, your password of today may become known.

The same does not apply to governments sending encrypted messages containing the names of agents, or details of alliances or agreements that they wish to keep secret for ever. They must clearly be wary of the fact that a foreign power may obtain and retain such encrypted messages, hoping that, some time in the future, once quantum computers are available, the dirty deeds of the past may be exposed for all to see.

In sketching Shor's algorithm, I shall say no more about discrete logarithms, but will discuss the relationship between factorization and period finding, which uses a subtle property of numbers.

The numbers we shall be considering will be products of two distinct odd numbers, so the smallest contender is 15. We shall start with this number which we call N, and we will use another number, which we shall call x. We take values for x from 2 up to $N-1$, though we shall leave out the actual factors of 15, that is 3 and 5. (It may seem strange that we are making use of these numbers, since these are, of course, precisely the numbers that our algorithm is supposed to find. The point is that in actual implementations of the scheme, numbers used would be enormously large, so accidentally alighting on one of the factors would just not happen.)

We now choose a value of x, say $x = 2$, and write down the sequence of numbers x, x^2, x^3, x^4, x^5,...This would give us 2, 4, 8, 16, 32, 64, 128....Let us now adjust this so that, instead of x^n, we write down $x^n(\text{mod } N)$ or here $x^n(\text{mod } 15)$. This will give us 2, 4, 8, 1, 2, 4, 8,..., and in fact these same four numbers will repeat indefinitely. We may say that the numbers form a periodic function of period 4, which we shall call f.

If we choose $x = 7$, the sequence we will obtain is 7, 4, 13, 1, 7,..., and the period, f, is again 4. However, if we choose $x = 4$, we obtain 4, 1, 4, 1,..., and the value of f is 2. For $x = 11$, the sequence is 11, 1, 11, 1,..., and again f is 2, and f is also 2 for $x = 14$, for which the sequence is just 14, 1, 14,....

A little algebra is now given, but please feel free to take this as read if you wish. If our working is (mod N), then we may write $x^{f+1} = x$ (mod N). This just means that we start with x, we multiply this by itself f times, and our answer (mod N) is just x; this follows from the definition of f.

Now we divide both sides by x and obtain $x^f = 1$ (mod N). (Technically here, because our arithmetic is modular, we must notice that x and N

cannot have any common factors, from the definition of x.) We now have the result that $x^f - 1 = 0 \pmod{N}$. In fact this means that $x^f - 1$ may be 0, N, $2N$, $3N$,.... The first possibility of 0 can easily be ruled out, for x must be greater than 1, and f is at least 1.

So let us try the next possibility, $x^f - 1 = N$. By the well-known mathematical rule that $a^2-b^2=(a+b)(a-b)$, which is always called 'the difference of two squares', we may obtain $x^f - 1 = (x^{f/2}+1)(x^{f/2}-1) = N$. Provided f is even, this tells us that $x^{f/2}+1$ and $x^{f/2}-1$ are factors of N, which are, of course, exactly what we are after. For the case $x = 2$, for which $f = 4$, these are $2^2 + 1$ and 2^2-1, or just 5 and 3, which is exactly the answers we hope for!

However, in other cases, $x^f - 1$ will be a multiple of N. For example, for the case $x = 11$, we found that f was equal to 2, so $x^f - 1$ is equal to 120, which is $8N$. In this case $x^{f/2}+1$ and $x^{f/2}-1$ are equal to 12 and 10, respectively. These will usually be multiples of the desired factors, and indeed here 12 is a multiple of 3, and 10 is a multiple of 5.

How, though, can we obtain the required factors, 3 and 5, from the numbers we have produced, 12 and 10? We can easily note that the highest (or maximum) common factor (HCF or MCF) of 12 and 15 is 3, while the HCF of 10 and 15 is 5. But such trivial manipulation will not be possible when we are dealing with the large numbers of an actual application.

Rather we may use a well-established method of Euclid. To find the HCF of 10 and 15, we divide the larger number by the smaller, and keep the remainder, which is 5. We then divide the smaller original number, 10, by 5 and find there is no remainder. This tells us that 5 is the HCF of 10 and 15. Similarly we may show that 3 is the HCF of 12 and 15.

This has established the connection between factorization and period finding, but unfortunately there will be cases where the method does not work. The value of f may be odd, or, for example, we may take $x = 14$, which, as we have shown, gives $f = 2$. Then $x^{f/2}+1$ and $x^{f/2}-1$ are 15 and 13. Unfortunately this is just not helpful! But, for an arbitrarily chosen value of x, the probability of success is greater than ¾, and if one value leads to failure, others may be tried.

Overall we may see that, while Shor's work was a massive step forward, and although it was somewhat complicated, it did not require any dramatically new concepts or any advanced mathematics, though admittedly the well-established mathematics that it did use is not particularly well known to most people today!

The actual quantum circuit necessary to implement Shor's algorithm is also quite complicated, but in fact it is broadly analogous to that required for implementation of the Deutsch algorithm. We start with the number to be factorized, N, and a value of x is chosen. First one must check by Euclid's method that it has no common factor with N, though of course in real cases it would be enormously improbable that we happened to pick a value of x that *is* a factor.

Then Hadamard gates are used to set the first of two quantum registers into a superposition of states representing all numbers a up to some limit. This limit is set by the number of qubits available. As for the Deutsch algorithm, there is a gate U, which in this case calculates $x^a \pmod N$, so that the second register holds the values of this for all values of a up to the same limit. The states of the two registers are entangled.

A Fourier transform is then applied to the first register by a series of logic gates. Because of the entanglement of the two registers, this Fourier transform produces the values of the function in the second register, and, if the values of the qubits on the first register are measured, there is a very good chance, as explained above, that the factors of N will be discovered.

Grover's algorithm

The other important and useful quantum algorithm discovered so far is due to Lov Grover, yet another employee of Bell Labs. This method is often described as searching a telephone directory by number [9] rather than by name. Clearly in a telephone directory, to search for a name is very easy because the names are in alphabetical order. However, if we are given a number, we must search bodily through the whole directory. If there are N names in the directory, we maybe have to search through all or nearly all N, but we may be lucky and find the number almost immediately; on average we will take $N/2$ steps. The great power of Grover's algorithm is that it reduces the number to about $\sqrt{N}$.

Since in real cases N must be an extremely large number this is clearly an enormous saving of effort. In a search through something like a million items, the number of steps required is reduced from something like a million to something like a thousand. It should be said that both N and $\sqrt{N}$, which we may write as $N^{1/2}$, are both polynomial or of complexity class P, so, unlike Shor's algorithm, Grover's algorithm does not change the complexity class of the problem. Nevertheless the saving of time will be substantial.

The areas of application go far beyond telephone directories, or any other types of directory. Any computer search engine is, of course, just searching through large numbers of pieces of information, and Grover's algorithm would make such processes much faster. It may also assist cryptography, searching through lists of keys, if examples of encrypted and unencrypted versions of the same text are available; or through all possible strategies in chess; or any problem where finding a solution is hard but checking a solution is easy, and Grover's algorithm may allow one to search through all possible solutions. It may be said that Grover's algorithm is less spectacular than Shor's algorithm, but probably of much wider potential use.

We now explain in fairly general terms how Grover's algorithm works. As always we attempt to utilize the quantum nature of the computation by examining all terms together, and manipulating them all together. In the algorithm, we start with a superposition of all the various items of information, but then manipulate this superposition very slightly but repeatedly so as to increase the contribution of the term we are searching for; we shall call this term $|i\rangle$.

The superposition is formed in a way we have become used to, by using a Hadamard gate. We then apply an operator that picks out $|i\rangle$ from the superposition and reverses its sign. Another Hadamard gate follows, then an operation that reverses the sign of all the terms in the superposition *apart from* $|i\rangle$, then yet another Hadamard gate. We are broadly back to where we started, apart form the fact that $|i\rangle$ has been treated very slightly differently from every other term in the superposition, and, in fact, the scheme is designed so that the coefficient of $|i\rangle$ increases very, very slightly.

We must repeat the process many times, and the coefficient of $|i\rangle$ continues increasing, but not indefinitely. It reaches a maximum and then starts decreasing again. Steane has quoted Karen Fuchs as likening the process to baking a soufflé. The desired result rises slowly but surely, but if the process is not caught at the right point, it, like the soufflé, will collapse again.

We shall now discuss this in a little more detail. Following the first Hadamard gate, all the N items in the directory are represented equally, and the laws of quantum theory tell us that the coefficient of each will be $\sqrt{1/N}$, a very small number of course. If we take a reading, we must obtain a single result, and this tells us that the possibility of obtaining any result, including, of course, $|i\rangle$, is equal to $1/N$.

As we shall see, it is mathematically convenient to write $\sqrt{1/N}$ as $\sin\theta$, where θ is an angle in radians. (In radians, $\pi/2$ radians = 90°.) Since $\sqrt{1/N}$ is very small, $\sin\theta$ is small, and it is a property of radians that, if an angle is small, $\sin\theta$ is very close to θ itself; the two may be treated as being equal.

We shall say that the coefficient of $|i\rangle$ as we work our way through the algorithm is $\sin\alpha$, so at the beginning α is just equal to θ. However, the first use of the Grover operations that we outlined above increases α to 3θ, so the coefficient of $|i\rangle$ is now $\sin 3\theta$, very nearly 3 times what it was before. As further Grover operations are used, α increase steadily to 5θ, 7θ... and so on. Thus the coefficient of $|i\rangle$ becomes successively $\sin 5\theta$, $\sin\theta$ and so on. After n operations, this is $\sin(2n-1)\theta$.

At first these successive values are larger and larger, but elementary trigonometry tells us that the sine function reaches a maximum value of 1, and then starts decreasing again. Clearly we must catch it at this maximum, when the coefficient of $|i\rangle$ will be 1, so a measurement is certain to give the particular answer we need.

It will reach this maximum value when $\sin(2n-1)\theta = 1$, but since $\sin \pi/2 = 1$, this means that $\sin(2n-1)\theta = \pi/2$. Since n must be large, we can approximate $(2n-1)$ by $2n$, and we thus obtain $n = \pi/(4\theta)$, but, remembering that $1/\sqrt{N} = \theta$, we finally reach the relation $n = \sqrt{N}(\pi/4)$. Remember that N is the number of items in the directory, and n is the number of runs required, so we see very clearly the result that we announced at the beginning of the section.

Decoherence and quantum error correction

The Shor and Grover algorithms should have convinced us that, if quantum computers can be built, they will definitely find important uses. Yet it must certainly be clear that computers with sufficient qubits and sufficient gates to use these algorithms will be extremely difficult to build. Indeed, at just about the time when these two algorithms were exciting everybody, some eminent quantum information scientists took the opportunity to express total pessimism about the possibility of them ever being implemented!

The reason is that, from several points of view, the operation of a quantum computer must be highly sensitive. The first reason is that in some ways quantum computers share the disadvantages of analog computers as compared with digital computers. A physical system representing a bit on a classical digital computer is analogous to a light switch. It requires just two positions of equilibrium, corresponding to 0 and 1. If the system moves away from one of these equilibrium positions, there is a strong restoring force to take it back. There will also be dissipation of energy as the system moves towards one of these equilibrium positions, so that it will come to rest rather than oscillating indefinitely.

Analog computers do not have these equilibrium positions, and thus do not have these positive features to prevent errors developing when the computer is in operation. This largely explains why digital computers and recording devices are more reliable than their analog equivalents, and why they have replaced them.

For a quantum computer, bit is replaced by qubit, and this means that, though for a classical computer, 0 and 1 are special, for a quantum computer, the states $|0\rangle$ and $|1\rangle$ are not. Any superposition of $|0\rangle$ and $|1\rangle$ has just as much (or as little) right to be regarded as fundamental. Thus there can be no stable potential minima. Also, because quantum computers will be reversible, there can be no dissipation of energy to settle the values of qubits.

Another massive problem for quantum computers is that their behaviour depends crucially on the maintenance of superpositions of states, including entangled states. This is what we have called coherence. As we discussed in Chapter 13, interaction with the environment will cause decoherence.

Suppose we have a state of a qubit given by $a_0|0\rangle+a_1|1\rangle$, which is coherent in the sense that $|0\rangle$ and $|1\rangle$ may interfere with each other. Interaction with the environment may lead to a state such as $a_0|0\rangle|e_0\rangle+b_1|1\rangle|e_1\rangle$. Here $|e_0\rangle$ and $|e_1\rangle$ are different environment states, and since these are macroscopic, not only do $|e_0\rangle$ and $|e_1\rangle$ fail to interfere, their presence prevents $|0\rangle$ and $|1\rangle$ from interfering. Though as we stressed in Chapter 13, there is no collapse of wave-function of the full system of qubit plus environment, when we restrict ourselves to considering the qubit on its own, it now behaves as a mixture of $|0\rangle$ and $|1\rangle$ rather than the superposition; this is why we talk of decoherence. This must certainly completely ruin the performance of the quantum computer.

The amount of decoherence may be reduced by clever techniques in the construction of the computer, but it will not be possible to eliminate it, and the gate in the quantum computer that suffers most must inevitably be the controlled-*NOT* gate, because it necessarily involves couplings between different spins, so inevitably may bring in other undesired couplings as well. Martin Plenio and Peter Knight have estimated that, on every use of this gate, the probability of decoherence occurring must be at least 10^{-6}.

This may seem remarkably low, but it must be remembered that, to be useful for demonstrating Shor's algorithm by factorizing numbers of perhaps 200 digits, thousands of qubits and billions of gates will be needed. A simple calculation shows that the amount of decoherence is still too high by a factor of about 10^7. As Steane says, it is easier to imagine increasing the speed of a classical computer by this factor than reducing the decoherence of a quantum computer by the same amount!

What it is hoped will save the day is quantum error correction, and a very large amount of successful work on providing the necessary codes has been done by Shor [10] and Steane [11] in particular. The codes for quantum error correction were broadly based on analogous codes for correcting errors in classical computation and communication, though we shall see that the quantum case brings in several extra problems. The classical techniques were very important in the early days of computers, when the computers used valves, which were prone to error. They became less necessary when the more reliable transistors were used. In classical communication, though, as distinct from classical computation, error codes have always been important as the channels will almost certainly be noisy.

To give an example of an error-correcting code quite away from computers or communication, we first mention the ISBN, or International Standard Book Number, which is at the front of every book. We briefly discuss the scheme used before 2007, after which date it became a little more complicated. In the ISBN, the first nine digits give information on the country of publication, the publishers and the individual book. The tenth digit, though, is a check digit. It is chosen so that if we calculate ten times the first digit, plus nine times the second digit,...plus twice times the ninth digit, plus the last digit, which is the actual check digit, the sum must be divisible by 11. (If 10 is required for this last digit, it is written as X.)

For example, if the first nine numbers are 0-521-67102, the sum so far gives 147. To make the complete sum divisible by 11, it must be made up to 154, or in other words, the check digit will be 7, and indeed the full ISBN is 0-521-67102-7. The presence of the check digit enables the presence of either of the two most common types of mistake, altering a digit or transposing two adjacent digits, to be made obvious.

To turn to the quantum computation case, in this section we do not want to give too much technical detail about the codes, but it is appropriate to explain the main principles of quantum error correction because it is such an important part of quantum computation.

The simplest check on a message is to add a parity bit. If the number of 1s is even, the bit is 0; if it is odd, the bit is 1. So 1011011 becomes 10110111. So if there is a single error in transmitting the message, whether in the original bits or the parity bit, this error is easily spotted, and the message must be sent again.

This technique can clearly only cope with a single error. A combination of two errors, either both in the original message or one in the message and one in the parity bit, will not be detected. If there is a significant probability of two errors occurring, additional check digits must be included. Another limitation is that the technique does not tell us where the error has occurred, so it cannot be automatically corrected; as has been said, the message must just be resent.

A more elaborate scheme does allow automatic correction of an error in most cases. In this scheme, we must have a number of lines of text, each with the same number of bits. We then add a parity bit for each line, and another parity bit for each column. Now imagine a single error in the whole block, though not, for the moment in any parity bit. Clearly this will cause an error in the row of the erroneous bit, and also in its column—parity bits for both row and column will fail to match the contents of their actual row or column. We will thus easily be able to trace the error and correct it.

If the parity bit of a row does not match up with its row, but there is no compensating mismatch for a column, it is clear that it is the parity bit itself that is in error (and of course the same with rows and columns interchanged). It is even clear that a number of errors may be spotted in this way, provided, of course, that there are not more than one error in any individual row or column.

We now introduce the Hamming code for classical computing, which can correct data with a higher proportion of errors than the schemes we have met so far. It is also the basis of Steane's system of error correction for quantum computers. We first note that there are 16 (or 2^4) possible 4-bit messages. In what we shall call the {7, 4, 3}

Hamming code, each of these is represented by a 7-bit codeword, and it is arranged that each of these codewords differs from all the others in at least three places. For example, 0000 is represented by 0000000, 0001 by 1010101, 1001 by 0101010 and so on.

We can note that the codewords for 0000 and 1001 differ in exactly three places. If a single error is made in each bit, 0000000 might become 0100000, and 0101010 might become 0100010, but they are still different in one place, and, as long as there is just the one error in each codeword, it will always be absolutely obvious what the correct bit should be.

We now turn to the actual quantum codes, and we must note a problem that is missing from classical computation but is important here; we must be careful never to perform a measurement on any qubit we are actually using for computation, because, in doing so, we would certainly disturb it—we would initiate a collapse of state or a decoherence.

Another difficulty for quantum computation that does not occur in the classical case is the so-called 'no-cloning' theorem. In the classical case, one way of tackling the possibility of error occurrence is to take a number of copies of a particular message or set of data. Then, even if an error occurs in one copy, the rest will give the correct result. However, the no-cloning theorem, which was proved independently in 1982 by Wooters and Zurek, and by Dieks, tells us that making such copies is impossible in the quantum case. It is not possible to find an operator U that will clone an arbitrary state that it receives.

The proof is simple. Such an operator must work as follows. It must cause the following change: $U|\alpha,0\rangle \rightarrow |\alpha,\alpha\rangle$ where the general state presented to the would-be cloner may be written as $|\alpha\rangle$, and the state of the second qubit, which is initially $|0\rangle$, becomes $|\alpha\rangle$ as a result of the cloning. Similarly it must cause $U|\beta,0\rangle \rightarrow |\beta,\beta\rangle$, where $|\beta\rangle$ is a different state.

Let us now imagine the same operator acting on the state with the first qubit in the state $\left(1/\sqrt{2}\right)\left(|\alpha\rangle+|\beta\rangle\right)$, and the second as always in the state $|0\rangle$. The laws of quantum theory tell us that we can add up the two previous equations, and insert the factor of $\left(1/\sqrt{2}\right)$. The final state of the system will then be $\left(1/\sqrt{2}\right)\left(|\alpha,\alpha\rangle+|\beta,\beta\rangle\right)$. But this is not a cloning of the original state. For a cloning, both qubits would have to end up in the state $\left(1/\sqrt{2}\right)\left(|\alpha\rangle+|\beta\rangle\right)$, and the combined state would have to be $\frac{1}{2}\left(|\alpha,\alpha\rangle+|\beta,\alpha\rangle+|\alpha,\beta\rangle+|\beta,\beta\rangle\right)$. This short argument makes it quite clear that, if we imagine such a cloning device, we will encounter inconsistencies, or, in other words, that such a device cannot exist.

Quantum error correction aims to solve both problems—the difficulties with measurement and the lack of cloning—by including extra qubits, called the *ancilla*, which is a Latin word for maidservant. These qubits take part in the computation process, but they do not actually contribute directly to the final answer, so there is no restriction on their measurement. In some ways it may be said that they play a role analogous to the copies of the original qubit that would be made if cloning were allowed.

To explain their use, let us first note that errors to single qubits may be complicated, but they can all be built up out of three basic types:

$$c_0|0\rangle + c_1|1\rangle \to c_0|1\rangle + c_1|0\rangle$$
$$c_0|0\rangle + c_1|1\rangle \to c_0|0\rangle - c_1|1\rangle.$$
$$c_0|0\rangle + c_1|1\rangle \to c_0|1\rangle - c_1|0\rangle$$

The first two are called the 'bit flip' and the 'phase flip', respectively, while the third is a combination of both.

There are schemes to cope with a single error using as few as five qubits for each qubit of information, but we shall sketch Shot's scheme, which was the first and is probably the simplest to describe. Shor's scheme uses nine qubits for each qubit of information.

First we shall see how to correct for bit flips during the transmission of information. This requires two extras qubits in an ancilla for each qubit that carries information. At the beginning of the process, both the qubits of the ancilla are in state $|0\rangle$, and in an encoder, two controlled-*NOT* gates are used to entangle each ancilla qubit with the information qubit. In this process, the ancilla qubits are flipped if but only if the information qubit is $|1\rangle$, so, if the initial state of the information qubit is $c_0|0\rangle + c_1|1\rangle$, the final entangled state of the combined system is $c_0|000\rangle + c_1|111\rangle$.

We now allow the qubits to propagate and then study how to determine what errors may have occurred and how to correct them. To do this we pass the qubits through a decoder, which, just like the encoder, consists of two controlled-*NOT* gates. Again each of the ancilla qubits flips if the information qubit is $|1\rangle$ but not otherwise.

First let us suppose that no error has occurred. In this case the decoder will change the entangled state $c_0|000\rangle + c_1|111\rangle$ back to $c_0|000\rangle + c_1|100\rangle$. This is a separable state; the state of the information qubit is $c_0|0\rangle + c_1|1\rangle$, and that of each ancilla qubit is $|0\rangle$. It is clear that the decoder has exactly undone the effect of the encoder.

Now let us see what happens if a bit flip has occurred in the second qubit (the first of the two ancilla qubits) in the transmission process. In this case the state of the combined system before decoding will be $c_0|010\rangle + c_1|101\rangle$, and this will be decoded as $c_0|010\rangle + c_1|110\rangle$. Again the state is separable; the state of each qubit is $c_0|0\rangle + c_1|1\rangle$, $|1\rangle$ and $|0\rangle$, respectively. Similarly, if the bit flip has occurred in the third qubit (the other ancilla qubit), the situation is as just described but with the two qubits of the ancilla interchanged.

If, though, the bit flip has occurred in the information qubit, the state of the combined system before decoding is $c_0|100\rangle + c_1|011\rangle$, and after decoding this will become $c_0|111\rangle + c_1|011\rangle$. Yet again the final state is separable, the information qubit being in the state $c_0|1\rangle + c_1|0\rangle$, and the two qubits of the ancilla both being in the state $|1\rangle$.

It is crucially important that in all cases the final state is, as we have stressed, unentangled. The next step we will take is to measure the qubits of the ancilla, and if there were any entanglement between these qubits and the information qubit, a measurement of the ancilla would disturb both, and this, of course, is exactly what the ancilla is there to avoid.

When we measure these qubits of the ancilla, we see that there are three possible results. It may be that both are $|0\rangle$, and that tells us that the information qubit has not suffered an error during the transmission process. It may be that one of them is $|0\rangle$ and the other $|1\rangle$; this also tells us that the information qubit is still correct; a bit flip has taken place in one of the ancilla qubits, and this is unimportant. Only if *both* ancilla qubits are found to be $|1\rangle$ is there an indication that the information has suffered a bit flip. Only in this case, therefore, do we apply a bit flip to the information qubit, which will restore it to the correct state $c_0|0\rangle + c_1|1\rangle$. In all cases the information qubit will now be in this state.

We have described this scheme in detail to give some flavour of how error codes work. Clearly it merely tackles one very small part of an enormous task. For a start it only tackles bit flips, but Shor was able to show how phase flips could also be corrected for: in his enhanced scheme a Hadamard gate is added to each output of the encoder and each input of the decoder. Three qubits are required for correction just of phase flips, and for correction of all the single errors for single qubits, nine qubits are required in Shor's method.

Even so we have only discussed correction of single errors in the transmission process. We have not discussed errors occurring in the actual computation process. For this we may use a technique called *fault-tolerant computation*.

And we should never assume that correction of single errors will be enough. In any working quantum computer, error rates are likely to be much higher. The techniques discussed in this section may be extended to cover more errors, but only at the expense of adding many more qubits, 23 to correct for three simultaneous errors, and as many as 87 to correct for seven.

It is natural to wonder whether things are not getting totally out of hand—does not the addition of more and more qubits merely add more and more possibility for errors? In fact the whole scheme *does* work, and extra qubits included to reduce overall error do succeed in doing so, provided the error rate for each component is kept low enough, less than something like 10^{-5}.

Overall there is a great improvement! Without error correction, the probability of decoherence every time a gate is used would have to be as low as 10^{-13} for quantum computation to be possible. This, Steane asserts, would be utterly impossible, purely on physical grounds. The figure of 10^{-5}, on the other hand, he describes as 'daunting but possible'. However, the downside is that, with error correction, the total number of operations may increase by a factor of thousands.

To conclude this section, we mention an entirely different approach to avoiding errors caused by decoherence. In Chapter 13, we met the quantum Zeno effect, which may be described as a way of inhibiting, or even totally suppressing, evolution of a quantum state. It is natural to consider that the effect might be used to suppress exactly the kind of decoherence that degrades the action of the quantum computer, and quite a lot of study of this possibility has taken place.

References

1. D. Deutsch, 'Quantum Theory, the Church-Turing Principle and the Universal Quantum Computer', *Proceedings of the Royal Society A* **400** (1985), 97–117.
2. J. Stolze and D. Suter, Quantum Computing: AShort Course from Theory to Experiment, 2nd edn (Weinheim: Wiley-VCH, 2008).
3. M. Le Bellac, Quantum Information and Quantum Computation (Cambridge: Cambridge University Press, 2006).
4. M. Nakahara and T. Ohmi, Quantum Computing: From Linear Algebra to Physical Realizations (Boca Raton, FL: CRC Press, 2008).
5. A. M. Steane, 'Quantum Computing', *Reports on Progress in Physics* **61** (1998), 117–73.
6. A. Ekert and R. Jozsa, 'Quantum Computation and Shor's Factoring Algorithm', *Reviews of Modern Physics* **68** (1996), 733–53.
7. S. Singh, *The Code Book: The Science of Secrecy from Ancient Egypt to Quantum Cryptography* (London: Fourth Estate, 1999).
8. R.J. Hughes, 'Quantum Computation', in: A. J. G. Hey (ed.), *Feynman and Computation* (Reading, MA: Perseus, 1999). pp. 191–221.
9. G. Brassard, 'Searching a Quantum Phonebook', *Science* **275** (1997), 627–8.
10. P. W. Shor, 'Scheme for Reducing Decoherence in Quantum Computer Memory', *Physical Review A* **52** (1995), R2493–6.
11. A. M. Steane, 'Error Correcting Codes in Quantum Theory', *Physical Review Letters* **77** (1996), 793–7.

18
Constructing a quantum computer

Requirements for a quantum computer

From the last section of the previous chapter, it is quite clear that constructing a useful quantum computer, one that could, for example, factorize numbers completely beyond the capability of classical computers, or sort through very large data sets with the advantage of Grover's algorithm, must be an enormous task. Very large numbers of qubits and enormous numbers of gates will be required, and the demand must be for a remarkably small amount of decoherence and other errors.

A smaller immediate challenge, though, from the mid-1990s on, was for a demonstration that quantum computation actually worked, leaving aside the question of whether it was useful. For this purpose, construction of a single controlled-*NOT* gate was important, and factorizing of the number 15 would be a triumph. (Fifteen is the smallest number that is a product of two prime numbers; here we are not including 2, which is highly unusual because it is the only even prime number.) This stage would require only two or three qubits, and a few hundred gates.

Even for this very much lesser problem, the challenge was extremely large, but on the other hand, it might be said that the minimum requirement even to *talk* about quantum computation was essentially a system with at least two energy levels and a way of causing transitions between them. It was probably in that spirit that it became almost *de rigueur* in the early years of the present millennium for any experimental physicist writing a research proposal requesting financial support, or presenting results, to add in the claim that the work could be used for quantum computation. It was seldom explained how, at least in any detail!

Rather more seriously, a very large number of physical implementations of a quantum computer have been suggested. They have included nuclear magnetic resonance (NMR), trapped ions, cavity quantum electrodynamics with atoms, optical photons, polymers, superconductors, quantum dots, Josephson junctions, electrons on a liquid helium surface and many more.

Progress overall has been steady but perhaps not dramatic. One important point is that the techniques that can achieve the first stage most quickly are in general not the ones most likely to be successful in the longer haul, when extension to large numbers of gates and qubits is required. The problem is called that of *scaling up*. The titles of two successive books edited by Samuel Braunstein are interesting. In 1999 he edited a book called *Quantum Computing: Where Do We Want to Go Tomorrow?* [1], but two years later, the book he edited with H. K. Lo was called *Scalable Quantum Computers: Paving the Way to Realization* [2].

In the latter book, David DiVincenzo included a list of what he called the five basic criteria for effective quantum computation, and we shall briefly discuss these, and also some comments on them by Mikio Nakahara and Tetsuo Ohmi [3].

The first criterion is that we need a system with well-characterized qubits, and that the system must be scalable. Each qubit must be separately addressable, and simultaneous use of different types of qubits corresponding to the same quantum system makes the quantum computation particularly efficient. For example, in the case of the ion trap, we can have qubits corresponding to (i) low-energy transitions between different sub-levels of the ground state; (ii) higher energy transitions between the ground state and an excited state; and (iii) oscillations of the ion.

The second criterion is that there must be a way of setting all the qubits, including all those to be used in quantum error correction, to $|0\rangle$. Anybody who has ever programmed a classical computer must remember the importance of setting all the various variables to 0, or *initializing* them, before writing any of the rest of the program. It is absolutely no good adding to the computer representation of your 'wealth' all the various pieces of income you may accumulate, and assuming that the final value of the variable is yours to spend, if you omit to ensure that the initial value has been put equal to 0. Such is the road to bankruptcy!

In many implementations, all that is required is to lower the temperature by cooling; this will ensure that all the systems will be in their ground state, which is our $|0\rangle$. In other cases, we can perform a measurement that will project the qubit into the desired state. In the following section, we shall find that NMR, although certainly the easiest system to use in the construction of quantum computers, in many ways, including this one, rather fails to obey the rules.

The third criterion is one of the most obvious. It relates to the decoherence time, which obviously we want to be as long as possible. Nakahara and Ohmi say that the equivalent time for a classical computer is just the length of time for which the hardware lasts, and it can be upwards of 10 years. The greatest contrast between classical and quantum computers is that the decoherence time of a quantum computer will often be about 10^{-6} s.

However, it is not the time as such that is crucial, but the ratio of the decoherence time to the gate operation time. The latter is determined by such factors as the strength of couplings between qubits, and atomic oscillation periods, and the good news is that it may often be in picoseconds (ps) or multiples of 10^{-12} s. Thus the ratio of decoherence and gate operation times may be as high as 10^6, or in other words, during the time in which a gate operates, only about one in a million qubits will dephase. We remember that Steane's requirement in the previous chapter was that, even with all the quantum error correction available, this number should be at least as large as 10^5, so we see that quantum computation should be possible, but there is little or no room for slackness in the system—everything must be practically perfect!

The fourth criterion is for a universal set of reversible gates. As we have seen the actual number of different gates to form a universal set is small—just single-qubit gates and the controlled-*NOT* double-qubit gate, but as we have stressed, the latter is extremely difficult to design and construct in such a way that the decoherence time is not much too short.

In addition, the interaction comprising the gate must be switched on and off at precise times. There may be two cases to be considered here. The first case is where the interaction may be provided by nature, but it may not be straightforward to switch it off. The second case is where there is no such interaction in nature, and a so-called 'bus qubit' will be required to interact with each qubit in turn; this bus qubit itself would seem certain to be a prime vehicle for decoherence.

It can be mentioned that although quantum computers do not *need* to include gates working on more than 2 qubits, in fact the implementation can be more efficient if reversible gates working with 3 qubits, such as the Toffoli or the Fredkin gate, are included in the circuit. And as one last point on gates, it can be mentioned that there are some implementations where the reversible gates can be replaced by irreversible gates generated by measurements.

The last of the five criteria is that it must be possible to measure the qubits. Here again, an NMR implementation does not rigorously fulfill this requirement as we shall shortly see. Any measurement will fail to be 100% efficient or accurate, so it may be necessary to repeat each computation several times.

Over the decade since DiVincenzo introduced his criteria, they have been used a great deal to study the progress made with each putative implementation of quantum computation. Back in 2001, DiVincenzo himself suggested that it was not possible to know which method would be optimal, and indeed that it was probably pointless even to ask the question. Instead many implementations should be attempted, and progress should be assessed as new ideas and techniques emerge. Nine years later things have probably not changed that much.

In the rest of this chapter we shall discuss some of the schemes [4, 5] that either have achieved most so far or seem capable of progressing most in the long run.

The NMR quantum computer

Nuclear magnetic resonance is perhaps the technique that has made the greatest steps towards demonstrating quantum computation, mostly because many of the manipulations of the spin-state that provides the qubit are exactly those that have been carried out routinely in NMR from the very beginning of use of the technique. However, it does not seem likely that NMR will be useful in pushing quantum computation out to the region where it can outperform classical computers. Also, rather strangely, it is possible to argue that the work done in this area is actually not quantum computation at all! We shall see this later.

Nuclear magnetic resonance was invented by Felix Bloch and Edward Purcell in 1945, and they were to be awarded the Nobel Prize for physics in 1952. The invention was in the period in which physicists, who had been using radio waves and other novel aspects of physics during the Second World War, returned to using these techniques for peaceful research.

The NMR technique relies on the fact that the neutron and the proton, like the electron, are spin-½ particles. This implies that they have two quantum states, the first with $s_z = \hbar/2$, and the second with $s_z = -\hbar/2$, or we can say that the states have the quantum number m_s equal to ½ and –½ respectively. Again we may call them spin-up and spin-down.

Larger nuclei are, of course, composed on protons and neutrons. For each pair of protons, one will have $m_s = \frac{1}{2}$, and the other $m_s = -\frac{1}{2}$, so the total spin angular momentum will be zero, and the same will be true for each pair of neutrons. It should thus be clear that nuclei with even numbers of both protons and neutrons will have no total spin, but all other nuclei will have a non-zero spin. In classical and in quantum mechanics there is always a close relationship between angular momentum, of which spin is one type, and magnetic moment, and so these nuclei will also have magnetic moments.

Now we imagine applying a strong steady magnetic field along the z-axis, to a sample of liquid containing many such nuclei. We will find that for each nucleus there are a number of different energy levels, each corresponding to a different z-component of spin, and hence to a different z-component of angular momentum.

We now apply an alternating field at right angles to this steady field. Let us suppose that the difference between two of the energy levels we mentioned in the previous paragraph is E, and let us also suppose that frequency of the alternating field is f. Then if $hf = E$, the nuclear spins will make transitions from the lower level to the upper level, and energy will be absorbed. This is, of course, absolutely typical quantum mechanical behaviour. It is resonant behaviour (hence the R in NMR), because if one sweeps through the frequency of the alternating field, the absorption of energy occurs in a very small range around a precise value of the frequency, f, which is know as the resonant frequency.

Nuclear magnetic resonance has many uses. As we have said, it is a technique invented by physicists, but used greatly by chemists and other scientists, and of course in recent years it has become extremely important in medicine. The most obvious use is to search for a particular type of nucleus. Each nucleus has its own characteristic frequency, and if a sweep of energy is made, and energy is absorbed at that frequency, that shows that that particular species is present. The process may be made quantitative—we may be able to discover how much of each type of nucleus is present by study of the height of the various peaks.

However, NMR can achieve vastly more than this. To make the discussion simple, we will restrict it to the case where the nucleus in question is hydrogen, so the nuclei are just protons. This is the only case we shall be concerned with from now on. In this case, as we have said, the nucleus has spin-½, so in the magnetic field there are two energy levels, and the transitions are between these levels.

So far we have been thinking of the steady magnetic field experienced by the actual nucleus as identical to that applied to the sample. In practice, though, this will not be exactly the case, because the magnetic moment of each nucleus produces a small field that affects its neighbours. Thus nuclei of the same species but in different chemical environments will experience slightly different magnetic fields, and so they will have slightly different resonant frequencies. (It can be said that here we are assuming the NMR is performed in a liquid; in a solid the interactions between nuclei are very direction-dependent and we wish to avoid that here.)

The standard example is ethyl alcohol. The chemical formula for this is CH_3CH_2OH, and there are three types of hydrogen nuclei in different chemical sites, those in the CH_3, those in the CH_2 and the one in the hydroxyl group OH. And indeed when the distribution with frequency of the NMR absorption of energy is studied, there are three peaks at slightly different frequencies and with heights in the ratio 3:2:1.

This approach is the basis of the use of NMR in medicine, where it is re-christened MRI—magnetic resonance imaging—to avoid frightening those being investigated by the terrible N-word. In this technique, the applied magnetic field is arranged to vary across the human body, and the signal produced gives detailed information on the chemical environment of protons through the body, which is extremely useful for an understanding of the local biochemistry.

Now let us turn to quantum computing with NMR. We must first stress a massive awkwardness. As we have described quantum computing so far, an individual system must be in either one of two states, $|0\rangle$ and $|1\rangle$, and as it reaches a gate, may, for example, make a transition from one state to the other. In NMR we are in a completely different type of environment. We never interact with individual spins at all. Rather in essence we look at the whole of the system, perhaps consisting of 10^{20} nuclei.

We must also appreciate that, for any realistic value of the steady magnetic field, the energy difference between the two energy levels is extremely small. To discuss this, we need to return to k, the famous Boltzmann's constant, and indeed the combination kT that we met so often in Chapter 16. Let us suppose that the energy difference between the lowest energy level of a system, that corresponding to the ground state, and the next highest level, that of the first excited state, is much greater than kT. It then follows that virtually all the systems will be in the ground state—in practice we will just say that they all are. This will be the case for samples consisting of many atoms and at reasonable temperatures; of course if T increases enough, the relationship must break down and other levels will become populated.

However, the NMR case is entirely the opposite. The energy difference between the two states for the case of the proton depends on the size of the magnetic field as well as the temperature, but for reasonable values of both, the energy difference is mush less than kT. This means that, although there will still be a larger number of nuclei in the lower energy level than the higher—such will always be the case—the difference is very small. The difference in the number of spins pointing in the two directions is tiny, perhaps one in a million of the total number of spins present.

When the alternating field is imposed at the resonant frequency, this causes further nuclei to make transitions to the higher level. However, there will also be decay of nuclei from the higher to the lower level, and a new equilibrium will be reached in which again there will be more nuclei in the lower level than the upper, but the difference between the numbers will be even smaller than before.

So, as we have said, far from considering one individual spin as a qubit, we must essentially treat the whole system as one large qubit. However, as we start to explain how NMR quantum computation works, we shall temporarily ignore this issue.

In quantum computing, a proton (we can say) is used as a qubit, so a molecule with two protons may act as two qubits. The reason why it can be said that NMR quantum computation got off to a flying start is that many of the operations necessary for quantum computation can be investigated using a conventional NMR spectrometer.

Let us start by describing the energy-level diagram of the pair of protons in the absence of any spin–spin interaction between them. This is in Figure 18.1a. For the purpose of this diagram, energies are written as $\hbar\omega$, where $\hbar$, as always, is $h/2\pi$, and ω is what is known as the angular frequency; it is the actual frequency, f, multiplied by 2π. Thus energies may be written as hf or $\hbar\omega$, and it is usual to drop the $\hbar$ and talk of energies just in terms of ω.

On the left-hand-side of this figure, the spins are assumed to be in different chemical environments but to behave independently. Thus the energy difference between $|00\rangle$ and $|10\rangle$, and that between $|01\rangle$ and $|11\rangle$ are the

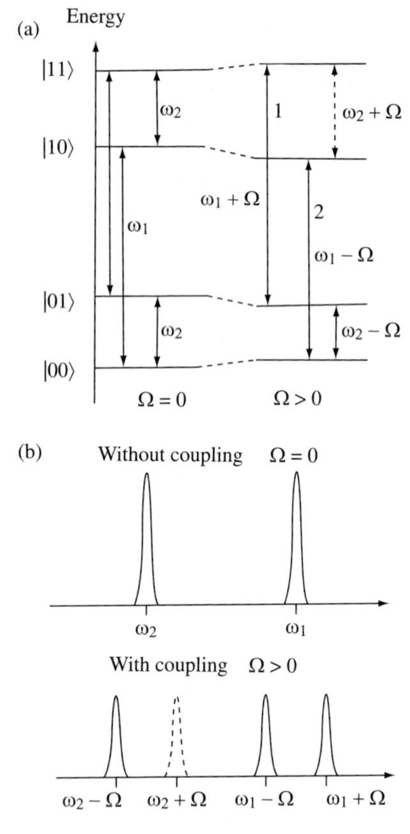

Fig. 18.1 (a) Energies of the states of the two spins without coupling ($\Omega=0$) and with coupling ($\Omega>0$). (b) Resonance frequencies of the two spins without and with coupling. The dotted transition corresponds to the *CNOT* gate.

same, $\hbar\,\omega_1$, or for convenience just ω_1 and each difference relates only to the first spin. Similarly the energy difference between $|00\rangle$ and $|01\rangle$, and that between $|10\rangle$ and $|11\rangle$ are the same, and they relate only to the second spin. The transitions marked on the diagram are those likely to occur; they are the ones for which only one spins changes its state.

On the right-hand-side of Figure 18.1a, the interaction between the spins is included. This interaction gives a positive contribution to the energy if the spins are in the same direction, that is to say for $|11\rangle$ and $|00\rangle$, and an equal and opposite negative contribution if the spins are in opposite directions, that is to say, for $|10\rangle$ and $|01\rangle$. It is handy to call the change in the energy $\pm\hbar\Omega/2$ so that the energies of the various transitions become $\hbar(\omega_1+\Omega)$, $\hbar(\omega_1-\Omega)$, $\hbar(\omega_2+\Omega)$ and $\hbar(\omega_2-\Omega)$. To obtain the corresponding frequencies we merely remove the $\hbar$ in each case, and so, unless there should be a coincidence, all the frequencies will be different, as in Figure 18.1b.

Now let us imagine an alternating field imposed at frequency, for example, $(\omega_2 + \Omega)$. The diagram shows us that the only transition this will cause is the one from $|10\rangle$ to $|11\rangle$. But—and this is the crucial point—in the language of quantum computation, this is precisely a controlled-*NOT* gate; spin 2 changes its state if but only if spin 1 is in state $|1\rangle$. So the controlled-*NOT* gate, which in almost every other implementation of quantum computation is extremely susceptible to decoherence, is a routine operation in NMR.

So NMR has many good features for quantum computation. Let us first consider the DiVincenzo criteria for up to about 10 qubits. We can define well-characterized qubits, which may be individually addressed. Initializing at the start of the process is fairly straightforward; one just lets the system come to the equilibrium with steady numbers in each of the two energy levels.

Decoherence times are long, maybe as high as 10^2–10^3 s, because the system has only very weak links to the outside. Single-qubit gate operating times are quite short, around 10^{-5} s, but those for two-qubit gates are much longer, 10^{-2} to 10^{-1} s. For a factorization of the number 21, which has been achieved by this method, around 10^5 gate operations are required, of which around 10^4 are for two-qubit gates. Putting the numbers together it can be seen that the decoherence times are sufficient, but maybe only just! As we have just seen, gates are one of the easiest features of NMR computation, and the fifth criterion, that of measurement, also taps into techniques well known in NMR for well over half a century.

Overall things look good for this case of a small number of qubits. Unfortunately this changes as the number of qubits increases much beyond 10. As the number of qubits goes up, the number of quantum states available increases, but the number useful for quantum computation does not change. The result is that addressing individual states, the initialization and the measurement stage all become more and more difficult, and we may say that the utility of the method decreases exponentially with the number of qubits. In particular the signal strength decreases as 2^{-n}, where n is the number of qubits. Overall it must be admitted that, promising as NMR is to demonstrate quantum effects, it is not scalable; it will almost certainly not be the technique that enables us to factorize really large numbers.

The last point to be made is that there have even been suggestions that the computation carried out as we have described it is not quantum mechanical at all! Because the scheme works with very large numbers of spins, and because there are only very small differences between the populations of the different energy levels, the state of the system can be decomposed into separable states. In this way the behaviour can be described without using the idea of entanglement, and it can actually be modelled classically, although the classical models can be rather contrived. As we suggested at the outset, NMR quantum computation presents a lot of nice answers but also some very awkward questions!

The ion trap quantum computer

We met the basic ideas of the ion trap in Chapter 11. As the concept of quantum computation moved to centre stage in the early to mid-1990s, the idea of the ion trap being the basis of a quantum computer was a fairly obvious one. Indeed use of the ion trap has been one of the very most successful approaches to achieving a quantum computer, and it is even possible that the technique could be scalable and so a useful device could be built, capable of delivering Grover's or Shor's algorithm for cases beyond the capability of a classical computer. The latter achievement, of course, would take a tremendous amount of work yet!

In this technique, ions are confined in a linear trap at a pressure of about 10^{-8} Pa. The ions are separated by about 2×10^{-5} m by their electrostatic repulsion. Thus the ions are lined up rather like the elements in a highly technologically advanced abacus!

Different types of qubit can be used. Two atomic states can be used, but it is essential to avoid the most common situation where the spontaneous decay time from the higher state to the lower is short, maybe of order 10^{-6} s, which would inevitably lead to a short decoherence time. One possibility is that the higher state is a so-called 'metastable' state, where, by the rules of quantum theory, decay does not take place by any of the stronger decay mechanisms that are likely to be fast. There will always be a small amount of decay by other weaker decay mechanisms, but these can be slower by many orders of magnitude. A calcium ion, Ca^+, is most used for this type of qubit.

The other possibility is to use a pair of what are called 'hyperfine' levels, which are essentially sub-levels of the atomic ground state, corresponding to different nuclear levels. The spontaneous decay-rate between such levels is quite slow, so that the lifetime of the state with the higher energy will be long as required for a qubit in quantum computation. The beryllium ion Be^+ is most commonly used in this case. In fact, though, in some cases both types of qubit are used in the same system.

Another type of qubit that is used relates to a quantum of vibrational or sound energy in the string of ions. In physics this is called a 'phonon' in analogy with the photon, which is a quantum of light energy. In some ion trap quantum computational routines, this qubit is used as a 'bus qubit', used to couple together different atomic qubits. The two states will be $|0\rangle$ and $|1\rangle$, states with 0 and 1 phonons, respectively.

Virtually every operation of the trapped ion quantum computer is achieved by interaction with lasers. This includes initializing, use of gates and final read-out. For gate operation, a laser beam is split by beam-splitters, and one beam-pair is used to illuminate each ion. Each laser beam drives transitions between the two states of a single qubit. If we call the two states $|g\rangle$ and $|e\rangle$, for 'ground' and excited', then transitions such as $|e\rangle \rightarrow |g\rangle$ and $|g\rangle \rightarrow |e\rangle$ are examples of single-qubit gates.

Also required is the controlled-*NOT* gate, which, for nearly every implementation of quantum computation (with the exception, as we have seen of NMR), will be much less straightforward to achieve. In fact it was the highly ingenious technique proposed for performing this operation, due to Juan Cirac and Peter Zoller [6] in 1995, that encouraged the development of the ion trap method of quantum computation. It may even be correct to go further and suggest that it was this invention, together with Shor's invention of his algorithm, which was the main stimulus for the intense generation of interest in the whole topic of quantum information theory around 1994–5.

In the Cirac–Zoller scheme, the laser performs yet another task, that of stimulating a phonon in the string of ions. This phonon allows interaction between an ion and its nearest neighbour, which is central in the controlled-*NOT* gate. The scheme consist of four operations: (i) a 180° pulse on ion m; (ii) a 360° pulse on its nearest neighbour, ion n; (iii) a second 180° pulse on ion m; and (iv) a 90° pulse on ion n. In stage (i), a phonon is also excited, but only if ion m is initially in state $|e\rangle$; in any case, if it is excited in stage (i), it will disappear in stage (iii), and in all cases the only change between the beginning and the end of the process is that, for ion n, states $|g\rangle$ and $|e\rangle$ are interchanged but only if ion m starts in state $|e\rangle$. This is exactly what is required for a controlled-*NOT* gate.

Progress in implementation of the ion trap quantum computer has been unsurprisingly slow, but perhaps also encouragingly steady. Many of the advances have been made by David Wineland and his group at Boulder, Colorado. As early as 1995, they reported the operation of a controlled-*NOT* gate using a single barium ion Ba^+. The control qubit was a phonon state, while the other qubit, which we may call the target qubit, was formed from two hyperfine levels of the ion. They were also able to produce the phonon mode for two Ba^+ ions, and to produce entangled states of two trapped ions in a deterministic fashion, referring to the latter achievement as providing entangled states *on demand*.

In 2003 the Cirac–Zoller controlled-*NOT* gate using two trapped Ca^+ ions was constructed by members of Rainer Blatt's group in Innsbruck. They used an electronic transition of one ion and the collective vibrational motion of two ions. Among other achievements, it is interesting that a small-scale Grover-type search algorithm has been implemented.

We shall now turn to the DiVincenzo criteria, and we find that the ion trap method in general does very well for modest numbers of qubits. It has well-characterized qubits that can be individually addressed by focused laser beams. A linear trap can accommodate up to something like 100 ions, though a greater number will probably lead to instability.

The achievement of eight ions in a trap surpasses the number of qubits achieved in any other implementation. The Cirac–Zoller scheme had neighbouring ions interacting by means of a phonon, but in principle distant ions may also interact in this way. This will be highly useful, saving many gates, when it is necessary to operate a two-qubit gate on two distant qubits.

Initialization may be achieved by suitable use of laser beams. When we turn to decoherence times, those for ionic qubits may be quite long, between seconds and even thousands of seconds, and limited by unwanted transitions. There are a number of factors that may cause decoherence of vibrational qubits, such as noise in the trapping potential and collisions with residual gas molecules. Achievement of the criterion that decoherence time should be much longer than gate operation time depends on the type of ion and the quality of the trap, but experience so far is good. On the last criterion, measurement can also be achieved by use of laser beams.

It can be said—so far, so good! But as we have indicated, problems are inevitable if we were to think of increasing the number of ions in a trap into the hundreds. As stray electric fields penetrate into the trap, heating will occur, and will increase as the number of ions goes up. Again as the number increases, it will be more and more difficult to couple the ions themselves and the phonons, as there will be more and more phonon modes.

Methods for overcoming this limitation have been suggested, though. One method is to use a so-called quantum charge-coupled device (QCCD), a microfabricated array of electrodes that trap the ions and shift them around between 'interaction regions' and 'memory regions'. A related approach is to set up several ion traps and to couple them via photons. Sufficient progress has been made to maintain the hope that, with these embellishments, the ion trap quantum computer may be genuinely scalable.

Computing with quantum dots

The enormous success of the semiconductor fabrication industry has involved building devices that have structure in the nanometre range, and the ability to control such properties of the system as the conductivity, the electric potential and the semiconductor bandgap (the energy gap between the lower valence band and the upper conduction band which determines much of the behaviour of the semiconductor). It is not surprising that such structures have been the basis of a very large number of different proposals for implementation of quantum computation.

Many of these have utilized the idea of *quantum dots*, a quantum dot being a three-dimensional structure in which electrons may be confined, so that the energy of the electron in the dot has discrete levels. Typical sizes of quantum dots may be between 5 and 50 nm (or between 5×10^{-9} and 5×10^{-8} m).

Quantum dots may form spontaneously when one semiconductor material is deposited or absorbed onto a substrate of a different semiconductor. There are two factors involved in this process. First, the different macroscopic electrical properties of the two materials may lead to the flow of electric charge perpendicular to the surface, and this can cause a so-called *inversion layer*, or a potential minimum in a region parallel to and in the vicinity of the boundary between the two semiconductor species. Then the mismatch in the lattice constants (the interatomic spacings) of the two materials leads to movement of the atoms parallel to the surface in the vicinity of this boundary, and this can cause localized potential differences to develop within the inversion layer itself. The net result is the creation of potential minima in all three dimensions, or just the required quantum dots.

Two of the most promising suggestions for use of quantum dots in quantum computation have been the *charge qubit* and the *spin qubit*. The first uses the double quantum dot, which consists of two neighbouring quantum dots. The two states of the qubit, $|0\rangle$ and $|1\rangle$, correspond to which of the two quantum dots is occupied; we can alternatively write them as $|L\rangle$ and $|R\rangle$, for whether the left- or right-hand quantum dot contains an extra electron. The qubit also includes five control gate electrodes to control the overall charge of the qubit, and the current into and out of the qubit.

On the DiVincenzo criteria, the charge quantum dot scheme can provide well-characterized qubits, and a rather more complicated scheme involving a GaAs/Al/GaAs heterostructure is potentially scalable using currently available semiconductor lithography technology. (Here GaAs is gallium arsenide, a much-used semiconductor, while Al is aluminium.) Already a two-qubit system has been fabricated and coupling has been demonstrated between the qubits.

Also initialization is easy, as one can merely inject electrons, but decoherence times are unfortunately short because charges fluctuate and phonons may be emitted. On the question of gates, one-qubit gates have been constructed, but no two-qubit gates have been produced so far. For the last criterion, measurement is relatively straightforward through investigation of the current tunneling from the qubit. Overall implementation of quantum computation by charge qubits is a promising scheme, but clearly substantial advances are still required for several of these criteria.

We now consider use of the spin qubit, which involves using the electron spin to encode the qubits. The technology to take advantage of this idea is given the rather nice name of *spintronics*. It is natural to compare the technique with that using NMR, which also uses spins, though, of course, nuclear ones. Spin-½ systems are always excellent candidates for qubits, as there are, of course, always exactly the two states. In atomic systems, for example, though our qubit may be formed from the two lowest energy levels, there are always many more actually in existence, and there will be some leakage of electrons to and from these levels.

Another advantage of spin qubits, particularly as compared with charge qubits, is that decoherence times can be as high as a few millionths of a second, because the spins are coupled only rather weakly to the environment. Also electron spins are coupled to magnetic fields much more strongly than nuclear spins, because although the charges of nuclei and electrons are of the same magnitude, the mass of the electron is only about $\frac{1}{1000}$ that of the nucleus, and coupling to magnetic field is inversely proportional to mass. This means that gate operations can be much faster than for NMR.

Actually it is possible to combine the advantages of fast gates for electron spins and long decoherence times for nuclear spins by storing the information with nuclear spins, but changing it to electronic spins for processing.

Let us now turn to the DiVincenzo criteria for spin qubits. The technique is expected to be scalable with current semiconductor lithographic technology, and initialization may also be achieved by electron injection. Unfortunately addressing of individual spins has not been demonstrated yet. As we have just said, a great advantage of the scheme is that decoherence times may be much greater than gate operation times. The last criterion of measurement can also be achieved by measurement of tunnelling current.

However, the provision of a universal set of quantum gates as yet presents serious difficulties, mainly because of the difficulty of addressing individual qubits. So, as with charge qubits, though some of the DiVincenzo criteria are met extremely well, others still prevent difficulties yet to be resolved. In fact that same comment might be extended to the whole field of quantum computation using solids, where there have been far more approaches than have been described here. With all the promise of the field, it is disappointing that, as yet, no clear way has been found to making at least substantial progress towards quantum computation.

Quantum computing with superconductors

In Chapter 13 we introduced superconductivity, the Josephson junction and the SQUID ring very briefly. Here we should add one more concept, the Cooper pair. This is named after Leon Cooper, one of the three discovers of the so-called BCS or Bardeen–Cooper–Schrieffer theory of 1957, which finally explained the existence of superconductivity, 46 years after it had been discovered by Kamerlingh Onnes in 1911. The Cooper pair was an early and a crucial part of this theory. Two electrons will, of course, normally repel each other because of the Coulomb interaction between their electric charges, but, when one electron is situated between two positively charged ions in a metal lattice, it brings these ions closer, and overall there is a net attraction for a second electron. These two electrons have their spins oppositely oriented and form the Cooper pair.

The three men, John Bardeen, Cooper and John Schrieffer, all then at the University of Illinois, were then able to develop the theory using Cooper pairs as the basis; they were able to show how a new cooperative state of all the conduction electrons could be built up. The momentum of the Cooper pairs is not affected by random scattering of individual electrons, so there is zero electrical resistance, which, of course, constitutes superconductivity.

This was a great theoretical achievement, and it is at first sight a little surprising that the three had to wait until 1972 to be awarded the Nobel Prize for physics. The problem may have been that Bardeen had already shared the 1956 prize with William Shockley and Walter Brattain for the invention of the transistor. (In fact, when Bardeen left for Stockholm to receive the 1956 prize, Schrieffer had come to feel that superconductivity was too difficult a problem, and was thinking of changing field. Bardeen asked him to keep going with superconductivity for the month that he would be away, and in that month Schrieffer made considerable progress.)

It had always been assumed that, if one has won one Nobel Prize, one would in effect be ineligible for another prize in the same discipline. For example, Einstein was awarded his prize for the photoelectric effect, and was never even subsequently nominated for general relativity, even though an unknown scientist producing the latter theory would certainly have received the prize. In the end, though, it must have been felt that to apply the policy to Bardeen would have been grossly unfair to Cooper and Schrieffer, while Bardeen himself could not have been left out of the later award because he was the team leader. He remains the only person to have shared the Nobel Prize twice in the same subject.

There have been three suggestions for quantum computation using superconductivity—charge qubits, flux qubits and phase qubits. All are based on Josephson junctions, and the qubits are carried, not by individual microscopic quantum objects, but by macroscopic currents in superconducting circuits containing one or more of these junctions. An obvious positive point is that the zero resistance of superconductors implies absence of the dissipation that makes quantum computation extremely difficult.

The Josephson charge qubit consists of a small 'box' of superconducting material in contact with a reservoir of Cooper pairs through a Josephson junction. A control electrode is used to determine the potential of the box, and thus the number of Cooper pairs, n, in the box, and the two states of the qubit may be written as $|n\rangle$ and $|n+1\rangle$, where the value of n can be chosen for experimental convenience.

The flux qubit is associated with two distinct states of magnetic flux through a superconducting ring. Compared to charge qubits, flux qubits offer longer decoherence times, because they are not subject to electrostatic couplings to stray charges. The basic element of the flux qubit is a superconducting ring with a Josephson junction. For suitable parameters, the total energy of the system has a double minimum, and these two minima give the two states of the qubit.

The phase qubit uses the SQUID ring as discussed in connection with Leggett's work in Chapter 13. Again it is found the potential energy has two minima, and this provides two states as the basis of the qubit.

For all three types of qubit, single-qubit gate operations may be achieved by changing control parameters at a speed much faster than that of the natural evolution of the system, or by applying pulses at frequencies suitable for exciting transitions between different states of the qubit. Two-qubit gates, of course, require coupling between qubits, and this may be provided by the Coulomb interaction or by inductive coupling. For larger systems it may be preferable not to use direct coupling but to couple each qubit to a common degree of freedom such as an LC oscillator (a circuit consisting of an inductance and a capacitance that oscillates at a characteristic frequency). This may be regarded as a bus qubit, receiving the interaction from one qubit and passing it on to the other one.

Let us turn now to the DiVincenzo criteria for all types of Josephson junction qubits. Initial good news is that the qubits are expected to be scalable within current lithographic technology developed by the semiconductor industry. To date, however, no register has been developed with more than a few qubits. For initialization, superconducting qubits work at very low temperature and they will usually be found in the required ground state if left for a long enough time. Alternatively it is quite easy to measure the state of the qubit and to flip it to $|0\rangle$ if it is found in $|1\rangle$.

The decoherence time may be as long as 10^{-6} s, while the gate operation may take only around 10^{-10} s. The ratio of 10^4 seems quite high but it will be remembered that, for an effective quantum computer, it still needs to be increased by a further order of magnitude. In contrast good progress has been made on the construction of both single- and two-qubit gates. On the last criterion, there are several schemes for measuring the state of the qubit, though not all are expected to have high efficiency.

Overall use of Josephson junction qubits appears to be one of the more promising techniques for eventually producing a genuinely useful quantum computer, although there is still, of course, a very long way to go. But the field is still quite wide open. As was said earlier, in his article that initially put forward his criteria, DiVincenzo said that it was not only impossible to decide which implementation of quantum computation would be successful in meeting them, but it might be counterproductive even to ask it. It was genuinely the best guarantee of eventual success that a large number of schemes should be tried, and progress would be made as new ideas and opportunities emerged. It was impossible to predict in which field these ideas and opportunities would lie, and probably this is still the state of affairs today.

References

1. S. L. Braunstein (ed.), *Scalable Computing; Where Do We Want to Go Tomorrow* (Berlin: Wiley-VCH, 1999).
2. S. L. Braunstein and H.-k. Lo (eds), *Scalable Quantum Computers: Paving the Way to Realization* (Berlin: Wiley-VCH, 2001).
3. M. Nakahara and T. Ohmi, *Quantum Computing: From Linear Algebra to Physical Realizations* (Boca Raton, FL: CRC Press, 2008).
4. J. Stolze and D. Suter, *Quantum Computing: A Short Course from Theory to Experiment*, 2nd edn (Weinheim: Wiley-VCH, 2008).
5. M. Le Bellac, *Quantum Information and Quantum Computation* (Cambridge: Cambridge University Press, 2006).
6. J. I. Cirac and P. Zoller, 'Quantum Computation with Cold Trapped Ions', *Physical Review Letters* **74** (1995), 4091–4.

19
More techniques in quantum information theory

Quantum cryptography

Quantum computation is the aspect of quantum information theory that everybody knows about. It would indeed be a great prize to be able to do what would comparatively recently have seemed completely impossible—to demonstrate the gap in Turing's analysis in a piece of working hardware, and to outdo comprehensively any classical computer. But of course we still expect the route to that point to be long and difficult.

There are other techniques in quantum information theory that seem to be less newsworthy than quantum computation, but which are conceptually just as significant, and which have the advantage of being possible today. The first we shall look at is quantum cryptography.

When we looked at Shor's algorithm in Chapter 17, we learned that if this algorithm were to be implemented successfully, public-key cryptography would be undermined because it would become straightforward to factorize large numbers.

Yet strangely quantum information in a sense provided the antidote for that difficulty. Also in Chapter 17 we saw that *private*-key cryptography, or the use of the one-time pad, is totally secure in principle, but suffers from the immense problem in practice that key distribution is an administrative and security nightmare. Quantum cryptography can make key distribution totally secure. Indeed a more down-to-earth name for quantum cryptography is just quantum key distribution.

The earliest ideas on the topic were invented by Stephen Wiesner in the 1970s, but unfortunately it seems that they were thought of as little more than amusements, instructive examples of how quantum theory might work perhaps, but of no real significance, conceptually or practically.

The first set of ideas to gain general interest was produced by Charles Bennett and Giles Brassard and collaborators, who included Wiesner, in 1982, and the famous BB84 protocol was published by Bennett and Brassard [1] two years later. We met Bennett in connection with Maxwell's demon and the reversible computer earlier, and we shall meet him again later in this chapter in connection with quantum teleportation. He is certainly a hero of quantum information theory.

To give some idea of the BB84 protocol, we shall briefly revise some elements of quantum physics and quantum measurement with polarized photons. We may start by considering an analyser of polarized light oriented so that it may distinguish between photons with vertical polarization and those with horizontal polarization. This means that if a stream of photons, each with either vertical or horizontal polarization, reaches the polarizer in this orientation, a complete list of the polarization of each photon may be built up.

However, let us now consider the analyser still in this orientation, but a photon with polarization bisecting the horizontal and vertical axes incident on the analyser. In fact we may note that there are actually two such directions of polarization, one bisecting the $+x$- and $+y$-axes, and the other at right angles to this direction. We can call these directions +45° and –45° respectively. When a photon polarized in either of these directions meets the analyser, it will record either a horizontal or a vertical polarization at random; for a stream of many such photons, the numbers of horizontal and vertical measurement results will be roughly equal.

Alternatively we can rotate the analyser so that it can record correctly and distinguish between photons with polarizations in the +45° and –45° directions. However, this will record random results when photons with horizontal or vertical polarization reach it.

Now we shall discuss the BB84 protocol. Alice and Bob are attempting to establish a shared set of private keys. They are able to communicate on a private quantum channel such as an optical fibre, but will also need to use a public channel, which can just be a telephone line.

Alice has four polarizers available, each of which can polarize photons along one of the horizontal, vertical, +45° and –45° directions. Bob has two analysers, one that can distinguish between vertically and horizontally polarized photons, and a second that distinguishes between photons polarized in the +45° and –45° directions.

Alice now sends a stream of photons to Bob, in each case choosing one of her four polarizers at random and recording which polarizer she chooses. Bob detects each photon with one of his analysers chosen at random. For each photon he records which analyser he chooses and the result that he obtains.

In half the cases there will be a mismatch between the choices made by Alice and Bob. For example, Alice may have sent a photon polarized in the +45° direction, but Bob has used the analyser that distinguishes between photons polarized horizontally and those polarized vertically. Bob then has a 50% chance of obtaining either result, so that the particular result that he does obtain is meaningless.

In order to remove these meaningless results from the record, Bob now sends Alice on the public channel a list of the analysers he used in each case. She replies by sending Bob a list of the numbers of the occasions when the polarizer and the analyser were compatible. This provides the two of them with common knowledge in the form of Alice's polarizer settings and Bob's results. This is called the *raw quantum transmission* (RQT), and it can be used to generate a shared key, at least provided Eve has not been at work.

Let us now examine what Eve will do if she is able to intercept the data being transmitted on the quantum channel. Actually we shall look only at her most straightforward strategy. She may be a little more sophisticated, but this does not change the general nature of the conclusions we shall reach. We assume that she intercepts and analyses each photon with her own analyser, and the direction of this analyser will be selected at random between distinguishing horizontally and vertically polarized photons, and distinguishing photons polarized along the +45° and –45° directions (just the same choices as Bob). We shall consider only those cases for which the choices of Alice and Bob were compatible, so the event has entered the RQT.

In 50% of the cases, Eve's choice of analyser is compatible with those of Alice and Bob. Her measurement result will tell her Alice's choice of polarizer. Her measurement does not interfere with the photon, so she sends it on to Bob in the same state as it was delivered by Alice, and he, of course, analyses it in the normal way. The event enters the RQT in an absolutely correct form, but, unfortunately for Alice and Bob, and unbeknown to them, Eve knows it as well.

However, in 50% of the cases, Eve's polarizer is incompatible with the choices of Alice and Bob. It may be that Alice has sent a photon polarized in the +45° direction, but Eve uses the analyser that distinguishes between vertical and horizontal polarization. In this case Eve will get a random result. Whichever result she obtains, she will send on to Bob a photon in a state corresponding to the result of her measurement. So Bob will receive a photon polarized in either the horizontal or vertical direction. Since we are considering an event that has entered the RQT, Bob uses his analyser that distinguishes between photons polarized in the +45° and −45° directions, and his result will be random. There is a 50% probability that he will register a result of +45%, which is what Alice had actually sent, but an equal probability that the result he registers will be −45°, which is obviously incorrect.

So let us sum up. There is a 50% probability that Eve makes the right choice of analyser, and does not disturb the passage of the photon from Alice to Bob. There is a 25% probability that she uses the wrong analyser but nevertheless the result measured by Bob corresponds to what was sent by Alice. But there is also a 25% probability that, as a result of her using the wrong analyser, the result measured by Bob has been affected by Eve, and in the RQT there is an error.

To check this, Alice and Bob use the public channel to check a section of the RQT, which must obviously then be discarded. According to our analysis above, it may, of course, be that their versions of the checked section agree totally, and this shows that there has been no eavesdropper. However, it may be that, for about a quarter of the events in the RQT, Alice's record is different from that of Bob. This is clearly the sign that Eve has been at work, and the whole RQT must be abandoned.

In practice things will probably be a little complicated. As we have said, Eve may use rather more sophisticated strategies. There may also be some noise in the circuit. And, of course, Eve may only have intercepted some part of the quantum signal. The good news, and it is really the central point of quantum cryptography, is that, however sophisticated Eve may be, if the channel is noiseless, Eve must always leave some trace of her activities. So if there is no such trace, Alice and Bob know that they have a secure key.

If the channel is noisy, things are obviously more complicated, because there will be mismatches between Alice's and Bob's version of the RQT, even in the absence of eavesdropping. Alice and Bob may use error correction, as sketched for quantum computation in Chapter 17, and they may then apply a series of operations called *privacy amplification* that cut down further any possible information Eve may have obtained. This is achieved, though, only at the expense of cutting down further the proportion of the RQT that is usable. In the end, it will never be possible to say that Eve has obtained no information, only to limit the amount of information she might have obtained. The whole process is known as *key distillation*.

It is important to stress how central the no-cloning theorem is to the fact that Eve will leave some trace of her misdeeds. If she had a cloning device, she could clone the state of the photon that she receives from Alice, and send the original directly on to Bob without having performed any measurement on it. Thus her presence could not be detected. She could also, in fact, clone many copies of the original state for her own benefit, and hence determine its polarization. If, for example, using the vertical/horizontal analyser always gives the result as horizontal polarization, but the +45°/–45° analyser gives either result randomly, it would be clear that the polarization is in the horizontal direction. But, as has been said, this relies on the availability of a cloning device, which is impossible.

The BB84 protocol is unusual among applications of quantum information theory in that it does not make use of entanglement or the ideas of Bell. So it is interesting that a different protocol was proposed by Artur Ekert in 1991 [2], and this was based totally on Bell's theorem. In this scheme, the first member of an EPR pair is sent to Alice, the second to Bob. They each make a measurement along the x-, y- or z-axis, their choices of axis being independent and random. Later, rather as in the BB84 protocol, they ascertain the case where they made the same choice of measurement direction using the public channel, and this will give them an RQT.

The remaining measurements are used to test Bell's inequality on the EPR system. If violation is found, they will conclude that there has been no eavesdropping, and so the key is secure. However, if Bell's inequalities are obeyed, that is a sign that there has been an eavesdropper. The point is that the result of eavesdropping effectively creates a hidden variable, because it collapses a superposition of different states into a single state, which corresponds to the value of the hidden variable, and which in turn causes Bell's inequalities to be obeyed.

We now turn to practical implementation of the BB84 scheme, and the first such implementation was by Bennett and Brassard [3] themselves in 1989. Weak polarized pulses were transmitted through free space over a distance of about 30 cm. The bit rate was slow and system errors were about 4%, but the key distillation worked efficiently. As Brassard later admitted, any potential Eve would have had a tremendous advantage; the devices used to put the photons into different polarization states gave very different noises depending on the state involved! But, despite the practical limitations, this was actually an extremely important achievement for quantum information theory. As Deutsch remarked, it was the very first time that a device of any type had been produced with capabilities exceeding those of a Turing machine.

However, it is in any case clear that, while achievement of the useful quantum computer may be decades away, making quantum cryptography useful imposes comparatively few demands on the experimenter. It is relatively straightforward to make the scheme work in a laboratory, and the significant step has been to make it work in a practical and challenging application.

The group that has performed the most work to this end is that of Gisin, and this work has followed on from that done on checking Bell's inequalities described in Chapter 11 in a very natural way. In 1996 [4], a quantum key was shared by users 23 km apart across Lake Geneva. The connector was standard Swiss Telecom optical fibre, and the normal telecommunication wavelength of 1.3×10^{-6} m was used. The key was encoded in very weak laser pulses, each pulse carrying on average only 0.12 photons.

In other interesting work, Richard Hughes' group at Los Alamos has demonstrated quantum cryptography in free space over a distance of about 1.0 km in broad daylight [5]. Possible problems from a high background of photons from the Sun and a turbulent atmosphere were overcome by a careful choice of experimental parameters. The atmosphere has a high transmission window for photons around 7.7×10^{-7} m, and it has little effect on the polarization states of photons, while the periods involved in atmospheric turbulence are long enough that its effect on Hughes' experiment could be compensated for. This is a very exciting development because it implies that perfectly secure communication between the Earth and a satellite may be possible.

Meanwhile on Earth itself, quantum cryptography is gradually coming into use for civic and commercial security. In October 2007, for example, during the Swiss National Elections, Gisin took charge of the process, made secure by quantum cryptography, by which the Geneva canon counted the votes and communicated the result between the polling station and the counting office [6].

Quantum teleportation

Quantum computation and quantum cryptography take areas of classical science and technology, and show how quantum techniques may be superior; they perform the various tasks faster, more efficiently or more securely. In contrast, quantum teleportation is not a development of a classical technique, or a more powerful way of doing something that could be done classically. It is an independent achievement, totally reliant on quantum technology. One may expect that, in time, further such possibilities will be discovered or invented by visionary quantum information scientists.

Though, as we have said, quantum teleportation is new to science, teleportation is extremely well known in science fiction. When there is trouble at the other end of the Universe, it is a natural time-saver for the hero, rather than having to travel there in a spaceship, to be teleported there, to be disassembled at point A and to be reassembled at point B, totally unchanged, without travelling through the intervening portion of space.

Quantum teleportation does achieve much the same thing, though, as yet at least, only for the state of a single elementary particle, such as an electron or a photon. In one way it obeys the implicit laws of science fiction teleportation, but in one way it violates them. The agreement is in the point that the hero disappears at one point, reappears at the other point. In fiction this is a natural requirement; one certainly does not want two versions of the hero. In quantum teleportation it is a requirement of the no cloning theorem. One cannot have the original electron or photon, and a teleported version, which would have to be a clone of the original.

Where, however, the fictional image differs from the quantum reality is that, in a film or on television, the impression is decidedly given that the teleportation process is instantaneous, or, at the very least, reappearance at B follows disappearance at A in an extremely short interval, far shorter than it would take a ray of light to travel right across the Universe from A to B. In fictional teleportation, it is generally implied that special relativity is violated.

In fact, as we saw in Chapter 7, quantum theory may well be in violation of locality, but it respects parameter independence, so it does not conflict with special relativity; it does not allow information to be sent at speeds greater than that of light. We shall find that quantum teleportation, like quantum cryptography, requires use of a public classical channel as well as a quantum channel. Since the classical channel certainly is restricted by the laws of special relativity to speeds no greater than that of light, it is obvious that quantum teleportation itself must be subject to this restriction.

Anyway, we now turn to quantum teleportation itself. It was invented by a group of what can be thought of as the great and the good of quantum information theory including Bennett and Brassard [7]. Quantum teleportation involves Alice transmitting the state of a spin-½ particle to Bob, without this state itself moving from Alice to Bob. It is important to recognize that what is transmitted is the state itself, not knowledge of the state. Alice herself may happen to know the state of the particle, but she does not need to do so, and, although at the end of the process, Bob is in possession of the identical state, he definitely does not know this state, and, of course, any attempt on his part to determine this state, would fail; it would almost certainly change it.

To understand what is involved in quantum teleportation, we need one further piece of quantum theory. This is the idea of what are called the Bell states. There are four of these, and we are already very familiar with one of them, the EPR–Bohm state, which we may write as $\left(1/\sqrt{2}\right)\left(|+_1 -_2\rangle - |-_1 +_2\rangle\right)$. Let us just remember what this means. The + sign refers to the spin-up state, $|s_z = \hbar/2\rangle$, and the – sign to the spin-down state, $|s_z = \hbar-/2\rangle$. The subscripts $_1$ and $_2$ refer to the first and second particles in this combined two-particle state, so $|+_1 -_2\rangle$ tells us that particle 1 is in the $|+\rangle$ state, and particle 2 in the $|-\rangle$ state. But, of course, in $|-_1 +_2\rangle$, the particles are interchanged between the states. In other words it is an entangled state, and, of course this was exactly the important point of the original EPR argument. As always, the $\left(1/\sqrt{2}\right)$ factor is just for normalization or bookkeeping.

We may ask *how* entangled the state is. In general it is far from easy to quantify the degree of entanglement of complicated combined states of a number of particles; many people have had their own different ideas of how to do this. But the EPR–Bohm state is so simple that it seems inevitable to say that its degree of entanglement could not be higher; we can speak of it as *maximally* entangled.

It is fairly obvious, though, that it is not the only maximally entangled state of these two particles in these two states. There are four such maximally entangled states and we call these the Bell states. We shall call the EPR–Bohm state, in fact, the second Bell state, and we shall write it as $|\psi_b\rangle$. Because it is the state of particles 1 and 2, we shall write it more specifically as $|\psi_{b12}\rangle$.

What we call the first Bell state is only slightly different from this EPR–Bohm state. It is written as $|\psi_{a12}\rangle$, and it is $\left(1/\sqrt{2}\right)\left(|+_1 -_2\rangle + |-_1 +_2\rangle\right)$. As compared to the EPR–Bohm state, there is a plus sign rather than a minus sign between the two constituent states of the two particles. This can be described as a relative phase factor, and it really makes very little difference to the fundamental description of the state, but it is very important for the detailed calculations we shall need to do while studying quantum teleportation.

The other two Bell states, though, are very different in nature from the first two. For $|\psi_{a12}\rangle$ and $|\psi_{b12}\rangle$, the two particles are always in different individual states; if one is in state $|+\rangle$, the other must be in state $|-\rangle$. However, in $|\psi_{c12}\rangle$ and $|\psi_{d12}\rangle$, the two particles are always in the same state. They are respectively $\left(1/\sqrt{2}\right)\left(|+_1+_2\rangle+|-_1-_2\rangle\right)$ and $\left(1/\sqrt{2}\right)\left(|+_1+_2\rangle-|-_1-_2\rangle\right)$.

We may say that the Bell states are a *complete* set of maximally entangled states for the particles in the sense that any state of the two particles in states $|+\rangle$ and $|-\rangle$ can be written as a sum of terms, each of which is a particular Bell state with its own coefficient.

We now come to the actual method of quantum teleportation. The particle whose state is to be teleported is particle 1 and is in the keeping of Alice. Its state may be completely general and we shall write it as $|\varphi_1\rangle$. Particles 2 and 3 are put into an EPR–Bohm state, and since this is also the second Bell state, we may write their state as $|\psi_{b23}\rangle$. Thus the combined state of the system is $|\varphi_1\psi_{b23}\rangle$.

Now of the EPR–Bohm pair, particle 2 is sent to Alice and particle 3 to Bob. So Alice is in command now of particles 1 and 2, Bob of particle 3, but the essential thing is that the state of the overall system is very much entangled. When Alice makes a measurement on her particles, it will also affect Bob's particle.

Alice, in fact, makes what we shall call a Bell-state measurement on particles 1 and 2. What this means is that she collapses the state of these particles down to one or other of these four states. And detailed analysis shows that the probability of her collapsing it to each of the states is equal, so each must be equal to ¼. The most pleasing part of the analysis is that, if she measures particles 1 and 2 to in $|\psi_{b12}\rangle$, which, as we have said, is the EPR–Bohm state, particle 3 is left in state $|\varphi_3\rangle$. This means that, in this particular case, the state $|\varphi\rangle$, which was the original state of particle 1, in the care of Alice, is now the state of particle of particle 3, which is in the care of Bob. This seems to be exactly what we require—quantum teleportation.

If this was all there was to it, on the one hand things would be excellent. A small worry might be that we would have violated special relativity because the state $|\varphi\rangle$ would seem to have travelled from particle 1 to particle 3 instantaneously.

Certainly things are not as simple as this. If Alice, instead of measuring the system of particles 1 and 2 to be in state $|\psi_b\rangle$, finds it in one of the other Bell states, $|\psi_a\rangle$, $|\psi_c\rangle$or $|\psi_d\rangle$, rather than being left in state $|\varphi\rangle$, particle 3 is left in different states, which we may call $|\varphi'\rangle$, $|\varphi''\rangle$ or $|\varphi'''\rangle$, respectively. However, there is still good news; the states $|\varphi'\rangle$, $|\varphi''\rangle$ and $|\varphi'''\rangle$ may each be transformed to $|\varphi\rangle$ by a very simple operation, the operation consisting of a rotation of 180° about the *x*-, *y*- or *z*-axis, respectively.

So all Bob must do is to make one of the three rotations, or alternatively to do nothing if Alice has determined that particles 1 and 2 are in state $|\psi_b\rangle$, so particle 3 is already in state $|\varphi\rangle$. But the question is, of course, which of these four actions should he perform? He knows that, just as Alice has an equal chance of measuring her particles to be in each of the Bell states, each of the actions has an equal chance of being appropriate, but which one should he choose?

The only way for him to find out is for Alice to let him know over the public channel which result she obtained in her Bell-state measurement. Once she does that, because there is a one-to-one correspondence between her result and the action Bob must take, this immediately tells him what he must do. Once he performs the appropriate rotation, or in one case does nothing, he then has particle 3 in state $|\varphi\rangle$. In other words, we can say that state $|\varphi\rangle$ has been transferred from particle 1, where Alice was in charge of it at the beginning of the teleportation, to particle 3, where Bob is in charge of it at the end.

We can now see clearly the reason for several points we mentioned earlier. We said that Bob is in command of the state, but he does not know what it is. We mentioned that the law of special relativity was not violated, and we now see that this is because of the necessity for the passage of information over the public channel discussed in the previous paragraph, which clearly takes place at less than the speed of light. Finally there is, of course, no cloning because at the end of the process, particle 3 is in state $|\varphi\rangle$, but particle 1 certainly isn't; it is by then part of the Bell-state measured by Alice.

Yet one may feel uneasy about the use of the public channel. Certainly it is true that this was necessary for Bob to know which of the four states he was left with following Alice's Bell measurement, but one might feel that Bob's knowledge should not necessarily be the central issue. Surely, it might be felt, the fact is that 'something' has moved instantaneously when Alice performs the Bell-state measurement. Whatever state particle 3 is in at this point, it is closely connected to $|\varphi\rangle$, the original state of particle 1.

It is absolutely true that Bob can deduce absolutely nothing useful without the call over the public channel, but surely, we may feel, somehow physics itself is violating special relatively even though physicists are not able to take advantage of this. In a sense this goes right back to the earliest discussions of entanglement; it was immediately clear that we cannot use an experiment such as EPR to violate special relativity, but not nearly so clear that nature itself is not doing exactly that.

Let us anyway turn to experimental validation of the original theoretical idea. Within five years, three groups of physicists, led by Zeilinger [8], Francesco De Martini [9] and Jeff Kimble [10] had all claimed to have demonstrated quantum teleportation, though in rather different types of system.

Dirk Bouwmeester and colleagues in Zeilinger's group used polarization states of photons, as in so many experiments before. Many of the same techniques were used in the same group's later study of GHZ systems described in Chapter 12. This is another example of the way in which ideas and experimental methods used in the study of Bell's inequalities and those used in studies in quantum information theory were often very closely connected.

This experiment relies on production and measurement of entangled states. In fact at the outset, two entangled pairs are produced by parametric down-conversion. One of the pairs consists of photons 2 and 3 as they were called above; the other pair is photon 1 and another photon that acts as a trigger telling us that photon 1 is on its way.

We must remember that photon 2 is sent to Alice, who already has photon 1. These two are now subjected to a Bell-state measurement. It is an important aspect of the method that the individual times of arrival within a pulse are sufficiently imprecise that they may be regarded as indistinguishable. In 25% of cases, two detectors, which we may call f_1 and f_2, register. This is the sign that we have created the EPR–Bohm state, which we have also called $|\psi_b\rangle$, one of the Bell states. Actually this is the only one of the Bell states able to be used in this experiment; extra polarizers would be required to bring $|\psi_a\rangle$ into play, while $|\psi_c\rangle$ and $|\psi_d\rangle$ would require quantum logic gates.

Now let us focus on photon 3. This is sent by Bob through a polarizing beam-splitter and in general experiments may proceed to one of two detectors, which we may call d_1 and d_2. However, the proof that quantum teleportation has occurred is that whenever Alice's detectors f_1 and f_2 register, Bob's particle 3 is always detected at d_2, never at d_1.

In De Martini's experiment, on the other hand, only two photons are used. This is a departure from the detailed scheme as originally published, but it makes the experiments simpler, and enables all four Bell states to be used. The EPR state is produced by entanglement of path rather than of polarization, and the two measurements are on the path and the polarization of the same photon. In Kimble's experiment, optical coherent states are teleported, and in this way a higher fidelity is reported than is possible without the use of entanglement. The authors report this as the first realization of quantum teleportation in which every state entering the system is teleported.

It can be mentioned that there has been controversy over which of the experiments demonstrates quantum teleportation most 'genuinely'. (Rather tellingly, for example, the paper of Kimble' group is called 'Unconditional Quantum Teleportation'.) It may be better just to acknowledge that all three experiments show interesting and important uses of entanglement.

In the decade since these experiments, quite a lot of further progress has been made. As with studies of Bell's inequalities and quantum cryptography, Gisin and his group [11] at Geneva have led the way in increasing the distance between Alice and Bob. They have performed experiments in which quantum teleportation has been demonstrated between laboratories separated 55 m in space, but separated by 2 km of standard telecommunications fibre. Apart from its own intrinsic importance, this work may help to extend the scope of quantum cryptography to larger distances.

Another important application of quantum teleportation will be to quantum computation. It will often be the case that the output of one quantum computer will be a quantum state, and this state will be the input to another quantum computer some distance away. Clearly the most convenient way of transporting the state from one computer to the other would be quantum teleportation.

Still looking into the future, we may ask whether the scope of quantum teleportation may be extended beyond single particles such as photons. Could it be extended right the way up to human beings as in the television programmes we all love so much? Unfortunately not; Sam Braunstein has estimated that it would take a hundred million centuries to transmit the information required to teleport a human being along a single channel.

Entanglement swapping

A technique very closely related to quantum teleportation is called *entanglement swapping*. Indeed it is also referred to [12] as *teleporting entanglement*, since that is exactly what it does. The idea for the technique was due to Marek Żukowski, Zeilinger, Horne and Ekert [13], and dates from 1993, only just after the invention of quantum teleportation itself, though a paper written by Bernard Yurke and David Stoler [14] the year before did suggest the basic idea.

Up to this point, it had been assumed that for two particles to be entangled, they must either have emerged from the same source or have become entangled through an interaction. The technique of entanglement swapping showed that particles that have no common past at all may become entangled.

One way of describing the scheme is that Alice and Bob share an entangled pair of particles. Bob then teleports the state of his particle to Carol, and since, at the end of the teleportation process, Carol's particle is in the same state that Bob's was at the beginning, Alice's and Carol's particles are themselves entangled, though they have never interacted or even met.

A more symmetric way of describing what happens is to say that Alice and Bob start off with an entangled pair, and Bob and Carol also share an entangled pair. Bob then performs a Bell-state measurement on his two particles, and, as a result, Alice's and Carol's particles have become entangled.

So far we have not needed a classical signal, but if we require that the entangled state of Alice and Carol's particles at the end of the experiment is identical of that of Alice and Bob's particles at the beginning, then a classical signal from Bob to Carol telling her the result of his Bell-state measurement is required, and Carol must perform one of the four operations, just as in quantum teleportation.

Thus the scheme should be of great use in a whole host of procedures in quantum information theory where provision of pairs of separated particles in entangled states is all-important. Indeed the final entangled state may even have a greater fidelity than a state obtained by relative motion of an initially entangled pair. Thus entanglement swapping will be an essential part of increasing the distances over which quantum cryptography and quantum communication will be carried out, and will also be an important component of any quantum computing network.

The first experiments on entanglement swapping were performed by members of Zeilinger's group shortly after the experiments on quantum teleportation in 1998, and three years later, the apparatus had been improved to the extent that Alice and Carol's particles could be used to demonstrate the violation of Bell's inequalities [15].

Further advances took place in the first decade of the twenty-first century. As so often before, Gisin's group [16] was able to demonstrate that the results were unchanged when Alice's and Carol's photons were separated by 2 km of optical fibre. More recently the same group [17] has been able to carry out an experiment in which continuous-wave sources rather than pulsed sources may be used. The advantage is that, whereas for pulsed sources there are substantial problems involving synchronization of the two initially entangled pairs, these problems do not exist for continuous-wave sources, and so they allow the use of completely autonomous sources, an essential step towards real quantum networks.

Lastly we mention recent work using an ion trap by Rainer Blatt and colleagues [18] at the University of Innsbruck. All the operations were carried out by the use of lasers. In the experiment, four ions were lined up in an electromagnetic trap, and entangled in two pairs. A Bell-state measurement was carried out, and, based on the result of the experiment, the ions were manipulated to produce a chosen entangled state. The authors call this work 'deterministic' in the sense that entangled swapping is achieved in every single experimental run, and, as just explained, a well-defined entangled state is produced. The technique would clearly be of great importance, in particular, for ion-trap computation.

Super-dense coding

When Alice transmits a classical bit to Bob, she naturally transmits either 0 or 1, and this is exactly and trivially what Bob will be able to read. She has clearly sent Bob 1 bit of information.

When we move to the quantum case, but at first not including any entanglement, we might wonder whether we could do better, but this is not the case. Suppose for example, Alice sends a vertically or horizontally polarized photon $|v\rangle$ or $|h\rangle$. Bob may observe the photon using a polarizer that distinguishes between $|v\rangle$ and $|h\rangle$, and, of course, he then discovers which polarization he had been sent; he gets 1 bit of information.

He could, of course, use a different polarizer, maybe one that can distinguish between photons polarized in the +45° and those polarized in the –45° direction. In this case, and whichever polarization Alice had chosen, he will get each of his directions of polarization with equal probabilities. Certainly he does not get any more information in this way. The often useful property of superposition does not add to his capabilities for this type of situation.

However, perhaps surprisingly, entanglement does so. To show this, we may yet again suppose that Alice and Bob share a pair of entangled particles. Let us suppose that this pair of particles is entangled in one of the Bell states, in fact $|\psi_{a12}\rangle$, using our notation from earlier in the chapter.

Now Alice behaves rather as Bob does when he receives the qubit in quantum teleportation. She may do one of four things, *a, b, c* or *d*. She may do nothing, in which case the entangled pair obviously remain as $|\psi_{a12}\rangle$. Or she may perform a rotation about the *z*-, *x*- or *y*-axis, and she will produce $|\psi_{b12}\rangle$, $|\psi_{c12}\rangle$ or $|\psi_{d12}\rangle$. Then she sends the particle that is in her possession to Bob—a single qubit!

Now Bob is in possession of both qubits, and he may perform a Bell-state measurement. This will enable to discover whether Alice had chosen *a*, *b*, *c* or *d*. So Alice has been able to send Bob one of four possible pieces of information, and Bob has discovered which she sent. These four possible answers clearly correspond to two classical bits of information. In other words, two bits of classical information may be transmitted by sending a single qubit from Alice to Bob.

The concept of super-dense coding was thought up by Bennett and Wiesner in 1992 [19], and the first experiments, using entangled photons, were performed by Klaus Mattle, Harald Weinfurter, Paul Kwiat and Zeilinger at Innsbruck in 1995.

References

1. C. H. Bennett and G. Brassard, 'Quantum Cryptography: Public Key-Distribution and Coin Tossing', *Proceedings of 1984 IEEE International Conference on Computers, Systems and Signal Processing* (New York: IEEE, 1984), pp. 175–9.
2. A.K. Ekert, 'Quantum Cryptography Based on Bell's Theorem', *Physical Review Letters* **67** (1991), 661–3.

3. C. H. Bennett and G. Brassard, 'The Dawn of a New Era for Quantum Cryptography: The Experimental Prototype Is Working!', *Sigact News* **20** (1989), 78–82.
4. A. Muller, H. Zbinden and N. Gisin, 'Quantum Cryptography over 23 km in Installed Under-Lake Telecom Wire', *Europhysics News* **33** (1986), 335–9.
5. W. T. Butler, R. J. Hughes, P. G. Kwiat, S. K. Lamoreaux, G. G. Luther, G. L. Morgan, J. E. Nordholt, C. G. Peterson, and C. M. Simmons, 'Practical Free-Space Quantum Key Distribution over 1 km', Physical Review Letters 81 (1998), 3283–6.
6. A. Shields and Z. Yuan, 'Key to the Quantum Industry', *Physics World* **30** (3) (2007), 24–9.
7. C. H. Bennett, G. Brassard, C. Crepeau, R. Jozsa, A. Peres and W. E. Wootters, 'Teleporting an Unknown Quantum State via Dual Classical and Einstein–Podolsky–Rosen Channels', *Physical Review Letters* **70** (1993), 1895–9.
8. D. Bouwmeester, J.-W. Pan, K. Mattle, M. Eibl, H. Weinfurter and A. Zeilinger, 'Experimental Quantum Teleportation', *Nature* **390** (1997), 575–9.
9. D. Boschi, S. Branca, F. De Martini, L. Hardy and S. Popescu, 'Experimental Realization of Teleporting an Unknown Pure Channel via Dual Classical and Einstein–Podolsky–Rosen Channels', *Physical Review Letters* **80** (1998), 1121–5.
10. A. Furusawa, J. L. Sorenen, S. L. Braunstein, C.A. Fuchs, H.J. Kimble and E. S. Polzik, 'Unconditional Quantum Teleportation', *Science* **282** (1998), 706–9.
11. I. Marcikic, H. de Riedmatten, W. Tittel, H. Zbinden and N. Gisin, 'Long-Distance Teleportation of Qubits at Telecommunication Wavelengths', *Nature* **421** (2003), 509–13.
12. A. Zeilinger, *Dance of the Photons: From Einstein to Quantum Teleportation* (New York: Farrar, Strauss and Giroux, 2010).
13. M. Żukowski, A. Zeilinger, M.A. Horne and A. K. Ekert, ' "Event-ready Detectors" Bell Experiment via Entanglement Swapping', *Physical Review Letters* **71** (1993), 4287–90.
14. B. Yurke and D. Stoler, 'Bell's Inequality Experiments Using Independent-Particle Sources', *Physical Review A* **46** (1992), 2229–34.
15. T. Jennewein, G. Weihs, J.-W. Pan and A. Zeilinger, 'Experimental Nonlocality Proof of Quantum Teleportation and Entanglement Swapping', *Physical Review Letters* **88** (2002), 017903.
16. H. de Riedmatten, I. Marcikic, J.A.W. van Houwelingen, W. Tittel, H. Zbinden and N. Gisin, 'Long-Distance Entanglement Swapping with Photons from Separated Sources', *Physical Review A* **71** (2005), 050302.
17. M. Halder, A. Beveratos, N. Gisin, V. Scarani, C. Simon and H. Zbinden, 'Entangling Independent Photons by Time Measurement', *Nature Physics* **3** (2007), 692–5.

18. M. Riebe, T. Monz, K. Kim, A.S. Villar, P. Schindler, M. Chwalla, M. Hennrich and R. Blatt, 'Deterministic Entanglement Swapping with an Ion-Trap Quantum Computer', *Nature Physics* **4** (2008), 839–42.
19. C. Bennett and S. J. Wiesner, 'Communication by One-and Two-Particle Operators on Einstein–Podolsky–Rosen States', *Physical Review* **69** (1992), 2881–4.

Conclusions

As I write these conclusions, work on both quantum foundations and quantum information continues at full pace. As we were reminded by Smolin in Chapter 15, although studies following from Bell's theorem have established much exceedingly important information about quantum theory, they have not actually succeeded in making the theory more comprehensible, or making sense of it, as he puts it. Much work goes on to appreciate more and more facets of the behaviour of this remarkable theory.

On the quantum information side, while quantum cryptography has made huge progress in a relatively short time, quantum computation remains an ideal. Yet week by week journals contain research papers that solve or contribute to the solution of one small part of the task, perhaps a way of achieving one component of the task of constructing one particular implementation of a quantum computer.

Increase in knowledge and understanding has led to the possibility of dramatic technological advance. But we certainly should not be surprised at this. Einstein was well aware that this was how progress would be achieved. Much of his demand for realism came from the rather pragmatic viewpoint that, if one addressed the interesting questions head-on, rather than avoiding them or declaring them rather arbitrarily solved, there would follow genuine conceptual advance, followed in turn, at least in some cases, by the possibility of technical innovation.

The story of quantum theory over the past hundred and ten or so years has reminded us that the frequent cry of the politician to forget theory and to leave aside thinking about fundamentals in order to concentrate on 'applications' must always be self-defeating. It took about fifty years for entanglement to move from being a theoretical idea, and rather an unwelcome one to many, to being a resource for highly novel technological advance.

Those who have spearheaded the advance, and their deeds have been described in some detail in this book, deserve enormous credit for their courage, their vision and their skill. History will certainly show them to be worthy of standing with the most celebrated scientists of the past.

Index

References to illustrations are in **bold**. References to boxes in Part 1 are in **bold** with a + sign added.